COMPUTUS AND ITS CULTURAL CONTEXT IN THE LATIN WEST, AD 300–1200

STUDIA TRADITIONIS THEOLOGIAE

Explorations in Early and Medieval Theology

5

Series Editor: Thomas O'Loughlin,
Professor of Historical Theology
in the University of Nottingham

COMPUTUS AND ITS CULTURAL CONTEXT IN THE LATIN WEST, AD 300–1200

Proceedings of the 1st International Conference on the Science of Computus in Ireland and Europe
Galway, 14–16 July, 2006

Edited by
Immo Warntjes
&
Dáibhí Ó Cróinín

BREPOLS

© 2010, Brepols Publishers n.v., Turnhout, Belgium

All rights reserved. No part of this publication may be reproduced, stored in a retrieval system, or transmitted, in any form or by any means, electronic, mechanical, photocopying, recording, or otherwise, without the prior permission of the publisher.

D/2010/0095/5
ISBN 978-2-503-53317-9

For Leofranc Holford-Strevens,
mirae doctrinae uir

C

CONTENTS

ABBREVIATIONS	IX
FOREWORD	XI
MAX LEJBOWICZ, Les Pâques baptismales d'Augustin d'Hippone, une étape contournée dans l'unification des pratiques computistes latines	1
IMMO WARNTJES, The *argumenta* of Dionysius Exiguus and their early recensions	40
ERIC GRAFF, The recension of two Sirmond texts: *Disputatio Morini* and *De diuisionibus temporum*	112
LEOFRANC HOLFORD-STREVENS, Marital discord in Northumbria: Lent and Easter, his and hers	143
DANIEL MC CARTHY, Bede's primary source for the Vulgate chronology in his chronicles in *De temporibus* and *De temporum ratione*	159
MASAKO OHASHI, The *Annus Domini* and the *sexta aetas*: problems in the transmission of Bede's *De temporibus*	190
KERSTIN SPRINGSFELD, Eine Beschreibung der Handschrift St. Gallen, Stiftsbibliothek, 225	204
BRIGITTE ENGLISCH, Karolingische Reformkalender und die Fixierung der christlichen Zeitrechnung	238
DAVID HOWLETT, Computus in Hiberno-Latin literature	259
DÁIBHÍ Ó CRÓINÍN, The continuity of the Irish computistical tradition	324
BIBLIOGRAPHY	349
INDICES	369

ABBREVIATIONS

RB	*Revue Bénédictine.*
CCCM	*Corpus Christianorum, Continuatio Medievalis.*
CCSL	*Corpus Christianorum, Series Latina.*
CLA	*Codices Latini Antiquiores.*
DACL	*Dictionnaire d'archéologie chrétienne et de liturgie.*
DSp	*Dictionnaire de spiritualité.*
DTC	*Dictionnaire de Théologie Catholique.*
LM	*Lexikon des Mittelalters.*
MGH	*Monumenta Germaniae Historica.*
Auct. ant.	*Auctores antiquissimi.*
Conc.	*Concilia.*
Epp.	*Epistolae (in Quart).*
Epp. sel.	*Epistolae selectae in usum scholarum seperatim editae.*
LL	*Leges (in Folio).*
Poetae	*Poetae Latini medii aevi.*
SS rer. Germ.	*Scriptores rerum Germanicarum in usum scholarum separatim editi.*
SS rer. Merov.	*Scriptores rerum Merovingicarum.*
NPNF	*Nicene and Post-Nicene Fathers.*
PG	*Patrologia Graeca.*
PL	*Patrologia Latina.*
PRIA	*Proceedings of the Royal Irish Academy.*
PW	*Paulys Realencyclopädie der Classischen Altertumswissenschaft.*

RÉA	*Revue des études augustiniennes*
RHE	*Revue d'histoire ecclésiastique*
SE	*Sacris Erudiri.*
WS	*Wiener Studien.*
ZCP	*Zeitschrift fur celtische Philologie.*
ZKG	*Zeitschrift für Kirchengeschichte.*
ZKT	*Zeitschrift für katholische Theologie.*
ZSK	*Zeitschrift für schweizerische Kirchengeschichte.*

FOREWORD

In 2002, a research proposal, entitled *FOUNDATIONS OF IRISH CULTURE*, was accepted by the Irish government for funding under the *Programme for Research in Third-Level Institutions* (PRTLI) 3, based at the National University of Ireland, Galway. Part of that general umbrella project involved prolegomena towards the compilation of a *Catalogue of Irish Manuscripts containing Scientific Texts in Latin, c.AD 600–800*, including works on computus, astronomy, and related subjects.

To mark the mid-point of our work, the project hosted an international conference of scholars with an interest in these subjects, in the period *c.*AD 400–850, which took place in Galway in July 2006, with speakers and auditors from Belgium, Canada, England, France, Germany, Ireland, Italy, Japan, the Netherlands, Spain, and the USA. This volume brings together the papers presented at that conference, here presented in the recognized international conference languages (English, French, German, with editorial principles generally used in English publications being consistently applied throughout).

The papers offered at the Galway conference were wide-ranging in chronological scope and varied in theme, and were intended to reflect the areas of scholarly interest that underlay the *Catalogue of Scientific Manuscripts* project that provided the inspiration for the event. The FOUNDATIONS OF IRISH CULTURE programme, of which the *Catalogue* was a part, was especially concerned with the cross-over between the Humanities and Science in the period AD 400–800 and with the modern study of medieval mathematics and science. Of particular importance were:

- The Irish role in the development of Computistical Mathematics,
- The transmission of Late Antique Mathematical Knowledge in Ireland & Europe,

- The development of Astronomy in Early Medieval Ireland & Europe,
- The Irish contribution to the development of the European Science Curriculum in the period AD 500–850.

'Since in the seventh century the leading experts on the computus were the Irish' (the verdict of Leofranc Holford-Strevens, *The History of Time, a very short introduction* (Oxford 2005), 56) it was felt to be entirely appropriate that this landmark conference should take place in Galway, which has become one of the leading centres for computistical studies in Western Europe. The conference brought together, for the first time, the leading scholars of computistics and related subjects from all over the world, and the resulting conference proceedings provide a panorama of Early Medieval scientific knowledge, both in Ireland and in the rest of Western Europe during the period of the so-called 'Dark Ages'.

Irish schools in the Early Middle Ages made no distinction between science and general culture, but regarded the two as essential aspects of the one educational system. The modern dichotomy between these two areas of knowledge would have been incomprehensible to Irish and European scholars of that time, and nowhere is that more obvious than in the Science of Computistics – the mathematics required to calculate the date of Easter and related topics (incl. astronomical observations and calculations). The Science of Computistics straddles the fields of mathematics and astronomy, biblical interpretation and cosmology, empirical astronomical observation, and the perennial quest to understand the concepts of Time and Time-Reckoning. It is the fundamental mechanism for bridging the gap between the Humanities and Science in the Early Medieval world (and gives the lie – if that were needed – to the tiresome cliché of the 'Dark Ages'. On the contrary, the papers presented at our conference impressively established the fact that the serious scientific study of time and the mathematics associated with time-reckoning flourished throughout this period AD 400–800 and even earlier, and the legacy of those cultural and scientific achievements was present right down to the twelfth century. The title of the present volume, *Computus and its cultural context in the Latin West, AD 300 to 1200*, was chosen for that reason, with the articles being ordered chronologically in order to reflect that continuity.

The Galway Conference provided a unique opportunity to bring together both the established authorities in computistical studies, and the next generation of young scholars in this field. We had hoped that the event would be marked in particular by the presence of Prof. Arno Borst, doyen in the field, whose monumental three-volume study of Carolingian computistical texts was scheduled to appear just after our event. It was with great sadness, therefore, that we learned of Professor Borst's death on 24 April 2007, following the brief illness that had precluded him from travelling. With his permission, however, and through the good offices of Prof. Dr Gerhard Schmitz (Monumenta Germaniae Historica, München, who published the volumes), we had the privilege of officially launching Prof. Borst's new publication as part of the Conference, and we know that he was particularly gratified by that. The fact that one of the editors of these proceedings, Dr Immo Warntjes (at that time completing his doctoral studies in Galway) had collaborated closely with Prof. Borst in the final stages of his work gave added significance to the Galway launch, and offered further proof of that fruitful contact between the generations that was mentioned above.

It remains only to thank all those whose encouragement and help made that 1st International Conference on the Science of Computus possible. They include the authorities in the National University of Ireland, which hosted the event and provided generous funding for it, the Director of the Moore Institute for the Humanities, where the talks took place, the Royal Irish Academy, and a group of generous private sponsors. Prof. Thomas O'Loughlin, University of Nottingham, kindly offered to 'host' the volume in the series of which he is General Editor, and the publishers, Brepols, who have been patient and supportive. Doctoral students and colleagues involved in the FOUNDATIONS OF IRISH CULTURE programme at NUI Galway were particularly supportive: Dr Eric Graff, Dr Mark Stansbury, Dr Pádraic Moran, and Dr Immo Warntjes. Maura Walsh (Ó Cróinín) designed the conference poster and programme, Eugene Jordan (Galway) and especially Daniel Frisch (Greifswald) assisted greatly with the electronic typesetting of these proceedings, and additional assistance was provided by Dr Jacopo Bisagni (Galway), Dr Catherine Emerson (Galway), and Morgan Riou (Brest).

To Immo, in particular, I owe special thanks for having taken on the major work of seeing the proceedings through to publication. The subject of his Galway Ph.D., an edition and study of the famous Munich Computus, will appear in print shortly and will undoubtedly represent a milestone in the modern study of medieval computistics. The volume of essays is dedicated to one person in particular, Dr

Leofranc Holford-Strevens, in appreciation of the extraordinary contribution that he has made to the field with his monumental study of calendars and chronology in *The Oxford Companion to the Year* (1999), co-authored with his wife, Bonnie Blackburn. By his presence at our Conferences, through his contributions to the field of computistics in his capacity as desk-editor with Oxford University Press, and by his unstinting help to younger scholars in the field, Leofranc has done more than most to pass on the torch of learning and meticulous scholarship from his generation to the next.

Finally, I would like to thank the following libraries for granting permission to reproduce photographic material under their copyright in this volume: the Biblioteca Apostolica Vaticana, for permission to reproduce the facsimiles on p. 106; the Bibliothèque Nationale de France, Paris, for permission to reproduce the facsimiles on pp. 107–9; the Bibliothèque Municipale, Nancy, for permission to reproduce the facsimiles on pp. 110–11; the Staatsbibliothek, Berlin, for permission to reproduce the facsimile on p. 193; and the Stiftsbibliothek, St. Gallen, for permission to reproduce the facsimiles on pp. 208–11, 214–5, 225, 228–9.

Dáibhí Ó Cróinín, Galway

MAX LEJBOWICZ

LES PÂQUES BAPTISMALES D'AUGUSTIN D'HIPPONE, UNE ÉTAPE CONTOURNÉE DANS L'UNIFICATION DES PRATIQUES COMPUTISTES LATINES

Abstract

At the end of AD 386, Ambrose of Milan sent a reply to the bishops of Emilia's queries concerning the correct date of Easter of the following year, which appeared to coincide with the date of Saint Augustine's baptism. This text contains the first evidence for the use of the 19-year luni-solar cycle among Latin computists. The arguments advanced by Krusch and Schwartz for this text being a forgery prove to be based on thin ground. Likewise, Charles W. Jones' theory of this text being original, but incorporating later interpolations, appears contestable. This letter as a whole should rather be regarded as originally Ambrosian, following the analysis of Zelzer, the editor of this work in CSEL – though many aspects of that analysis are worthy of criticism. Particularly the rhetorical structure of the letter deserves thorough examination; it shows parallels to other works from the pen of the bishop of Milan, while, at the same time, taking into account the unfamiliarity of the author of the letter with the subject treated by him. Additionally, the technical substratum of the letter allows for an early dating of the role played by the arians of *homoiousios* belief in the spread of the cycle in question, in the Greek East as well as in the Latin West. This, then, sheds more light on the legend of Pachomius, to whom an angel revealed the existence of the 19-year cycle.

Keywords

Ambrose of Milan, Constantius II, Auxentius of Milan, Athanasius of Alexandria, Georgios of Alexandria, Pachomius, Theodosius, Theophilus of Alexandria, Proterius of Alexandria, Dionysius Exiguus, *Chronicon paschale*, Charles W. Jones, Bruno Krusch, Eduard Schwartz, Georg Waitz, Michaela Zelzer, Easter, arianism, *homoiousios*, manichaeism 19-year cycle, 84-year Easter cycle, Lent, festal letters, sacred rhetoric, symbolism, Councils of Nicea, Sirmium, Ancyra, Sardica, Rimini, Constantinople, Aquileia.

I

Introduction

À la fin de l'année 386, douze ans après son accession à l'épiscopat, Ambroise de Milan répond au courrier que les évêques de la province d'Émilie viennent de lui adresser. Sa lettre appartient aux *Epistulae extra collectionem* de l'édition du corpus de Vienne commencée par Otto Faller et terminée par Michaela Zelzer. Comme leur désignation le signale et comme la tradition manuscrite le prescrit, ces *Epistulae* complètent le recueil épistolaire de l'évêque de Milan, qui a été réparti en dix livres. La lettre de 386 est la treizième du supplément, alors que l'édition des Mauristes Jacques de Friche et Denis-Nicolas Le Nourry (Paris, 1690), plus attentive aux indices chronologiques internes aux lettres qu'à la tradition manuscrite, lui avait assigné la vingt-troisième place dans l'ensemble de la correspondance ambrosienne considérée comme un tout continu.[1] La lettre envoyée par les évêques d'Émilie est aujourd'hui perdue mais sa substance se laisse deviner à partir de la réponse qu'elle a suscitée. Les prélats sont troublés par le désaccord qui règne entre les tables pascales pour la prochaine année; ils pressent leur métropolitain *de facto* de trancher le débat.[2] Pour un historien de la liturgie particulièrement averti: 'La réponse de saint Ambroise est un des documents les plus importants de l'âge patristique sur la signification même du mystère pascal dans la liturgie.'[3] L'historien du comput devrait être tout aussi comblé. La lettre aurait dû

---

[1] Ambroise de Milan, *Epistula extra collectionem* 13 (Maur. 23) (CSEL 82,3, 222–34). La patrologie latine de Migne reprend l'édition des Mauristes (PL 16, 1069–78). Je reviens dans la partie "Une mise au point salutaire mais incomplète et mal accompagnée" sur la formation du recueil épistolaire d'Ambroise. Sur la date de la lettre, outre l'éditrice (CSEL 82,3, CXVIII–CXIX), voir Palanque (1933), 176 et 518 et, plus récemment, McLynn (1994), 280.

[2] Sur la juridiction d'Ambroise, voir Gryson (1968), 155–8: 'Ambroise, "métropolitain" de l'Italie septentrionale'. Gryson (1968), 157 n 25 reprend, à partir de la *Clavis*, la thèse de Jones (1943): la lettre d'Ambroise est interpolée (voir *infra* les notes 27 et 28).

[3] Gy (1955), 8. Comme la plupart des langues romanes et à l'inverse de la plupart des langues germaniques, le français n'a pas deux mots pour distinguer les fêtes pascales selon qu'elles se rapportent à la religion juive, *Passover, Passahfest*, ou à la religion chrétienne, *Easter, Ostern*. Une convention existe toutefois pour parler au singulier de la Pâque (juive) et au pluriel des Pâques (chrétiennes). Elle n'a rien de systématique; j'essaierai malgré tout de l'appliquer dans les présentes pages même si cette décision a parfois des conséquences inattendues. La lecture chrétienne du

s'imposer comme un document capital dans l'histoire de la discipline: elle contient la première attestation, sous une plume latine, du cycle soli-lunaire de dix-neuf ans appliqué au calendrier ecclésiastique. Aegidius Bucherius l'avait fait entrer dans le champ des études computistes dès 1634: il l'édite sans préciser le manuscrit qu'il utilise.[4] En dépit de ce début prometteur et de la valeur intrinsèque du document, force est d'en rabattre. La lettre d'Ambroise n'a guère suscité un grand intérêt chez les historiens contemporains du comput. Les premiers à en avoir parlé l'ont même malmenée, mais en proposant une série d'interprétations dont je crois pouvoir démontrer l'inanité. Il importe de les épingler avec précision: elles rencontrent encore aujourd'hui la faveur de la majorité des historiens généralistes.

Une historiographie déficiente

Bruno Krusch, dans son grand livre de 1880, suspecte l'authenticité ambrosienne de la lettre. Il exprime ses doutes dans deux courtes notes de bas de page, sans renvoyer de l'une à l'autre, et bien qu'il opère différemment dans chacune. Dans la première, il se contente d'exprimer sa suspicion: 'Ich nehme hier den Brief des Ambrosius an die Bischöfe der Aemilia (gedr. Bucherius, De doctr. temp., p. 474 sqq.) aus, welchen ich deshalb nicht benützt habe, weil ich Bedenken gegen seine Echtheit hege.'[5] Il ne cherche pas à fonder son jugement. Il demande implicitement aux lecteurs de le croire. La légèreté de son propos a été par la suite relevée.[6] Il vaut la peine de la rappeler, tant la défaillance initiale de Krusch continue à avoir des conséquences déplorables et tant la juste critique de Zelzer s'est

---

Tanakh, l'appellation usuelle de la Bible hébraïque par les juifs, transforme celle-ci en Ancien Testament, de sorte que, pour un même texte qui appartient à ce Tanakh/Ancien Testament, il faut soit parler de Pâque si on l'aborde du point de vue juif, soit de Pâques si on l'aborde du point de vue chrétien. Quand j'évoque les fêtes pascales sans être obligé de distinguer leurs fidèles, j'écris Pâque(s).

[4] Bucherius (1634), 474–80.
[5] Krusch (1880), V n 1.
[6] Zelzer (1978), 190 n 10: 'Das ist keine brauchbare Methode, um mit Angaben, die nicht ins Konzept passen, fertig zu werden.'

accompagnée de supputations tendancieuses qui nuisent à sa part de vérité (j'essaie plus loin de trier le bon grain de l'ivraie). Dans sa seconde note, Krusch mentionne l'attribution à Jérôme, par un moine espagnol du VIIe siècle, d'une large partie de la lettre. Il est enclin à faire sienne une telle attribution, sans chercher à en tester la validité ni se poser de questions sur la formation du corpus épistolaire d'Ambroise.[7] Tout se passe comme si l'affaire était entendue avant d'être jugée. Devant ces manquements méthodologiques successifs perpétrés par un chercheur dont l'œuvre inspire le respect et l'estime,[8] l'historien s'interroge. Est-ce qu'il ne faudrait pas appliquer aux *Studien I* le jugement qu'un autre historien allemand qui domine sa discipline, Theodor Mommsen, a porté sans aménité sur un de ses livres majeurs, son *Römische Geschichte*? Une œuvre de jeunesse remplie d'erreurs, a lâché sur le tard le prix Nobel de littérature![9] Les quelque dix ans d'âge qui séparent le cadet de l'aîné lorsqu'ils étaient, chacun de son côté, attelés à leur chef-d'œuvre respectif, rend l'hypothèse plus crédible quand elle s'applique au plus jeune. À peine sorti de ses années de formation, Krusch a pu être plus réceptif qu'il n'aurait dû l'être à l'image que les historiens se faisaient d'Ambroise depuis plusieurs siècles et encore dans les années 1880: '[. . .] un grand homme d'action, un *Kirchenpolitiker*, qui aurait été par surcroît et comme par accident un écrivain prolixe, souvent ennuyeux, et un médiocre penseur.'[10] Un autre facteur a pu renforcer ses préventions. Il aurait été victime de la thèse qui,

---

[7] Krusch (1880), 302 n 2. Floëri and Nautin (1957), 30, considèrent que cette note contient 'des motifs sérieux' pour contester l'authenticité ambrosienne de la lettre. Ces 'motifs' sont si 'sérieux' qu'ils n'en rapportent aucun . . .

[8] Langlois and Seignobos (1898), 90, donne Krusch en exemple pour ses études sur les écrits hagiographiques de l'époque mérovingienne – postérieures il est vrai aux *Studien I* (1880).

[9] Nicolet (2003), 185. Krusch avait 23 ans quand ses *Studien I* (1880) sont parues; Mommsen a écrit sa volumineuse *Römische Geschichte* entre 32 et 39 ans, soit entre 1849 et 1856. Pour ces repères chronologiques, voir, outre le livre de Nicolet, le catalogue de l'exposition préparée par les MGH à l'occasion du centième anniversaire de la mort de Theodor Mommsen, Paris 24 avril–28 mai 2004 (Orthet. al. (2004)), accessible sur le site: www.mgh-bibliothek.de/mgh/mommsen/Katalog_Paris_5.1.pdf.

[10] Savon (1977), i 7, qui, de plus, commente pp. 7–8 la formation de cette conception d'Ambroise et donne, pp. 8–19, les étapes de sa destruction (voir Savon (1977), ii 11–8 pour les notes et la bibliographie); également, Savon (1993).

lancée en 1840 par Georg Waitz, a marqué pendant un siècle les travaux d'une bonne partie des historiens d'outre-Rhin: l'arianisme, dont Ambroise fut un des plus farouches adversaires,[11] est la version germanique du christianisme.[12] Otage de cette construction idéologique, l'évêque de Milan devient rien moins qu'un adversaire résolu de "la nation allemande" et ses écrits ne peuvent qu'inspirer aux patriotes la méfiance la plus extrême, jusqu'à l'aveuglement. À l'appui de ce second facteur, j'avance la manière dont Krusch interprète la remarque qui termine la présentation d'Agriustia dans l'anonyme *Computus Carthaginiensis* (II 9): *Ob quam rem hereticos omnes ad veritatis regulam docuit descivisse, et ut caeci iam facti ultra luminis ianuam non poterint invenire.* Il voit dans cet hérétique un arien, peut-être, mais en second lieu seulement, un donatiste.[13] Louis Duchesne en fait la remarque dans l'une des deux recensions qu'il consacre aux *Studien I* (= Krusch (1880)): un arien n'aurait certainement pas cherché à prolonger le cycle de quatre-vingt-quatre ans d'Augustalis; il se serait tout de suite réclamé du cycle de dix-neuf ans.[14] Tout se passe comme si la valorisation d'un arianisme défiguré écartait spontanément l'évidence d'un donatiste dans l'esprit d'un Krusch imprégné du germanisme répandu chez ses confrères historiens; et que, dans un deuxième temps, l'hypothèse raisonnable d'un donatiste se laissait néanmoins entrevoir.

Un historien postérieur a transformé en certitude les doutes de Krusch sur l'authenticité de la lettre d'Ambroise: c'est tout simplement un faux. C'est ainsi qu'Eduard Schwartz comprend vingt-cinq

---

[11] À l'idéalisation séduisante de cette page d'histoire par Jérôme, *Chronicon* (Jeanjean and Lançon (2006), 104): *Post Auxenti seram mortem Mediolanii Ambrosio episcopo constituo omnis ad fidem rectam Italia conuertitur*, on peut préférer l'approche réaliste et documentée du lutteur implacable et inventif que fut Ambroise: Williams (1995).

[12] Meslin (1967), 13–5, cité Waitz (1840). Georg Waitz a été, de 1875 à sa mort en 1886, le directeur de la Direction centrale des MGH (voir Orth et. al. (2004), 15, 20 et 51) où Krusch a fait paraître une bonne douzaine d'édition de chroniques et d'hagiographies.

[13] Krusch (1880), 291 (pour le texte) et 30 (pour l'assimilation à un arien). L'éventualité d'un donatiste est repoussée dans la n 3 de la page, bien que, pour l'avancer, Krusch s'appuie sur une documentation, alors qu'il se contente d'une affirmation pour celle de l'arien.

[14] Duchesne (1880a), 148. Je reviens plus loin sur le comput arien. Duchesne (1880b) a rédigé une autre recension des Krusch (1880), qui ne fait pas double emploi avec la précédente et qui est, elle aussi, d'une grande pertinence.

ans plus tard son prédécesseur: 'Aber der Brief ist, wie schon Krusch ausgesprochen hat [Studien, p. V], eine Fälschung.'[15] Il n'en avance pas moins cette fois une argumentation. La falsification date du début du VIe siècle: ce qu'il estime être un pseudépigraphe contient l'extrait d'une autre lettre, celle qu'un évêque d'Alexandrie, Protérius, a adressée au pape Léon à l'occasion des controverses pascales de 455.[16] Denys le Petit lui-même a révisé une ancienne traduction latine anonyme de la lettre de Protérius et évoque sa "traduction" dans son *Libellus de cyclo magno paschae* (525). Soixante-dix ans après sa rédaction, la lettre de Protérius était ainsi intégrée à ce qui allait devenir une des pièces majeures de la tradition computiste latine.[17] La lettre d'Ambroise n'a pas connu une pareille promotion. Elle a dû attendre douze siècles pour bénéficier, avec Bucherius, d'une insertion dans un classique du comput, et plus d'un millénaire et demi pour réapparaître avantageusement à nouveau, en dépit de Krusch et de Schwartz: August Strobel en fait la cinquième des onze pièces de son anthologie, en l'attribuant à l'évêque de Milan.[18] Encore s'agit-il dans les deux cas d'un livre d'historien, non du traité d'un computiste. Pour significative que soit l'initiative tardive de Strobel, elle laisse un goût d'inachevé: l'anthologiste ne s'est pas attaché à analyser et à repousser les positions de Krusch et de Schwartz. Affirmer une thèse sans invalider la thèse contraire pourtant répandue, n'a jamais fait bouger les lignes ni provoquer des adhésions. Le pas en avant de Strobel est resté en suspens. Il n'a pas mis fin aux vicissitudes de la lettre ni n'a ouvert à celle-ci une période de reconnaissance générale: aujourd'hui encore, vingt ans après, la paternité ambrosienne de l'*epistula extra collectionem 13* n'est pas unanimement reconnue par les historiens.

Curieusement, la version latine de la lettre de Protérius (l'original grec n'a pas été retrouvé), était bien connue de Krusch: il en donne

---

[15] Schwartz (1905), 54; Schwartz ne tient pas compte de la possible attribution de la lettre à Jérôme. Jones (1943), 35 donne un résumé plus fidèle et plus complet de la position de Krusch: 'Krusch (p. V, 302) hinted that the letter was falsely attributed to Ambrose.'

[16] Schwartz (1905), 54–5, qui cite la lettre de Protérius à partir de l'édition qu'en donne Krusch (voir *infra*, n 19).

[17] Krusch (1938), 67, ll. 17–20.

[18] Strobel (1984), 96–102 (traduction à partir de PL 16, 1069–78; voir cependant *infra*, n 34) et 102–6 (commentaire). Déjà Strobel (1977), 259–62 avait considéré cette lettre comme authentique, tout en faisant seulement état des positions, contraires à la sienne, de Krusch et de Schwartz.

une édition qu'il dote d'une étude préalable.[19] L'idée ne lui est pas venue de l'utiliser, à l'instar de Schwartz quelques années plus tard, pour étayer ses préventions: c'est dire le peu d'intérêt qu'il porte à cette version et combien ses préjugés à l'encontre d'Ambroise se confirment. Mais Schwartz n'est pas à l'abri de tout reproche. Il n'a pas pensé que l'emprunt qu'il avait décelé pouvait fonctionner en sens inverse, fût-ce par des voies détournées. De fait, Charles Jones souligne près d'une quarantaine d'années après, sans rappeler toutefois l'argumentation soutenue par Schwartz, que Protérius paraphrase explicitement le prologue dont un de ses prédécesseurs sur le siège épiscopal d'Alexandrie, Théophile, avait doté sa propre table pascale,[20] édités également par Krusch dans leur version latine.[21] Par conséquent, Protérius lui-même dépend d'un auteur, Théophile, qui a accédé à la fonction épiscopale près de onze ans après Ambroise, à la mi-385, alors qu'un écart d'environ seulement de cinq ans sépare leur naissance – 339/340 pour Ambroise et vers 345 pour Théophile.[22] De plus, l'aîné a eu par la suite l'occasion de solliciter l'aide de son cadet pour faire appliquer les décisions arrêtées par le concile de

---

[19] Krusch (1880), 266–9 (étude, avec la mention des éditions précédentes, celle de Jan (1718), étant reprise dans PL 57, 507–14) et 269–78 (édition).

[20] Jones (1943), 29 n 2 et 58–9. Protérius est le troisième successeur de Théophile sur le siège d'Alexandrie (Maraval (1997), 439), non le deuxième comme l'écrit Zelzer in CSEL 82,3, CXIX et CXXI.

[21] Krusch (1880), 221–6. De larges extraits du prologue subsistent en grec dans un autre prologue, celui d'un écrit anonyme de la première moitié du VIIe siècle, le *Chronicon paschale* (Dindorf (1832), i 3–31: 28–31); Krusch les a édités à son tour, en les disposant parallèlement au texte latin. Le prologue du *Chronicon* a été traduit et commenté en français par Beaucamp et al. (1979), 254–8 pour les extraits de Théophile. Sur l'importance de la table dressée par Théophile aujourd'hui disparue, voir: Krusch (1880), 84–8; *DACL* 13,2, 1523–74 (Leclercq, *s.v.* 'Pâques'): 1554 (avec des erreurs); Jones, (1943), 29–33; Grumel (1958), 37–8; Beaucamp et al. (1979), 297–301 (qui ne distingue pas le prologue de la table de la lettre que Théophile adresse à l'empereur Théodose pour accompagner l'envoi de sa table); et Declercq (2000a), 78–9. Plus généralement, l'étude de Théophile est rendue difficile par l'état déplorable de la tradition manuscrite de ses écrits et l'édition très insuffisante de ceux-ci dans PG 65, 33–68: Richard (1939). Les écrits de Théophile ont pâti des âpres assauts que leur auteur a menés contre Jean Chrysostome, dont l'autorité posthume n'a cessé de croître. La lecture successive des contributions de Delobel and Richard (*DTC* 15,1, 523–30 (*s.v.* 'Théophile d'Alexandrie')) et de Crouzel (*DSp* 15, 524–30 (*s.v.* 'Théophile d'Alexandrie')) est édifiante: le lecteur n'a pas l'impression d'avoir affaire au même auteur.

[22] Savon (1997), 29–31 pense devoir fixer la naissance d'Ambroise en 340, alors que McLynn (1994), 32–3 préfère 339. Pour Théophile: Favale (1958).

Capoue, qu'il avait lui-même présidé pendant l'hiver 391–392.[23] Cette postsynodale contient de surcroît des allusions à un échange de lettres, aujourd'hui disparues, entre les deux métropolitains.[24] Enfin, Théophile écrit à l'évêque d'Antioche pour lui demander de suivre l'exemple d'Ambroise; celui-ci n'a pas réordonné les clercs ariens qui se sont ralliés à lui, après son accession à l'épiscopat.[25] Ambroise et Théophile: deux fortes personnalités qui ont marqué l'histoire de leur siège, deux évêques qui se sont observés et qui, à l'occasion, se sont épaulés dans leurs charges ecclésiastiques. Or Jones date les tables de Théophile de la fin du règne de Théodose (†395);[26] et, dans la mesure où il considère qu'Ambroise a bien rédigé vers la fin de 386 sa lettre aux évêques d'Émilie, il est conduit à voir des interpolations dans les extraits de Théophile contenus dans cette lettre.[27] La *Clavis Patrum Latinorum* reprend à son compte la conjoncture de l'historien américain: elle lui confère l'autorité de la chose jugée, tout en lui assurant une large audience.[28] Autres expressions institutionnelles, favorables avec des nuances à la diffusion de l'interprétation de Jones: la version électronique d'Albert Blaise, *Dictionnaire latin-français des auteurs chrétiens*, revue et corrigée sous la direction de Paul Tombeur, la reprend textuellement; et l'*Index des citations et allusions bibliques dans la littérature patristique* affecte la lettre *extra collectionem 13* d'un point d'interrogation.[29]

Les raisons qui incitent Jones à attribuer une datation basse à la table de Théophile ne sont guère probantes. Je renvoie sur ce point à l'analyse que j'ai proposée ailleurs.[30] Si l'on admet une datation haute au point de

---

[23] Ambroise de Milan, *Epistula* 70 (Maur. 56) (CSEL 82,3, 3–6). Sur ce concile et le rôle qu'Ambroise y a tenu, voir, outre l'éditrice (CSEL 82,3, XXVI–XXIX), Palanque (1933), 254–62; McLynn (1994), 278–9 (qui ne parle pas de la lettre d'Ambroise à Théophile) et Savon (1997), 278–9.

[24] Ambroise de Milan, *Epistula* 70 (Maur. 56) (CSEL 82,3, 4, §3, ll. 20–24): *Cum ex his igitur aequissimis synodi constitutis speraremus iam remedium datum finemque allatum discordiae, scribit sanctitas tua iterum fratrem nostrum Flavianum ad precum auxilia et imperialium rescriptorum suffragia remeauisse.* et 5, §6, ll. 45–47: *Qua de re quoniam propriis texuisti litteris posse typum reperiri aliquem, quo possit auferri fratrum discordia.*

[25] Brooks (1902–4), i,2 343 (texte) et ii,2 304 (trad. anglaise), cité par Gryson, (1968), 232 n 14, que complète Savon (1997), 71 n 1.

[26] Jones (1943), 29–33.

[27] Jones (1943), 35 n 3 (où il s'en tient au seul témoignage de Protérius).

[28] *Clavis Patrum Latinorum*, n° 160 (je n'ai pu consulter l'édition suivante de 1995).

[29] http://www.brepols.net/publishers/pdf/Brepolis_DLD_FLyer_FR_2009.pdf; Allenbach et. al. (1995).

[30] Lejbowicz (2006), 59 n 158.

faire de cette table une des toutes premières productions de l'épiscopat de Théophile, rien n'empêche de voir en elle une des sources de la lettre d'Ambroise considérée, dès son origine, comme un ensemble d'un seul tenant. Mieux: le prologue de cette table s'avère l'unique source technique de la lettre (voir ma quatrième partie). Celle-ci relève donc dans sa totalité, sinon de la seule plume d'Ambroise, du moins d'une élaboration proprement ambrosienne. Elle est dans son intégralité un document authentique. Son auteur tire profit du meilleur texte computiste qui a été rédigé à son époque, juste quelques mois à peine avant que lui-même réponde aux évêques d'Émilie. Il suit l'actualité de la chrétienté gréco-latine, comme l'exige la charge épiscopale qu'il assume avec une absolue détermination dans l'une des capitales de l'empire[31] et comme sa connaissance du grec le lui permet aisément. Nous sommes loin de la thèse avancée naguère par Schwartz. En dépit des quelques quatre-vingts ans qui les séparent, Ambroise et Protérius utilisent, dans leur lettre respective, une même source: Théophile; autrement dit: le computiste le plus éclairé de son temps. Sa réputation en la matière est telle que Jérôme a traduit trois de ses lettres festales[32] et que deux siècles et demi plus tard ses compétences sont toujours sollicitées pour rappeler les règles qui président à la détermination des dates pascales (voir la note 21).

Une mise au point salutaire mais incomplète et mal accompagnée

Confrontée à la situation contrastée que je viens d'évoquer, la majorité des historiens a hésité à s'engager dans ce débat et à prendre parti en toute connaissance de cause.[33] Un seul d'entre eux lui a consacré une étude spécifique, Michaela Zelzer.[34] Elle l'a fait, tout en continuant les travaux d'Otto Faller sur la correspondance d'Ambroise interrompus par la mort en 1970[35] et en menant en vingt ans

[31] Capitale (1990), et plus spécialement les pages 79–89 et 415–22.
[32] Labourt (2002–3), lettres XCVI (début 401), v 8–32; XCVIII (début 402), v 35–67; C (404), v 68–91, qui se rapportent pour l'essentiel à la controverse origéniste. Sur les 25 lettres festales que Théophile aurait écrites, 14 seulement subsistent soit intégralement soit par fragment (Favale (1958), 6–9).
[33] Je renvoie à la typologie de ces attitudes esquissée dans Lejbowicz (2006), 50 n 134; ajouter l'analyse de Beaucamp et al. (1979), 298–9, qui va dans le même sens.
[34] Zelzer (1978); Strobel (1984), 103 a tiré parti de cette étude et 106, au moment de la lecture des épreuves, a eu recours à l'édition de la lettre dans le Corpus de Vienne.
[35] J'emprunte cette date à Savon (1995), 4.

l'entreprise à bonne fin. Elle a intégré la réponse aux évêques d'Émilie à l'édition en trois tomes des *Epistulae et Acta* ambrosiens du corpus de Vienne.³⁶ L'historienne n'a donc pas circonscrit son travail sur Ambroise à la seule lettre sous examen. Elle a embrassé la totalité de la correspondance de l'évêque de Milan, dont elle s'est attachée à défricher la tradition manuscrite dans la continuité de la démarche du grand ambrosien que fut Faller. L'entreprise est plus difficile qu'il n'y paraît. Vers la fin de sa vie, l'évêque de Milan a commencé à rassembler ses lettres en vue de leur édition raisonnée. Il n'a vraisemblablement qu'esquissé ce reclassement; et, de toute façon, un gouffre de quatre siècles propice à bien des accidents et réaménagements s'insère entre la mort d'Ambroise et les premiers témoins de la collection. Dans ces conditions, le principe du reclassement que l'épistolier avait adopté est toujours en discussion chez les spécialistes.³⁷ Les difficultés que l'éditeur de la correspondance d'Ambroise rencontre proviennent donc pour l'essentiel des conditions de formation du corpus. Je n'entrerai pas dans les discussions techniques que ces problèmes ont suscitées, et qui ne sont pas terminées: elles outrepassent le cas de l'*epistula extra collectionem 13*. Je me contente de renvoyer à la deuxième partie des *Prolegomena* du troisième tome de la *Pars X* des œuvres d'Ambroise dans le corpus de Vienne: les informations que Zelzer y donne, complétées par celles de son étude parues en 1978 dans les *Wiener Studien*, sont autant de présomptions en faveur de l'authenticité de la lettre.³⁸ Je serais disposé à conclure sans tarder en reprenant l'appréciation flatteuse que Dáibhí Ó Cróinín porte sur les travaux de Michaela Zelzer relatifs à cette lettre: 'Her arguments more than outweigh the flimsy objections of Krusch and Schwartz.'³⁹ J'en suis cependant empêché, ne serait-ce qu'en raison du peu d'écho rencontré par ces travaux en dehors des historiens spécialisés dans le comput.

Il semblerait à première vue que les travaux de Zelzer satisfont pleinement à l'un des principaux impératifs méthodologiques auxquels l'étude des textes antiques et médiévaux doit obéir. Jones l'avait rappelé, à propos du document en cause, quarante ans avant qu'il lui soit appliqué: 'No solution of the question [l'authenticité de la lettre d'Ambroise] can be satisfactory without a thorough study of the extant

---

³⁶ CSEL 82,1–3.
³⁷ Voir l'excellent article de synthèse de Savon cité à la n 35.
³⁸ CSEL 82,3, CXVII–CXXV, CXXXIII et CXLVI; Zelzer (1978).
³⁹ Ó Cróinín (1986), repr. in Ó Cróinín (2003), 31 n 14.

MSS.'⁴⁰ Or l'étude des manuscrits réalisée par Zelzer, pour complète et excellente qu'elle soit, se heurte malgré tout à deux écueils, d'ordre différent. L'un d'eux dérive de la date que Jones attribue à la table de Théophile. Zelzer ne semble pas connaître le fameux 'Development of the Latin Ecclesiastical Calendar' qui ouvre l'édition de 1943 de l'œuvre computiste de Bède ou, si elle le connaît, elle ne réfute pas directement l'une des remarques qui l'émaillent. L'étude de Jones contient non seulement l'extrait que je viens de donner, mais aussi, on l'a vu, la proposition d'une datation basse de la table de Théophile; elle exclut par conséquent l'utilisation par Ambroise du prologue qui précède cette table. Tant que la datation de Jones n'est pas explicitement remise en cause, la thèse de passages interpolés dans la lettre subsiste et, à partir d'elle, la suspicion gagne l'ensemble de la lettre. Dans son article de 1978 et ses *Prolegomena* de 1982, Zelzer a beau suggérer une date de rédaction de la table compatible avec celle de la rédaction de la lettre,⁴¹ elle n'aborde à aucun moment nommément la thèse de Jones. Elle laisse hors du champ de son étude une datation qui, émanant du meilleur historien du comput de sa génération, incite à douter de l'intégrité ambrosienne de la lettre. Par une singulière anomalie, en satisfaisant largement à l'impératif codicologique exprimé par l'historien américain, Zelzer se met aussi en position de faiblesse envers lui. Elle perd d'un côté ce qu'elle gagne d'un autre et prend le risque de ne pas faire avancer le débat, en dépit de la qualité du travail qu'elle a accompli sur les manuscrits ambrosiens.

Le deuxième écueil n'est pas moins important. La division en dix livres de la correspondance d'Ambroise que Faller et Zelzer posent à l'origine du regroupement thématique commencé par l'épistolier lui-même, aurait pour fondement, selon cette dernière, la correspondance de Pline le Jeune. Hervé Savon a fait justice de cette filiation: elle est pour le moins contestable.⁴² À parler crûment, elle est même historiquement insoutenable.

---

⁴⁰ Jones (1943), 35 n 3. On peut penser que, pour générale qu'elle est, cette remarque vise aussi les insuffisances de Krusch et de Schwartz. Celui-là doute de l'authenticité ambrosienne de la lettre en ayant simplement consulté: 1. l'édition de Bucherius (Krusch (1880), V n 1); 2. MS Köln, Dombibliothek, 83², 185v–187v (voir http://www.ceec.uni-koeln.de) qui n'en contient que des extraits, avec cette fois encore un renvoi à l'édition de Bucherius (Krusch (1880), 302 n 2). Quant à Schwartz, il n'élargit pas sur ce point les maigres informations codicologiques et bibliographiques données par Krusch.

⁴¹ Zelzer (1978), 202–3 et CSEL 82,3, CXX.

⁴² Savon (1995), qui donne les références de l'ensemble des études que Zelzer a consacrées à ce thème.

Je ne vois pas comment elle pourrait s'intégrer à l'univers mental d'Ambroise. Ce fils de Rome fait exploser la tradition littéraire latine par son imprégnation biblique et par son application à assimiler les Pères grecs et Philon d'Alexandrie. Il écrit en latin. Il agit en Romain. Mais il réfléchit, conçoit et décide en chrétien qui, à partir de son baptême reçu à l'âge de 34 ans, s'est délibérément imprégné de la culture grecque la plus favorable à l'approfondissement de sa foi. Croire qu'un évêque de combat au caractère bien trempé et aux convictions invincibles aurait pris Pline et ses coquetteries stylistiques comme un exemple d'épistolier et aurait organisé d'après lui sa correspondance, relève du contresens. C'est transformer un christianisme en cours d'institutionnalisation en ersatz d'une patrie romaine au faîte de sa gloire. Ambroise ne s'est jamais adressé à un empereur comme à Trajan un Pline 'consciencieux, mais timide et sans aucune initiative; sa bonté un peu flasque perd beaucoup en face du bon sens robuste et ferme de l'empereur.'[43] Cette conception relative à l'ensemble de la correspondance d'Ambroise jette une ombre sur celle, autrement plus solide, relative à l'*epistula extra collectionem 13* et à son authenticité. En la proposant et en y adhérant sans réticence, Zelzer se met, là encore, en position de faiblesse, même s'il n'est pas possible d'inférer d'une erreur sur la conception du tout à une erreur sur la conception d'une partie de ce tout.

Je voudrais terminer ce passage de mon étude en examinant plus rapidement trois autres aspects de la lecture de la lettre d'Ambroise par Michaela Zelzer. Moins fondamentaux que les deux précédents, ils peuvent cependant atténuer la portée des analyses conduites par l'historienne. Quand elle suggère de voir dans cette lettre la réponse à une demande possible de l'empereur Théodose, elle succombe à la surinterprétation.[44] Rien dans le texte n'appuie une telle initiative. Même si l'empereur et l'évêque ont très vraisemblablement en commun une aspiration à l'unification des dates liturgiques et ne peuvent que souhaiter une échéance pascale commune à tous les chrétiens, ils ne partagent pas des mobiles identiques. L'un, encore en 386 à la tête de la seule *pars orientis* de l'empire, privilégie un ordre public strictement entendu sur les territoires soumis à son autorité. L'autre, à la tête de l'Église de Milan, cherche à favoriser une célébration collective dont la date annuelle, quand elle est unanimement acceptée et universellement pratiquée, parachève l'unité de la vérité chrétienne, en rappelant au même moment l'unique fondement de son message: la crucifixion

---

[43] Bayet (1973), 402.

[44] Zelzer (1978), 203; même réserve dans McLynn (1994), 280 n 120, quoiqu'il ne l'argumente pas.

et la résurrection du fils de Dieu. Des trois lois du Code Théodosien qui traitent de la date des Pâques, une seule est cosignée par Théodose. Elle s'en prend à: 'tous ceux qui se rassemblent pour Pâques un autre jour que celui observé par la foi catholique.'[45] Elle vise les Quartodécimans, les Novatiens et les Montanistes, que fustige cinq ans plus tard l'auteur anonyme d'une homélie pascale anatolienne, partisan du cycle soli-lunaire de dix-neuf ans.[46] Et, tout comme cette homélie, elle ne cherche pas à régler le différent pascal qui oppose Rome, champion du cycle soli-lunaire de quatre-vingt-quatre ans, et Alexandrie, qui pratique celui de dix-neuf ans. Aucun des signataires de la loi, Gratien, Valentinien ou Théodose, ne s'aviserait de mettre les deux métropoles en dehors du catholicisme, alors que, deux ans plus tôt, ils les proposaient nommément en exemple d'orthodoxie.[47]

L'analyse de Zelzer mérite d'être nuancée sur un deuxième point. À suivre l'historienne, la démarche de Théophile est dans l'ordre des choses. Or, depuis l'ouverture de l'empire au christianisme, c'est la première fois qu'un computiste prend l'initiative d'envoyer à l'empereur, et plus précisément à l'empereur dont relève son évêché, une table pascale – de surcroît, une table vraisemblablement de son cru. Ce geste novateur survient vers la fin d'un siècle où, du haut au bas de l'échelle sociale et sans que le règne de Constantin ait marqué un changement substantiel, les éclats ininterrompus des controverses pascales n'ont pas cessé de résonner. Le prélat qui, au sujet des prévisions pascales, rompt avec des habitudes pluridécennales de distance vis-à-vis du pouvoir politique doit être animé par un puissant mobile. La lettre à l'empereur qu'il joint à sa table ouvre une piste de réflexion, qu'il faut emprunter pour tenter d'expliquer la nouveauté de sa démarche. Elle se réfère à deux reprises à l'Église d'Alexandrie, alors qu'elle n'occupe qu'une page dans l'édition de Krusch.[48] Elle est en revanche muette sur les décisions qui ont été prises à Nicée à propos de Pâques.[49] Indépendamment de son intérêt propre, la table de Théophile est destinée à rappeler à l'empereur l'apport des Alexandrins à la vie de la chrétienté

---

[45] *Codex Theodosianus* XVI 5, §9 (31 mars 382) (Magnou-Nortier (2002), 210): *quicumque in unum Paschae die non obsequenti religione convenerint*. Zelzer ne se réfère pas dans son article à cette documentation.

[46] Floëri and Nautin (1957), 116–21, qui donne tous les détails sur la détermination de la date pascale par ces différentes Églises.

[47] *Codex Theodosianus* XVI 1, §2 (28 février 380) (Magnou-Nortier (2002), 96).

[48] *Epistola Theophili* §§1 et 3 (Krusch (1880), 220–1).

[49] Sur l'enjeu pascal au concile de Nicée, voir l'analyse sur nouveaux frais que j'en propose: Lejbowicz (2006), 30–63.

dans une période de réorganisation des Églises peu favorable à celle de la capitale égyptienne. En 381, le concile œcuménique de Constantinople a réduit la zone d'intervention du métropolitain égyptien[50] et, dans la hiérarchie des sièges, a rétrogradé celui d'Alexandrie au profit de celui de la nouvelle Rome.[51] Ainsi située dans la conjoncture de la fin du IVe siècle, la table de Théophile acquiert une épaisseur historique. Elle témoigne d'une vie intellectuelle ancrée dans des rivalités d'Églises et des enjeux de pouvoir. Elle incarne le sursaut d'un siège qui, ayant subi un revers, cherche à rester dans les premiers rangs en affichant ses compétences.

Le dernier point que je souhaite examiner découle du précédent. Il porte sur le rôle du concile de Nicée. Théophile passe sous silence cet événement majeur dans la vie de l'Église ancienne. Ambroise pour sa part en parle à deux reprises dans sa lettre; et il le fait chaque fois en s'y référant comme à une autorité incontestable.[52] Il note aussi à trois reprises les apports purement techniques, donc à ses yeux secondaires, des Alexandrins au comput.[53] Théophile inscrit délibérément sa table dans la seule tradition de son siège; par l'intermédiaire de Théodose, il cherche bien à renforcer le prestige de son Église et à en maintenir le pouvoir d'intervention. En revanche, quand Ambroise attribue au bien commun de l'Église qu'instaure le concile de Nicée le cycle de dix-neuf ans qu'il emprunte à Théophile, il met au compte d'une tradition chrétienne indifférenciée ce qui revient aux seuls Alexandrins. Il facilite subrepticement l'assimilation latine d'un apport grec; il le minore par calcul parce qu'il est grec; mais il le valorise par la bande parce qu'il est porteur de vérité. Qu'il se soit senti forcé d'adopter un tel subterfuge ou qu'il l'ait librement accepté, dans tous les cas Ambroise se présente comme l'intermédiaire masqué du comput alexandrin.

Ayant parcouru le champ des études historiques de ces quelques derniers cent vingt-cinq ans pour y relever les principales traces de la présence de la lettre d'Ambroise, je voudrais avancer plus avant et traiter de la composition de cette lettre. D'une part, elle affiche une singularité dans le corpus de la littérature computiste: elle emprunte, à

---

[50] Voir le canon II du concile de Constantinople I dans Alberigo (1994), 86–9.

[51] Voir le canon III du concile de Constantinople I dans Alberigo (1994), 88–9.

[52] Ambroise de Milan, *Epistula extra collectionem* 13 §§3 et 16 (CSEL 82,3, 222, 230).

[53] Ambroise de Milan, *Epistula extra collectionem* 13 §§8, 14 et 15 (CSEL 82,3, 225, 228–9).

la fois, au traité technique et à l'homélie, sans satisfaire pleinement, dans ses parties les plus homogènes, l'un ou l'autre. D'autre part, si elle reste proche de l'écriture usuelle d'Ambroise, elle n'est pas sans donner au lecteur une impression de lourdeur, voire de confusion.[54] Il est possible de mettre ces travers au compte du manque de familiarité de l'auteur avec un sujet que sa notoriété l'a contraint de traiter et qu'il a essayé de maîtriser dans l'urgence. Étrange destin que celui de cette lettre: si la singularité de son écriture nuit à son insertion dans l'histoire du comput, ses maladresses sont parfois mises en avant pour suspecter son appartenance au corpus ambrosien. Ni typiquement ambrosien par sa facture ni à proprement parler computiste par son argumentation, elle est trop souvent rejetée dans un no man's land, où elle sommeille. Il convient de l'en sortir et de lui rendre sa vigueur, en respectant sa nature composite.

La lettre d'Ambroise comme structure rhétorique

Le mode de composition qu'Ambroise pratique est déroutant pour le lecteur non-prévenu. En prenant connaissance des écrits de l'évêque de Milan, il ressent un véritable malaise, que Gérard Nauroy a fort bien décrit:

> 'La plupart des traités d'Ambroise [et la lettre sous examen est en soi un petit traité] laissent, après une première lecture, le sentiment d'ouvrages composites, à la démarche diffuse et déconcertante, aux intentions disparates, à la terminologie incertaine, où la suite des idées, rarement explicite, semble dépendre du discours lui-même, du rapprochement fortuit de deux mots, de deux images, de deux citations scripturaires plutôt qu'obéir à un propos préétabli et ordonné selon des articulations apparentes.'[55]

Confronté à cette manière d'écrire si éloignée de la nôtre, le même historien propose d'en tirer malgré tout parti. Il esquisse un mode de

---

[54] L'expression la plus extrême de ces faiblesses se trouve sous la plume de van Wijk (1936), 13 n 4: la lettre d'Ambroise est: 'une curieuse décoction d'écrits dionysiens avec de l'homilétique baroque'.

[55] Nauroy (1974), repris dans le recueil de Nauroy (2003), 301–54: 303, à partir duquel je le cite.

décryptage qui est susceptible de faire entrer le lecteur dans l'intelligence de ce type d'écrits:

> 'Il importe donc de retrouver les transitions, souvent gommées à dessein par l'évêque de Milan, de sentir et de suivre ce discours fluctuant, cette technique de la surimpression des images profanes et scripturaires, ce tissage en un réseau dense et subtil des sources les plus diverses, pour faire apparaître une logique et un dynamisme de la pensée, révéler un usage réfléchi et critique de l'*imitatio*.'[56]

Cette prescription oriente la lecture de l'historien qui aborde en néophyte les textes d'Ambroise. De surcroît, le 'Tableau des correspondances thématiques'[57] que Nauroy dresse à la fin de son étude pour décrire la composition du *De Iacob et uita beata* est un précieux stimulant. J'ai à mon tour tâché de déceler le mode d'organisation de la lettre pour tenter de saisir le déroulement de la pensée de son auteur. Je me suis finalement arrêté au 'Schéma de la lettre' placé à la fin de ces pages (voir en Annexe). Il sert de support à la lecture que je propose.

La lettre s'organise autour de deux thèmes principaux: Choix d'échéances et Pâque(s), et de deux thèmes secondaires: Images et Jeûnes. J'aborde successivement les deux thèmes principaux et ne traite que marginalement des deux thèmes secondaires. L'un des thèmes secondaires demande toutefois une précision préliminaire. Le libellé de la première colonne, Images, ne forme pas avec celui des autres un ensemble homogène, tandis que les informations contenues sur les lignes appartenant à cette même colonne semblent faire double emploi dans l'économie générale du Schéma. Il s'avère qu'Ambroise lui-même fait un double usage du texte biblique. Il le considère comme la source exclusive de l'autorité; elle énonce des règles de conduite et pose des normes de pensée. Il la considère aussi comme un réservoir d'images, lesquelles ont leur consistance propre: elles stimulent la vie de l'âme et orientent celle de l'esprit. Cette dualité de conception d'un texte qui est à la fois normatif et suggestif, explique le double emploi que je viens d'évoquer: il manifeste le double usage qu'Ambroise fait spontanément des Écritures.[58]

---

[56] Nauroy (2003), 303; l'*imitatio* est l'équivalent latin du grec *mimésis* tel qu'il est compris par les grammairiens: la production par l'art d'effets analogues à ceux provoqués par la nature.

[57] Nauroy (2003), 339–51.

[58] Ambroise de Milan, *De officiis* I 49 (Testard (2002), 211): *seruemus imaginem ut ibi perueniamus ad ueritatem*. *Imago* est polysémique sous la plume d'Ambroise; voir le commentaire de Testard (2002), 274–5 n 21; je m'en tiens dans mon texte au sens de métaphore littéraire, sans préjuger de l'usage plus spécialisé qu'Ambroise en fait.

Je qualifie de non-confessionnel le thème Choix des échéances, encore qu'il puisse avoir des incidences confessionnelles. Quand cette éventualité se réalise, elle particularise une vérité qui est en son fond générale. Ce premier grand thème est résumé par le verset de Qo 3:1: 'Un temps pour chaque chose',[59] qui a valeur de maxime. On conviendra aisément qu'elle n'est pas spécialement juive ni chrétienne, bien qu'elle soit extraite de l'Ancien Testament. Ambroise lui-même remarque qu'elle s'applique dès la vie animale; il l'évoque à partir de ses manifestations ailées, qui sont communément considérées comme le plus facilement transposables à la vie spirituelle: 'La tourterelle et l'hirondelle, les moineaux des champs ont connu la saison de leur arrivée' (Jer 8:7).[60] Certaines croyances pervertissent cette vérité non-confessionnelle. L'apôtre Paul les fustige dans l'Épître aux Galates, ainsi que le §3 de la lettre le rappelle et que le §4 le détaille. En dépit de semblables dégringolades dans la superstition, la maxime n'en a pas moins acquis ses lettres de noblesse. Elle se rencontre sous différentes formes dans l'Écriture divine, comme Ambroise désigne la Bible. Elle est illustrée par plusieurs passages de l'Ancien Testament, mentionnés aux §4 (Ps 117:24 et Ps 88:29), 6 (en plus des versets Qo 3:1 et Jer 8:7 déjà cités, Ps 118:126, Is 1:3) et 7 (Is 49:8). Ce dernier paragraphe met aussi en avant le seul verset du Nouveau Testament qui est rapporté au premier thème: 'Voici maintenant le temps favorable, voici maintenant le jour du salut' (2 Cor 6:2b).[61] L'Écriture divine dans ces deux aspects, vétéro- et néotestamentaire, recommande par conséquent aux fidèles de choisir certaines échéances pour accomplir certaines tâches et pratiquer leurs dévotions. La recherche des dates pascales est ainsi légitimée aux yeux d'Ambroise, qui évoque par la suite, au §16, la *consuetudino peritorum* pour fixer l'équinoxe de printemps au 21 mars et pour définir le début et la fin du premier mois de l'année: il court du 21 mars au 21 avril. À ce stade, la maxime acquiert sa spécialisation chrétienne la plus nette et devient directement applicable à la fête chrétienne par excellence, Pâques.

Le deuxième thème domine la lettre plus nettement que le premier. Il suffit pour s'en convaincre de dénombrer dans le Schéma de la lettre,

---

[59] Ambroise de Milan, *Epistula extra collectionem* 13 §6 (CSEL 82,3, 225): *Omni rei tempus*; Vulgate: *Omnia tempus habent*.

[60] Ambroise de Milan, *Epistula extra collectionem* 13 §6 (CSEL 82,3, 225): *Turtur et hirundo agri passeres agnoverunt tempora introitus sui*; Vulgate: *Turtur et hirundo et ciconia custodierunt tempus adventus sui*. Chevalier and Gheerbrant (1969), *s. v.* 'Oiseau' avec renvoi à un certain nombre d'espèces.

[61] Ambroise de Milan, *Epistula extra collectionem* 13 §7 (CSEL 82,3, 225): *Ecce nunc tempus acceptabile, ecce nunc dies salutis*. Même texte dans la Vulgate.

le nombre de cellules consacrées à chacun d'eux: douze au premier, trente et une au second. L'écart est sans ambiguïté. Ce nouveau thème est exclusivement chrétien. Il lie la Passion et la résurrection de Jésus. Cette solidarité s'exprime dans la notion de *triduum sacrum*, la Passion, étant survenue un vendredi, l'ensevelissement ayant duré tout le samedi et la résurrection s'étant produite un dimanche. Selon les historiens de la liturgie, l'expression *triduum sacrum* est une création ambrosienne.[62] Elle est énoncée au §13, où elle est analysée dans les trois moments consécutifs qu'elle englobe (voir *infra*); mais elle est implicite au §10.[63]

Les développements assez tardifs du second thème ne doivent pas induire en erreur. Ambroise a pris la précaution de l'introduire dès la première ligne du §1: 'L'Écriture divine et la tradition des anciens nous enseignent que déterminer le jour de la célébration pascale ne relève pas d'un savoir ordinaire.'[64] Il signale ainsi le véritable thème de la lettre, même si pour le traiter, il doit faire un détour par celui du Choix des échéances. Il parle même, dans ce 1er §, du cycle soli-lunaire de dix-neuf ans, en utilisant la seule désignation grecque, *ennéadékaétéride* – ce qui est une manière de laisser deviner la langue usuelle de sa source technique, Théophile en l'occurrence. Il se réfère aussi, toujours dans ce 1er §, au concile de Nicée. Pareille concomitance dans l'évocation d'un cycle soli-lunaire et dans celle de la première grande réunion épiscopale de la jeune chrétienté n'a rien de fortuit. Elle est reprise au §16, sans que rien de nouveau n'ait été dit entre temps à son sujet, qu'il s'agisse de savoir astronomique ou de témoignages historiques. Ambroise juge important d'insister sur cette association, qui devient une des idées fortes de sa lettre. Or elle est à double titre étonnante.

L'évêque de Milan ne cite aucun texte, issu directement du concile de Nicée ou repris postérieurement d'un des participants, qui attesterait sans ambiguïté que les Pères conciliaires ont ne serait-ce qu'évoqué le cycle de dix-neuf ans – je ne dis pas qu'il est allé jusqu'à dire: les Pères conciliaires ont étudié les avantages et les inconvénients de ce cycle ou ont demandé à des experts de se livrer à une telle étude, comme il conviendrait de procéder pour parvenir à une décision rationnellement établie. Ambroise précise toutefois que le choix en

---

[62] Gy (1955), 7–8; Vogel (1981), 270 n 85; Jounel (1983), 59.

[63] Ambroise de Milan, *Epistula extra collectionem* 13 §10 (CSEL 82,3, 226): *sexta feria crucifixus est luna quinta decima, sabbato quoque magno illo sexta decima fuit ac per hoc septima decima luna resurrexit a mortuis.*

[64] Ambroise de Milan, *Epistula extra collectionem* 13 §1 (CSEL 82,3, 222): *Non mediocris esse sapientiae diem celebritatis definire paschalis et scriptura divina nos instruit et traditio maiorum.*

faveur du cycle de dix-neuf ans se révèle 'si on fait particulièrement attention'[65] aux délibérations du concile. Cette manière de certifier au lecteur qu'un texte fondamental existe et que son sens se dévoile à celui qui l'aborde *diligenter*, alors qu'aucun extrait n'en est donné, ni littéralement ni dans une paraphrase, ni dans un résumé, a valeur d'aveu.[66] Le décodage s'impose: le cycle de dix-neuf ans n'a pas été proposé par le concile pour régler le différent computiste qui oppose Rome et Alexandrie. Le constat d'une pareille réalité doit être fâcheux au plus haut point pour Ambroise dès lors que l'un des deux cycles en concurrence s'impose à sa clairvoyance et qu'il n'a pas reçu la formation nécessaire pour le défendre en spécialiste. Il est même trop fâcheux pour être avoué en toutes lettres. Pour ce grand Romain 'porte-parole d'une faction pro-nicéenne intransigeante',[67] le cycle de dix-neuf est bien d'origine alexandrine, et non pas nicéenne. Mais, pour en parler dignement et arriver à l'imposer aux Latins, il faut le soumettre à l'onction épiscopale la plus large, celle précisément du concile le plus fameux de la jeune chrétienté. Les quelques soixante ans qui le séparent de cet événement fondateur facilitent une interprétation assez libre, dès lors que le dogme n'est pas en cause et que l'unité des chrétiens est en jeu.

Le §1 provoque un second sujet d'étonnement, une fois la lettre intégralement lue. Le cycle de dix-neuf ans a beau y être abordé une deuxième fois, il ne fait à aucun moment l'objet d'une justification technique, ni n'intervient dans l'argumentation proprement dite. Il honore la lettre de sa présence sans que ses propriétés y soient établies et encore moins défendues. Seuls les compléments des cycles solilunaires–les termes lunaires et les termes pascals – sont fondés en droit sous la forme alexandrine (XV–XXI *lunae* et 22 mars–25 avril), en référence au déroulement de la Passion historique. Si l'exigence dont ils sont porteurs est intégrée à la détermination de la date pascale, ils ne reçoivent pourtant pas leur désignation technique, *termini lunae*,

---

[65] Ambroise de Milan, *Epistola extra collectionem* 13 §16 (CSEL 82,3, 230): *Unde et maiores nostri in tractatu concilii Nicaeni eundem* ἐννεαδεκαετηρίδα *si quis diligenter intendat statuendum putarunt.*

[66] Je rappelle que les actes du concile de Nicée 'n'ont probablement jamais été rédigés par des secrétaires', selon Alberigo (1994), 30. Le concile a produit une *Expositio fidei*, vingt canons et deux synodales, celle des Pères aux Égyptiens et celle de Constantin aux Églises d'Orient: dans cet ensemble de documents, le cycle de dix-neuf ans brille par son absence.

[67] Brown (1998), 155.

termini pascales. Les deux luminaires sont rapportés l'un à l'autre uniquement sous la forme d'une métaphore, qui est reprise des Psaumes: 'Son trône [celui du Fils de Dieu] reste en ma présence dans l'éternité, comme le soleil et comme la lune dans sa plénitude' (Ps 88:38).[68] Est-ce que la puissance suggestive de l'image ne cherche pas à faire oublier que le pasteur-exégète est ici victime de sa conception étriquée du savoir? C'est bien Ambroise qui reprend Cicéron et interroge: 'Qu'y a-t-il d'aussi obscur que de traiter d'astronomie et de géométrie, [. . .], et de mesurer les espaces de l'altitude éthérée, d'enfermer dans des nombres le ciel aussi et la mer?'[69] Cette interrogation oratoire est extraite du *De officiis*, dont la rédaction définitive est contemporaine de la lettre.[70] Une pareille proximité entre ces deux textes porte à voir en Ambroise un auteur qui, dans un même mouvement, dédaigne l'arithmétique et l'astronomie et cependant se préoccupe de comput. Il s'enferme dans un piège dont il ne peut sortir qu'en glissant sur les données computistes trop techniques. Par contrecoup, il se réclame, fut-ce abusivement, d'une tradition ecclésiastique, celle issue de Nicée. Il en impose à tout prix les caractéristiques sans en poser les fondements techniques, ni être à même d'en esquisser la véritable histoire.

Le pasteur-exégète retrouve son terrain de prédilection dès le §2, avec les versets que l'Évangile de Luc consacre à la célébration par Jésus et ses disciples de leurs dernières Pâques. Avant même de fixer des règles pour en déterminer la date annuelle, Ambroise soumet ces versets à une double allégorie au §3. D'abord, l'allégorie de la pièce à l'étage, celle-là même que l'habitation des personnes aisées met à la disposition des hôtes de passage, et que Jésus a envoyé Pierre et Jean réserver pour faire leurs Pâques.[71] Cette pièce à l'étage est l'image de la purification des sens qui doit s'accomplir durant les Pâques et, qui, seconde allégorie, est rendue possible par la *spiritali aqua fontis aeterni.*

---

[68] Ambroise de Milan, *Epistula extra collectionem* 13 §4 (CSEL 82,3, 224): *Sedes eius sicut sol in conspectu meo et sicut luna perfecta in aeternum manet*; la Vulgate, selon la Septante: *Et thronus eius sicut sol conspectu meo et sicut luna perfecta in aeternum.*

[69] Ambroise de Milan, *De officiis* I 26 (Testard (2002), 154): *Quid tam obscurum quam de astronomia et geometria tractare, [. . .], et profundi aeris spatia metiri, caelum quoque et mare numeris includere?*

[70] Pour l'éditeur du *De officiis* (Testard (2002), 44–9), la rédaction définitive du traité est postérieure au printemps 386; peut-être date-t-elle de la fin 388 ou en 389. Savon (1997), 246 n 32 accepte ces indications, que McLynn (1994), 272–3 semble faire siennes.

[71] Sur cette coutume proche-orientale, voir Lagrange (1930), 498–9.

L'allusion à la cérémonie du baptême couplée au IVe siècle, à la célébration de la résurrection, est manifeste[72].

Ambroise aborde le problème posé par la date pascale en rappelant dans le §4 la règle édictée dans l'Ancien Testament: les Pâques vétéro-testamentaires sont fixées à la quatorzième lune du premier mois. Il précisera par la suite, au §16, qu'il faut compter le premier mois à partir de l'équinoxe, sans jamais indiquer qu'il s'agit de celui du printemps; au §18, il apportera un complément d'information: le mois ainsi compris s'identifie à la lunaison. Le mode de composition qu'il pratique entraîne une dissémination des éléments techniques du calendrier ecclésiastique; il n'est manifestement pas un computiste dans l'âme.

La règle vétéro-testamentaire ayant été mentionnée, il convient de la transposer dans le monde néo-testamentaire, en insistant sur la révélation chrétienne, grâce à laquelle elle acquiert pour Ambroise toute sa vérité. Le §5 sert d'écrin à l'interpellation de Jésus: 'Père, l'heure est venue; glorifie ton fils afin que ton fils te glorifie' (Jn 17:1).[73] Le Père est partie prenante de la Passion. Les §6 et 7 présentent des variations sur la nécessité de choisir les échéances, avant d'en venir directement, avec le §9, au fond du problème qui, précise le §8, a été posé également 'même après les calculs effectués par les Égyptiens et les précisions apportées par l'Église d'Alexandrie, à la plupart des évêques de l'Église de Rome.'[74] Par *Aegyptiorum supputationes*, il faut sans doute comprendre une table pascale reposant sur le cycle de dix-neuf ans. Comme celle de Théophile est trop récente pour avoir déjà eu une large diffusion surtout chez les Latins non-hellénophones, il faut en poser une autre plus ancienne; à moins qu'il ne s'agisse que d'une clause de style: Ambroise donne à la table de Théophile une réputation qu'elle n'a pas encore acquise. Quant à la *Alexandrinae ecclesiae definitio*, elle renvoie vraisemblablement à la lettre festale annuelle du métropolitain d'Alexandrie. En dépit de ces élaborations orientales, la

---

[72] Ambroise de Milan, *De sacramentis* I 4, §12 (Botte (1980), 66): *qui per hunc fontem transit, hoc est, a terrenis ad caelestia – hic est enim transitus, ideo pascha, hoc est, transitus eius, transitus a peccato ad uitam, a culpa ad gratiam, ab inquinamento ad sanctificationem – qui per hunc fontem transit, non moritur sed resurgit.*

[73] Ambroise de Milan, *Epistula extra collectionem* 13 §5 (CSEL 82,3, 224): *Pater, venit hora, clarifica filium tuum ut filius clarificet te*; même texte dans la Vulgate.

[74] Ambroise de Milan, *Epistula extra collectionem* 13 §8 (CSEL 82,3, 225): *Unde necesse fuit, quia etiam post Aegyptiorum supputationes et Alexandrinae ecclesiae definitionem episcopi quoque Romanae ecclesiae per litteras plerique meam adhuc expectant sententiam, quid existimen scribere de die paschae.*

plupart des évêques de l'Église de Rome sont également dans l'incertitude. L'adverbe *quoque* de la dernière citation ajoute aux évêques d'Émilie, destinataires au premier chef de la lettre, certains autres de l'Italie suburbicaire, comme le laisse entendre le passage *episcopi quoque Romanae ecclesiae per litteras plerique meam adhuc expectant sententiam*. Ambroise précise ensuite que sa réponse vise à dépasser le cas des Pâques de 387; il veut apporter une norme pérenne.

S'il n'est pas un computiste confirmé, l'évêque de Milan est un organisateur aguerri, comme le montre la dernière remarque et comme le §9 le confirme. La règle de la quatorzième lune du premier mois est rappelée à l'aide cette fois de nombreuses références explicites (Ex 12:2; Lev 23:5 et Num 9:3 et 28:16) ou implicite (Dt 16:1). Le paragraphe suivant opère une transition de l'Ancien Testament, celui de la loi, au Nouveau, celui de la grâce et de la vérité (Jn 1:17 et Mt 5:17, effectivement cités mais, comme à l'ordinaire, sans donner les références). Il précise, en quantième lunaire et en quantième hebdomadaire, la chronologie qui conduit des Pâques célébrées par Jésus (quatorzième lune, un jeudi) à la crucifixion (quinzième lune, un vendredi), puis au sabbat (seizième lune) et, enfin, à la résurrection (dix-septième lune, un dimanche).

Avec le §11, un nouveau sujet de préoccupation apparaît, le jeûne. Il est repris immédiatement au §12 et plus loin au §20. Le jeûne permet de bien différencier les deux moments forts de la célébration pascale: il intervient durant la commémoration de la Passion, alors qu'au temps de la résurrection, c'est la fête, qu'un repas accompagne. Ambroise s'emporte contre les Manichéens: ils jeûnent le dimanche,[75] jour de joie et d'allégresse.[76] Mais même pour les fidèles catholiques – Ambroise s'en tient aux verbes utilisés à la première personne du pluriel – la situation n'est pas toujours facile. Lorsque la première quatorzième lune du printemps, jour de la commémoration de la crucifixion, tombe un

---

[75] Plusieurs rescrits impériaux antérieurs à 386 condamnent les Manichéens; aucun ne parle de la spécificité de leurs jeûnes: *Codex Theodosianus* XVI 5, §3 (2 mai 372), §7 (8 mai 381) et §9 (31 mai 382) (Magnou-Nortier (2002), 194–7, 202–7 et 208–11). Il reste que le jeûne est un des rites fondamentaux des fidèles de Mani: Puech (1979), 275–87: 278, où, à l'appui de ses analyses, il cite notamment la présente lettre d'Ambroise, dont il récuse cependant l'authenticité. L'évêque de Milan a pu lui-même observer cette coutume, à moins qu'il ne reprenne une remarque de Théophile (Krusch (1880), 224). Les Manichéens ne jeûnent pas spécialement le dimanche pascal; ils jeûnent tous les dimanches de l'année. Ils jeûnent aussi lors de la pleine lune équinoxiale, en référence à la Passion du Christ.

[76] Ambroise de Milan, *Epistula extra collectionem* 13 §20 (CSEL 82,3, 233): *Resurrectionis die exsultatio refectionis est atque laetitiae*.

dimanche, ce jour de fête dédié tout au long de l'année à la résurrection, les exigences festives de la semaine entrent en conflit avec celles pénitentielles de la lunaison. Le désaccord est réglé en donnant la priorité au rythme hebdomadaire au détriment de celui de la lunaison. Pour respecter la fête dominicale annoncée par les Psaumes: 'C'est le jour que le Seigneur a fait, nous nous réjouissons en Lui et nous sommes pleins d'allégresse' (Ps 117:24),[77] il faut dans ce cas précis repousser la célébration de la résurrection au dimanche suivant, jour de la vingt et unième lune; et de là remonter mentalement dans la semaine pour déterminer le quantième hebdomadaire de la crucifixion.

Ambroise accorde une large importance à l'argumentation sur le choix des jours de jeûne en période pascale. Il est en cela l'héritier d'une tradition, cette fois historiquement établie.[78] Ce qui pourrait s'apparenter à un déséquilibre entre la part dévolue dans sa lettre au jeûne et celle dévolue aux cycles soli-lunaires, dénote une tournure d'esprit ancrée dans la chrétienté des premiers siècles. Cette valorisation du jeûne a valeur de symptôme. Ambroise a tendance à ramener le comput à des mesures de discipline ecclésiastique. En dépit des allusions que les paragraphes 1 et 16 font au cycle de dix-neuf ans, l'astronomie n'est pour l'essentiel présente dans sa lettre qu'à partir des règles disciplinaires édictées dans l'Ancien Testament. Tout se passe comme si le soleil et la lune n'avaient pas de consistance propre. Ils sont détachés de leur réalité de corps célestes et envisagés à partir du statut que des textes dogmatiquement privilégiés leur accordent. On peut se demander si Ambroise est totalement convaincu de la permanence de l'objet et s'il ne souscrit pas à l'universalité de la foi plus aisément qu'à celle des objets célestes.[79]

---

[77] Ambroise de Milan, *Epistula extra collectionem* 13 §11 (CSEL 82,3, 227): *Hic est dies quem fecit dominus, exultemus et laetemur in eo*; Vulgate: *Haec est dies quam fecit dominus, exultemus et laetemur in ea.*

[78] Un témoignage d'Irénée de Lyon rapporté par Eusèbe de Césarée, *Historia ecclesiastica* V 24 (Traduction de Gustave Bardy revue par Neyrand (2003), 302), rappelle l'importance du jeûne dans la célébration des Pâques: 'Ce n'est pas seulement sur le jour que portent les divergences (pascales entre les Églises). Elles portent également sur la façon de jeûner.'

[79] Ambroise est le fils de son époque. Brown (1998), 34, illustre 'la loyauté viscérale qu'un notable municipal manifeste à l'égard de sa ville natale' en citant un recueil de plaisanteries du IIIe siècle: 'un petit garçon riche demande à son père s'il est vrai que la lune au-dessus de sa ville brille plus qu'au-dessus d'aucune autre!' À suivre le texte à la lettre, l'enfant interroge son père sur la validité d'une conception dont il a entendu parler, non sur la sienne. L'anecdote n'en appartient pas moins à un recueil de plaisanteries: la conception de la lune dont elle témoigne suscite des réserves chez ceux qui malgré tout y adhèrent.

Le §13 est consacré à l'un des fondements bibliques de la principale notion pascale déjà évoquée, celle du *triduum sacrum*. Elle repose sur la typologie de la destruction du Temple et sur sa reconstruction en trois jours: 'Détruisez ce temple et je le rétablirai en trois jours' (Jn 2:19).[80] À travers cette figure, Jésus annonce que sa mort et sa résurrection forment un tout indissociable, qu'il meurt et ressuscite sans changer d'identité et que la joie du dimanche doit suivre le deuil du vendredi. La leçon est double. Le chrétien est appelé à ressusciter dans la totalité de sa personne, corps et âme confondus. Quant au dimanche qui suit la quatorzième lune de la lunaison printanière, il est l'élément fixe de la liturgie; c'est bien sur lui que le vendredi de la crucifixion doit en quelque sorte se caler.

Pour différencier une dernière fois le jour de la commémoration de la Passion de celui de la fête de la résurrection,[81] Ambroise sollicite, aux paragraphes 19 et 20, et beaucoup plus longuement qu'il ne l'a fait dans le reste de la lettre pour d'autres extraits bibliques, les versets d'un même chapitre vétéro-testamentaire. Ces passages si largement cités (§19: Ex 12:5–8 et 11–14; §20: Ex 12:29, 31, 33, 34 et 39) font masse. Ils racontent une histoire, celle du départ d'Israël d'Égypte. Elle est envisagée d'un point de vue chrétien, comme le signale sans ambiguïté la référence à 1 Jn 2:18 placée au commencement du §20: 'Mes enfant, c'est l'heure ultime'.[82] Le sens de la destinée chrétienne se révèle dans cet épisode fondateur parce que, revisité, il métamorphose Pâque en Pâques.

Les deux derniers paragraphes (21 et 22) sont riches d'éléments concrets empruntés à la Bible (la mer et la nuée: 1 Cor 10:2-4; le pain et la boisson: Ex 16 et 17:1–7; la porte: Col 4:6; les lèvres: Ps 140:3; le parfum: 2 Cor 2:15; l'azyme: 1 Cor 5:8). La magie du verbe ambrosien hausse cet ensemble de réalités au rang d'images, où se laisse entrevoir la splendeur de la vie chrétienne. L'évêque de Milan a rédigé sa lettre en n'ayant qu'un objectif véritable à l'esprit: non pas tant préciser en

---

[80] Ambroise de Milan, *Epistula extra collectionem* 13 § 13 (CSEL 82,3, 235): *Solvite hoc templum et in triduo resuscitabo* (Vulgate: *excitabo*) *illud*; sur l'interprétation pascale de ce passage, voir Gy (1955), 9; l'autre typologie est celle de Jonas, dont les trois jours passés dans le ventre de la baleine annoncent le *triduum sacrum*. Absente dans la présente lettre, elle appartient à la liturgie milanaise, et Ambroise en a fait l'exégèse: Duval (1973), i 41–2 et 230–6.

[81] Ambroise de Milan, *Epistula extra collectionem* 13 §19 (CSEL 82,3, 235): *Alium diem passionis, alium resurrectionis celebrandum veteris quoque testamenti series ostendit*.

[82] Ambroise de Milan, *Epistula extra collectionem* 13 §20 (CSEL 82,3, 232): *Pueri* (Vulgate: *Filioli*), *novissima hora est*.

computiste la date des prochaines Pâques, mais découvrir en pasteur, et faire découvrir à ses correspondants, les fidèles rassemblés dans les lieux de culte, et tous, au même moment, 'chantant, grâce à la sainte doctrine, la gloire du Père et du Fils et la majesté indivisible de l'Esprit'.[83]

L'apport de la lettre à l'histoire du comput

La formule conclusive témoigne en faveur du Dieu Un et Trine et se constitue en doxologie; plus encore en une doxologie qui appelle le chant.[84] Elle couronne d'autant mieux la lettre qu'elle survient après qu'Ambroise a calmé l'inquiétude des évêques d'Émilie en répondant avec précision à leur demande ponctuelle. Il ne communique cependant l'information sollicitée que dans le §21, l'avant-dernier de la lettre, après l'avoir seulement évoquée au §17. En la plaçant si tard dans sa réponse, il montre que le choix d'une date pascale ne relève pas de la seule logique du calendrier, ni n'obéit aux seules contraintes d'un cycle soli-lunaire. Elle engage une conception du christianisme, qui doit être brossée avant que l'échéance soit arrêtée et diffusée. Mais si Ambroise pose tout au long de sa lettre les fondements de sa conception de la vraie foi, il mentionne aussi épisodiquement les quatre dernières années où la date des Pâques a été, selon lui, un objet de débats: au §14, en 373 et en 377; au §17, en 379, qui est présenté comme un modèle possible de 387; et au §21, qui est l'avant-dernier de la lettre, en 360. Ce sont autant de cas qui orientent son analyse et favorisent sa décision. Les trois premiers sont abordés en suivant une chronologie croissante, que le quatrième interrompt en opérant un retour en arrière; cette remontée dans le temps creuse un écart de dix-neuf ans par rapport au cas immédiatement précédent et de treize par rapport au cas avancé en premier. Est-ce qu'Ambroise obéit à un mobile particulier en ne respectant pas intégralement le parallélisme entre l'ordre d'exposition qu'il s'est fixé et l'ordre temporel des cas qui illustrent ses propos? La

---

[83] Ambroise de Milan, *Epistula extra collectionem* 13 §22 (CSEL 82,3, 234): *pia doctrina gloriam patris et filii et spiritus maiestatem individuam concinentes.*

[84] Je n'ai pas retrouvé dans les quatorze hymnes éditées dans *Ambroise de Milan, Hymnes*, éd. et tr. sous la dir. de Fontaine (1992), un passage qui puisse se rapprocher de la présente doxologie. Au moment de la rédaction de sa lettre, Ambroise n'a commencé que depuis quelques mois à s'illustrer en hymnodie; ses premiers essais datent du conflit des basiliques, pendant la première partie de l'année 386 (Fontaine (1992), 12).

réponse se laisse deviner lorsque sont connues les circonstances qui ont entouré le plus ancien d'entre eux, autrement dit le dernier de la série.[85]

En 360, Auxence, un homéen, occupe depuis cinq ans le siège épiscopal de Milan; il y a accédé grâce au soutien de l'empereur Constance II, de même obédience que lui, qui règne en Orient de 337 à 352, puis, à partir de 353 et jusqu'à sa mort en 361, sur tout l'empire.[86] En procédant à cette nomination, Constance a déposé et exilé en Arménie l'évêque

---

[85] Je remercie Leofranc Holford-Strevens d'avoir attiré mon attention sur l'environnement politico-religieux qui a entouré la détermination de la date de Pâques en 360. Je lui dédie cette quatrième partie, que je n'aurais probablement pas écrite sous cette forme sans son intervention amicale et documentée.

[86] Il est pour l'historien difficile d'aborder un sujet qui interfère avec d'intenses débats doctrinaux, alors qu'en dehors du cercle restreint des spécialistes, ils n'ont pas laissé de traces très nettes dans la mémoire collective. Telle est l'histoire du comput dans les années 330–390: elle est étroitement mêlée aux controverses sur les rapports entre Dieu le Père et son Fils, dont on admettra qu'ils sont fort complexes à concevoir dans un contexte monothéiste où le Fils est d'une totale humanité et, à l'instar du Père, pure transcendance. Pendant les deux tiers du IVe siècle, ces débats ont agité jusqu'à l'ébranler l'encadrement d'une chrétienté en pleine expansion. Ils ont bien souvent viré en affrontements violents: menaces, coups fourrés, anathèmes, exils, excommunications, et même échauffourées avec mort d'hommes y ont été monnaie courante. L'élément déclencheur a été l'arianisme.

Depuis sa première formulation par Arius (v. 260–336), l'arianisme, condamné au concile œcuménique de Nicée (325), s'est différencié. Chacune des thèses en présence s'est nourrie de la contestation qu'elle suscitait. Auxence et Constance II appartiennent au camp des homéens, le plus modéré parmi les ariens, sous réserve que certains de ses membres sont d'un tempérament irascible et n'ont pas manqué de le faire savoir sans aucune retenue (les camps adverses n'ont rien eu à lui envier sur ce plan). Les homéens pensent que le Fils est semblable (*homoios*) au Père en toutes choses comme les Saintes Écritures le disent et l'enseignent. Ils se distinguent, à la fois, des ariens radicaux, les anoméens, pour qui le Fils est différent (*anomoios*) du Père, et des nicéens, pour qui le Fils est de même substance (*homoousios*) que le Père; encore faut-il introduire chez ces derniers deux sous-catégories: les vieux-nicéens, attachés à la lettre de Nicée, et les néo-nicéens, sensibles aux élucidations doctrinales ultérieures au concile et notamment à l'apport des homéousiens. Ceux-ci, tard venus sur la scène (concile de Sirmium de 358), sont taxés de semi-ariens par les uns et d'antiariens par les autres; pour eux, le Fils est semblable au Père selon l'essence (*homoiousios*). Les troubles durent jusqu'au concile œcuménique de Constantinople (381), qui scelle l'accord des vieux- et des néo-nicéens et leur assure la victoire sur les homéens; voir l'*Expositio fidei* dans Alberigo (1994), 72–3. Mais sous la pression des polémiques, l'arsenal conceptuel de la métaphysique grecque (*phusis*, *prosôpon*, *hupostasis*) a été mis à profit et sert maintenant à définir une théologie trinitaire qui trouve son point d'équilibre au concile de Chalcédoine (451; voir l'*Expositio fidei* dans Alberigo (1994), 192–9: 198–9). La définition d'une orthodoxie s'est cependant accompagnée d'un schisme, toujours en cours; pour les Églises orientales, cette scission enlève au concile de Chalcédoine la qualité œcuménique que les Églises catholiques, orthodoxes, et protestantes lui reconnaissent.

nicéen en place, qui meurt peu après. Il a décidé ce changement au terme d'un concile convoqué précisément à Milan par le pape Libère. Le pontife est un nicéen, 'généreux et naïf',[87] dépourvu de cette ténacité qui forge les grandes figures lorsque ceux qui en bénéficient sont plongés dans l'adversité. De semblables substitutions d'évêques sont courantes durant ces décennies. Elles ne sont qu'une des applications des résolutions prises au cours de conciles successifs (Sirmium en 351, 357 et 358, Ancyre en 358, Séleucie et Rimini en 359 et Constantinople en 360 pour ne citer que les principaux), où la ligne homéenne s'élabore et finalement triomphe sous la conduite d'un empereur persuadé d'être mandaté par Dieu pour assurer l'unité doctrinale de l'Église et la paix dans l'empire.[88]

Or en 360, le siège épiscopal d'Alexandrie est occupé, depuis trois ans, par un homéen choisi, là encore, par Constance II. Le nouveau titulaire, Georges, a évincé Athanase, un nicéen ombrageux, taillé pour la lutte, fort de l'expérience d'un épiscopat mouvementé de plus de trente ans et qui, en homme averti, n'a pas attendu d'être mis devant le fait accompli. Il a pris ses précautions: avant l'annonce officielle de sa destitution, il s'est réfugié dans la clandestinité. Il vit en se cachant

---

Après la mort de Constance II, après la courte rupture de Julien (361–363) et le bref règne de Jovien (juin 363–février 364), l'homéisme militant de l'empereur est poursuivi en Orient par Valens (364–378) jusqu'à ce que sa fin dramatique à la bataille d'Andrinople jette la suspicion sur sa foi, tandis qu'en Gaules, Valentinien Ier (364–375) pratique la neutralité religieuse. Valentinien II (375–392, préfectures d'Illyrie, d'Italie et d'Afrique), initialement homéen sous l'influence de la cour (il accède à l'empire à l'âge de quatre ans), devient nicéen au contact d'Ambroise.

Ambroise appartient aux nicéens, sur une ligne où se retrouvent les empereurs Gratien (375–383, préfecture des Gaules) et Théodose (379–395, préfecture d'Orient, puis, à partir de 392 seul empereur), sous les règnes desquels se déroule la majeure partie de son épiscopat (374–396). Le concile de Constantinople à peine achevé, le 9 juillet 381, Ambroise dut affronter les homéens d'Italie (concile d'Aquilée, septembre 381), qui ne renoncèrent à la défense de leur foi qu'avec l'installation de Théodose à Milan, en 388.

Pour l'histoire de ces débats doctrinaux, voir Simonetti (1975), auquel j'emprunte l'essentiel de ma documentation, que j'ai complétée par les biographies d'Ambroise déjà citées et par Mayeur et al. (1995), plus spécialement par les chapitres III et IV de la Deuxième partie et le chapitre I de la Troisième, dus tous les trois à Charles Pietri: 'L'épanouissement du débat théologique et ses difficultés sous Constantin: Arius et le concile de Nicée', pp. 249–88; 'De la *partitio* de l'Empire chrétien à l'unité sous Constance: la querelle arienne et le premier 'césaropapisme'', pp. 289–336; et 'Les dernières résistances du subordinatianisme et le triomphe de l'orthodoxie nicéenne', pp. 357–98. Pour l'histoire des péripéties politiques, voir plutôt Piganiol (1972), 27–302. Pour une approche synthétique et à jour, voir Maraval (1997), 313–48.

[87] Pietri (1976), 25–8 et 237–68.
[88] Sansterre (1972).

dans la ville ou dans le désert proche pour mieux résister à l'assaut homéen et pour préparer sa contre-attaque.[89] Ces péripéties politico-religieuses ont une incidence directe sur l'histoire du comput. Georges met un terme à la politique de compromis computiste scellée pour cinquante ans au concile de Serdique (343); elle a été pratiquée plus ou moins assidûment, selon les aléas de la conjoncture, par un Athanase persuadé de la vérité de son comput mais disposé à nouer des alliances et à trouver des arrangements sur des points mineurs pour maintenir son influence et propager l'intégrité de sa foi.[90] En cas de différend pascal, les deux grandes métropoles de la chrétienté cherchent un compromis entre leurs cycle respectifs. Je les rappelle: à cette époque, Rome suit un cycle de quatre-vingt quatre ans, qui fixe l'équinoxe au 25 mars et dont les termes couvrent les périodes XVI–XXII *lunae* (sous l'influence de la chronologie johannique de la Passion) et 25 mars–21 avril;[91] et Alexandrie, un cycle de dix-neuf ans, qui fixe l'équinoxe au 21 mars et dont les termes couvrent les périodes XV–XXI *lunae* (sous l'influence de la chronologie synoptique de la Passion) et 22 mars–25 avril.[92] Le nouvel évêque homéen de la capitale égyptienne n'est pas homme à ménager une Rome nicéenne. 'Totalement dépourvu de scrupule et de morale', il n'en est pas moins un 'homme éclairé'.[93] La qualité de sa bibliothèque est fameuse et il excelle dans la dialectique.[94] Calcul politique pour assurer la réputation de son siège? Ou choix rationnel en

---

[89] Martin (1996), 474–540.

[90] Ce compromis n'est pas mentionné dans les actes officiels du concile; voir Hess (1958), 14; je n'ai pu avoir accès durant ce travail à la nouvelle monture du livre de Hess (2002). Il est mentionné dans l'Index qui sert de préambule aux Lettres festales d'Athanase connu seulement dans sa version syriaque: Martin et Albert (1985), 242–3 et 249 (texte) et 289 et 295 (notes).

[91] Le calendrier festif romain traditionnel impose la limite du 21 avril. À cette date, les Romains honorent annuellement la divinité protectrice du Palatin, Palès, et célèbrent le jour anniversaire de la fondation de Rome. Le dimanche pascal devant être précédé par une période d'abstinence qui débute le jeudi précédent, il est difficile de jeûner dans une ville livrée aux festivités, d'autant que la chrétienté du IVe siècle n'exclut pas le patriotisme romain; sur ce dernier thème, voir Inglebert (1996), 683–90. D'autre part, à une date que l'on situe dans la deuxième moitié du IVe siècle, Rome adopte un équinoxe au 21 mars (voir en dernier Declercq (2000a), 59 et 69).

[92] Sur les compromis passés entre Rome et Alexandrie au temps d'Athanase, les avis divergent. Schwartz (1905), 51–2 pense que les concessions ont été réciproques. Pour Declercq (2000a), 78, c'est Rome qui cède le plus souvent, sauf quand les dates alexandrines se situent en dehors des termes pascals romains. Richard (1974), 330–1 réserve les concessions à l'acceptation par Athanase des termes 25 mars–21 avril. L'Index syriaque confirme, me semble-t-il, l'interprétation de Richard.

[93] Martin et Albert (1985), 519–20.

[94] Martin et Albert (1985), 519.

faveur du vrai? Il s'en tient strictement au cycle de dix-neuf ans tel qu'il est entendu à Alexandrie.[95] Pour lui, la vérité du cycle qu'il pratique, et qu'il cherche à imposer sans ménagement à ses partenaires, s'accorde à la vérité de la christologie qu'il défend et illustre.

L'Index syriaque qui sert de préambule aux Lettres festales d'Athanase l'assure: en 360, Alexandrie a fêté Pâques le 23 avril, XXe lune.[96] Le quantième mensuel est pour Rome irrecevable. À cette date cependant, Georges avait fui depuis plus d'un an la ville dont il était l'évêque, tant sa gestion avait révolté la population. L'agglomération avait été aussitôt reprise en main par les hommes de l'empereur mais Georges n'y était pas revenu tout de suite. Il avait participé à la rédaction du document qui allait peu après être présenté aux évêques occidentaux réunis en concile à Rimini (fin mai–octobre 359), puis aux évêques orientaux réunis en concile à Séleucie (septembre–décembre 359); tous le signèrent, non sans réticence et après avoir obtenu quelques aménagements. Georges parfait ainsi sa figure d'évêque de cour et de piètre pasteur. Athanase qualifie ironiquement de Credo daté ce document conciliaire: il porte la date de sa rédaction préconciliaire, le 22 mai 359. Quelle que soit sa validité, il consacre officiellement l'homéisme, que couronne aussitôt le concile de Constantinople du début de 360.[97] De son côté, le pape Libère, qui avait été exilé en Thrace à l'été 356 pour avoir pris fait et cause pour Athanase, était revenu sur son soutien à l'Alexandrin. Après deux ans de relégation, il avait pu regagner Rome, où il dut affronter son ancien diacre consacré entre temps évêque par des ariens à la demande de Constance. Son autorité sort évidemment amoindrie de ces épreuves,[98] tandis qu'à Milan celle d'Auxence est renforcée d'autant.

Comment comprendre, en tenant compte des conditions qui viennent d'être rappelées, les informations qu'Ambroise livre sur les Pâques de 360?:

'Si la quatorzième lune (pascale) tombe le dimanche, une semaine supplémentaire doit être ajoutée, comme ce fut le cas lors que la soixante-seizième

---

[95] Richard (1974), 331: 'Cet évêque (Georges) n'a pas laissé bon souvenir. Il faut pourtant mettre à son crédit qu'il a probablement sauvé le comput de 19 ans. Le compromis accepté par Athanase sous la pression romaine, s'il s'était maintenu, aurait pu avoir les plus fâcheux résultats.'

[96] Martin et Albert (1985), 260–1.

[97] Martin et Albert (1985), 520–7.

[98] Pietri (1976), 255–63: 261: 'Le prélat [Libère] vaincu par l'exil ne pouvait plus représenter l'autorité d'une ferme tradition.'

année de l'ère de Dioclétien [360]. En effet, le 28 pharmouthi, qui correspond au 23 avril, nous avons célébré les Pâques du Seigneur sans aucune réserve de la part des anciens.'[99]

Le quantième mensuel des Pâques de 360 de la lettre est identique à celui de l'Index syriaque. Son quantième lunaire est cependant en excédent d'une unité: si la lune est à son vingtième jour le dimanche 23 avril, comme l'Index l'indique, elle était à son treizième le dimanche précédent, non à son quatorzième comme la lettre le prétend. Est-ce une erreur d'Ambroise? Ou est-ce qu'il utilise un biais, le quantième hebdomadaire de la lune du printemps, pour faire accepter à ses interlocuteurs l'alignement milanais sur le comput alexandrin aux mains, à cette époque, d'homéens au faîte de leur puissance, et quitte à délaisser un comput romain géré par des nicéens en mauvaise posture? Le *celebravimus* résonne curieusement: en utilisant à cette occasion la première personne du pluriel, Ambroise accepte l'héritage homéen de son siège et l'intègre à sa recherche sur la datation de Pâques, alors qu'il manifeste généralement la plus implacable opposition à la formule homéenne de Rimini.[100] Le *sine ulla dubitatione maiorum* résonne tout aussi curieusement, mais pour d'autres raisons. Le cycle pascal joint au fameux Calendrier de 354 et prolongé jusqu'en 411, date les Pâques de 360 du 16 avril sans préciser l'âge de la lune; la liste des Pâques célébrées à Rome contenue dans le manuscrit Vatican, Biblioteca Apostolica, Reg. Lat. 2077 leur attribue le 9 avril, XVIe lune;[101] et la table pascale jointe à la *Supputatio romana*, le 17 avril, XVIe lune.[102] Dans les trois cas, la limite du 21 avril, propre à Rome mais inutile à Milan, n'est pas dépassée. La nouvelle capitale impériale affirme une universalité que l'ancienne ne peut honorer. De plus, les quantièmes hebdomadaire et mensuel du Calendrier et de la liste sont cohérents avec ceux de l'Index et de la lettre, même si leurs résultats s'en démarquent, tandis que le quantième lunaire de la liste est, par rapport à l'Index et à la lettre, une

---

[99] Ambroise de Milan, *Epistula extra collectionem* 13 §21 (CSEL 82,3, 234): *si quarta decima luna in dominicam inciderit adiungendam ebdomadem alteram sicut et septuagesimo sexto anno ex die imperii Diocletiani factum est; nam tunc vicesimo octavo die Farmutii mensis qui est nonum kalendas Maias dominicam paschae celebravimus sine ulla dubitatione maiorum.*

[100] Duval (1969), 57–8, 92–4, 97 et 103.

[101] MGH Auct. ant. 9, 62–4: 63 et 739–43: 741; Krusch (1880), 80 pense que le second document est un meilleur témoin, alors que Schwartz (1905), 34 préfère le premier, tout comme, implicitement, à suivre son analyse pour cette année-là, Richard (1974), 329, 330 et 338; mais voir *infra*, n 105.

[102] Krusch (1880), 239.

anomalie que je n'explique pas. Anomalie également par rapport à l'Index et à la lettre les quantièmes mensuel et lunaire de la *Supputatio romana*. À prendre littéralement les témoignages du Calendrier, de la liste et de la *Supputatio romana*, sans évoquer d'éventuelles erreurs de scribe ou de transmission, ni chercher à déterminer le plus pertinent de trois témoignages, le constat s'impose: la cacophonie règne chez les tenants du comput romain, qui sont aussi obligés de respecter une coutume que le reste de l'empire ne pratique pas. La position de faiblesse du pontife pourrait expliquer partiellement ce désordre. Ambroise, chroniqueur occasionnel des dates pascales de son siège, serait confronté à l'alternative suivante: soit, au nom de la christologie nicéenne, partager les désordres pascals du cycle romain et conserver une règle sans grand intérêt pour la chrétienté; soit fermer provisoirement les yeux sur les errances doctrinales des homéens et se rattacher fermement à un comput alexandrin qui a fait ses preuves et n'est pas entaché d'une vaine restriction. L'ancien consulaire n'est pas homme à tergiverser: la rigueur du cadre temporel prévaut pour lui sur toute autre considération, dès lors que ce cadre n'est pas en lui-même la manifestation d'une erreur doctrinale et qu'il est compatible avec l'expression de la rectitude de la foi.[103]

En 373, Auxence occupe toujours le siège épiscopal de Milan. À Rome, à cette date, Damase a succédé depuis sept ans au malheureux Libère; il reconquiert peu à peu le terrain perdu par son prédécesseur et affirme la primauté romaine, au cours d'un pontificat exceptionnellement long (366–384).[104] Athanase a retrouvé officiellement son siège depuis neuf ans mais signe sa dernière lettre festale: il meurt quelques mois après. L'Index fixe la date de Pâques de cette année-là au 31 mars, à la XXIe lune.[105] La lettre d'Ambroise s'accorde cette fois complètement avec l'Index, tant pour le quantième mensuel que pour le quantième lunaire.[106] La discordance lunaire de 360 confirme ses mobiles politiques. À Rome, la situation est si confuse que l'historien du comput a du mal à s'y retrouver. Il vaut la peine de citer *in extenso* les réflexions désenchantées de Marcel Richard:

---

[103] Je me démarque de Zelzer (1978), 194–5, pour qui le *celebravimus* se rapporte à l'ensemble des Églises italiennes.

[104] Pietri (1976), 2e partie, livres 1 et 2, 403–884.

[105] Martin et Albert (1985), 276–7.

[106] Ambroise de Milan, *Epistula extra collectionem* 13 §14 (CSEL 82,3, 228): *cum quarta decima luna esset nonum kalendas Aprilis* [24 mars], *nos celebravimus pascha pridie kalendas Aprilis* [31 mars].

'Nous n'avons pas pu décider quelle confiance il convenait d'accorder aux sources historiques qui nous assurent que Rome a fêté Pâques le 24 mars, XVe lune, en 373. Cette année-là, le cycle de 84 ans mettait la XIVe lune au samedi 23 mars, donc Pâques au 31 mars,[107] date prescrite aussi par le calendrier alexandrin. Aucune explication de cette anomalie n'a pu être trouvée. L'hypothèse d'une distraction du computiste romain, qui aurait déplacé la XIVe lune du samedi 23 au vendredi 22, comme dans l'octaétéris, est vraiment une solution désespérée et pourtant paraît la seule possible. Si tel est le cas, Rome n'avait le choix qu'entre ce dimanche 24 mars, XVIe lune, celui du 31 mars, XXIIIe lune, et, avec l'embolisme, celui du 21 avril, XIVe lune, solution déplorable. On comprendrait donc le choix du 24 mars. Or cette date insolite est le seul exemple d'une célébration de Pâques à Rome antérieure au 25 mars de 312 à 381 pendant les trente ans d'usage de l'octaétéris et les quarante ans d'usage du cycle de 84 ans que nous avons pu étudier.'[108]

Les incertitudes computistes qui planent sur le siège de Pierre en 373 ne doivent pas dissimuler que, cette année-là encore, Milan, toujours dirigée par un homéen, se démarque vraisemblablement de Rome et marche de concert avec Alexandrie, que dirige maintenant un nicéen. Les affinités christologiques n'entraînent pas le ralliement au même comput. Aucune des trois métropoles n'a fait, dans les deux domaines en cause, un choix entièrement identique à ceux de deux autres: Milan homéenne suit le cycle de dix-neuf d'Alexandrie qui est redevenue nicéenne, tandis que Rome nicéenne fait bande à part avec le cycle de quatre-vingt-quatre ans. En commençant par cet exemple, et non par celui de 360, chronologiquement premier, Ambroise atténue le reproche qui pourrait lui être adressé: aligner son comput sur des pratiques homéennes.

Quatre ans plus tard, en 377, un chassé-croisé doctrinal se produit entre Milan, où le nicéen Ambroise a succédé à l'homéen Auxence, et Alexandrie, où l'homéen Lucius, un ancien prêtre de Georges, a évincé, grâce à l'appui de l'empereur Valens, le successeur nicéen d'Athanase, Pierre.[109] Ces va-et-vient n'empêchent pas que l'accord pascal des trois métropoles se réalise cette année-là: elles se réunissent pour fêter Pâques en même temps le dimanche 16 avril,

---

[107] La date est effectivement donnée par le MS Vatican, Biblioteca Apostolica, Reg. Lat. 2077: MGH Auct. ant. 9, 741, alors que le Calendrier de 354 donne le 15 avril: MGH Auct. ant. 9, 63.

[108] Richard (1974), 329.

[109] Martin (1996), 790–9; une coquille s'est glissée p 159 n 181: la fin de l'épiscopat alexandrin de Lucius est 378, non 375.

XXIe lune.[110] Ambroise traite dans le même paragraphe les Pâques de 373 et de 377, en les soumettant à la même problématique: la XIVe lune dominicale du printemps conduit à retenir le dimanche suivant pour célébrer les Pâques. Les témoignages historiques ne permettent d'établir ce rapprochement entre les quantièmes lunaire et hebdomadaire que pour Alexandrie et Milan. Rome obéit à une autre logique, fût-elle devenue insaisissable pour l'historien. En réunissant les trois métropoles sous un même chef, Ambroise obéit à un mobile politique: il noie l'héritage homéen de son siège dans un œcuménisme dont une Rome en cours de rétablissement est le symbole encore imparfait.

Une quasi unanimité doctrinale et computiste se réalise enfin en 379: les trois sièges, tous les trois confiés à des nicéens conquérants, fêtent leurs Pâques le dimanche 21 avril, XXIe lune.[111] Au-delà de cette concordance approximative, Ambroise est soucieux d'expliquer les Pâques tardives de cette année. Elles ménagent une transition en permettant de préparer ses correspondants à admettre celles, encore plus tardives, de 387, un 25 avril difficile à admettre pour des connaisseurs de l'histoire de Rome. La suite de la citation donnée à la note 110 est explicite.[112] Ambroise évite de parler des Parilia, qui est la croix du comput romain (voir la note 91). En 379, le dimanche pascal tombe en même temps qu'elles – ce qui est regrettable sans être totalement inconvenant: la joie de la résurrection se mêle aux festivités de la naissance de Rome.

---

[110] Ambroise de Milan, *Epistula extra collectionem* 13 §14 (CSEL 82,3, 228–9): *Rursus nonagesimo et tertio anno a die imperii Diocletiani* [377] *cum incidisset quarta decima luna in quartum decimum diem Farmutii mensis quae est quintum idus Aprilis* [9 avril] *quae erat dominica dies, celebrata est paschae dominica Farmutii vicesimo et primo die qui fuit secundum nos sextum decimum kalendas Maias* [16 avril]. Voir le Calendrier de 354 et le MS Vatican, Biblioteca Apostolica, Reg. Lat. 2077 in MGH Auct. ant. 9, 63 et 74] et la *Supputatio Romana* in Krusch (1880), 239.

[111] Ambroise de Milan, *Epistula extra collectionem* 13 §17 (CSEL 82,3, 230): *Sed cum ante sexennium* [379] *celebraverimus paschae dominicam undecimun kalendas Maias* [21 avril]. Calendrier de 354 et MS Vatican, Biblioteca Apostolica, Reg. Lat. 2077 in MGH Auct. ant. 9, 64 et 74]; Richard (1974), 329 et 338. Un bémol toutefois avec la *Supputatio Romana* in Krusch (1880), 239, qui donne le 23 avril, XXIe lune.

[112] Ambroise de Milan, *Epistula extra collectionem* 13 §17 (CSEL 82,3, 230–1): *hoc est tricesimo die mensis secundum nostram scilicet calculationem* [il s'agit d'un mois lunaire, dont, comme l'a rappelé le § 16, l'origine est l'équinoxe de printemps c'est-à-dire, précise-t-il, le 21 mars], *moveri non debemus si et proxime* [en 387] *tricesimo die Farmutii mensis* [21 avril] *celebraturi sumus paschae dominicam. Verum si quis mensem secundum id existimat quia post triduum completi mensis qui compleri videtur undecimum kalendas Maias* [25 avril] *paschae dominica erit, illud considerent quia quarta decima luna quae quaritur quartum decimum kalendas Maias* [18 avril] *erit hoc est intra praescriptum mensis numerum; lex autem diem passionis exigit intra primum novorum debere celebrari.*

En 387, l'ordre nicéen règne dans les trois métropoles avec Sirice à Rome, Théophile à Alexandrie et Ambroise à Milan. La date du dimanche de Pâques n'en est pas moins diverse. Marcel Richard a bien résumé la situation:

> 'En 387, elle [Rome] s'est retrouvée devant [. . .] le choix entre Pâques au 28 mars, XXIIIe lune ou au 18 avril, XIVe lune. Cette fois-ci la difficulté a été aggravée par la décision d'Alexandrie de célébrer Pâques le dimanche 25 avril, XXIe lune, solution tout à fait raisonnable, mais qui a fait quelque scandale.'[113]

Scandale à Rome du moins: les Parilia étant fixées, comme tous les ans, au 21 avril, elles interfèrent cette année-là avec le Carême, puisque le dimanche pascal tombe le 25 avril. Quant aux deux dates romaines, elles ne sont pas satisfaisantes, selon les critères même de Rome: la première va au-delà du terme pascal (mais Ambroise n'a pas abordé cet aspect); la seconde est irrecevable pour Ambroise lui-même (voir la citation donnée à la note 99). Le comput romain débouche sur une situation bloquée, que le comput alexandrin permet d'éviter. Entre une fidélité latine qui se heurte à des impossibilités et une ouverture au monde grec qui évite les apories, le choix d'Ambroise est fait. Le cycle de dix-neuf ans est le sésame qui fait accéder le computiste à la maîtrise de sa discipline.

Ambroise s'est attaché aux Pâques de 360, 373, 377 et 379 par attention au futur; son regard rétrospectif doit lui permettre d'établir quelques règles pour fixer la date des Pâques de 387 et, au-delà, pour déterminer à coup sûr la date de toutes les Pâques à venir. S'appliquant à sa tâche de pasteur, il pressent, plus qu'il ne démontre, les ressources du cycle alexandrin; il le légitime en prétendant qu'il est issu des travaux du concile de Nicée. Sa formation initiale est celle d'un agent de la haute administration romaine. Il voit spontanément dans les règles qui organisent le corps social en général et l'Église en particulier l'expression d'une assemblée de sages, où l'apport de l'adversaire, homéen ou autre, est passé par pertes et profits. Il ne pense pas que ces règles aient une valeur en soi et qu'elles adaptent éventuellement aux réalités humaines des lois physiques: une pareille problématique se place en dehors de son champ de vision; elle lui est absolument étrangère. Sa socialisation des règles de conduite s'accompagne d'une sorte de 'déréalisation' des lois physiques. Il en fait la manifestation d'une réunion d'hommes responsables, que rehausse leur statut ecclésial. Encore quelques temps, et Ambroise sera bientôt débordé sur sa

---

[113] Richard (1974), 332–3, qui renvoie à l'exposé détaillé de Floëri et Nautin (1957), 18–42; voir aussi Declercq (2000a), 78–9.

droite: une légende établit que le cycle de dix-neuf ans a été révélé par un ange à la grande figure du cénobitisme du IVe siècle, Pachôme.[114] Ambroise avait éliminé de l'histoire du cycle le rôle des homéens (ce qui est de bonne guerre) et des astronomes (ce qui est plus difficile à concevoir), au profit du concile de Nicée. Le Pachôme hagiographique est plus radical: il en élimine les hommes, au profit d'un ange. Une pareille élaboration est connotée. Elle s'enracine dans le milieu égyptien, à une époque, le Ve siècle, où l'Église locale abandonne peu à peu ses ambitions dans la direction de l'Église universelle et se replie progressivement sur ses traditions en cultivant la langue vernaculaire. Elle est d'autant plus facile à concevoir qu'elle se produit dans le milieu monastique égyptien, où les visions occupent une place éminente aux moments décisifs des existences – les moines étant des 'théodidactes' selon l'enseignement de Jean (Jn 6:45) repris des prophètes (Is 54:13 et Jer 31:22–24). Enfin, les évêques d'Alexandrie, les promoteurs véritables du cycle, ont entretenu d'étroits rapports avec le mouvement monastique: il est aisé d'attribuer à celui-ci ce qui revient à ceux-là.

L'hagiographie pachômienne force un trait déjà présent chez Ambroise, l'accentuation extrême de la signification des choses et des événements, au détriment d'une approche raisonnée des données physiques et des faits historiques. Est-ce que cette prééminence de la signification est un trait proprement pachômien et ambrosien, monastique et épiscopal? L'échéance pascale dont traite l'*epistula extra collectionem 13* appartient à l'année où Augustin a été baptisé à Milan même, par Ambroise en personne. J'ai eu beau chercher: aucune des principales biographies consacrées à l'évêque d'Hippone ni les principales études consacrées aux épisodes majeurs de sa vie, celles de Pierre Courcelle, d'Henri-Irénée Marrou, d'Aimé Solignac, de Peter Brown, d'Agostoni Trapé, de Serge Lancel, ne mentionne les problèmes computistes que les responsables ecclésiastiques ont eu à affronter cette année-là. À l'instar d'Ambroise, des historiens insistent sur les enjeux existentiels et doctrinaux de ce baptême; et, encore plus nettement que lui, ils délaissent la technique computiste, au point de l'oublier totalement. Le cycle de dix-neuf est bien entré avec Ambroise dans le monde latin nicéen, mais par la petite porte, par un passage on ne peut plus étroit que peu d'historiens ont cherché à emprunter et, au besoin, à élargir. Leur réserve n'enlève rien à cette concomitance: le baptême d'Augustin a correspondu à la promotion latine du cycle de dix-neuf conduite par un nicéen de haut rang, plus soucieux d'ordre – un ordre sacralisé – que de vérité.

---

[114] Jones (1943a), repris dans Jones (1994) VII. Sur le personnage: Rousseau (1999).

ANNEXE: SCHÉMA DE LA LETTRE D'AMBROISE DE MILAN AUX ÉVÊQUES D'ÉMILIE (*Epist. extra coll.* 13 (23 Maur.))

Thèmes §	Images	Choix d'échéances			Pâque(s)			Jeûnes
		Païens	chrétiens		chrétiennes		juive	
			vétérotestamentaires	néotestamentaires	vétérotestamentaires	néotestamentaires		
1					Le concile de Nicée et l'ennéadékaétéride			
2					Pâques de Jésus, Lk 27:7-12			
3	La pièce à l'étage, Lk 22:7-22 La source [Jn 4:14b]	Critique dans Gal 4:10-11						
4	Le soleil et la lune, Ps 88:38	Lunaison, Calendes, [...], Jours égyptiens	Ps 117:24 Ps 88:38		[Ex 12:6; Num 9:3] : le premier mois à la quatorzième lune			
5						Jn 17:1; Lk 13:32: rôle de Jésus		
6	La tourterelle & l'hirondelle, Jer 8:7 Le bœuf & l'âne, Is 1:3		Ps 118:126 Qo 3:1; Jer 8:7 Is 1:3					

7	Is 49:8				
8		2 Cor 6:2b	Interventions d'Alexandrie, de Rome et d'Ambroise		
9*			[Dt 16:1]; Ex 12:2; Lev 23:5 et Num 9:3 et 28:16: le premier mois à la quatorzième lune		
10			Jn 1:17; [Mt 5:17] : accomplissement de la loi dans la grâce et la vérité. Le triduum pascal		
11*			[Ex 12:8] Jour de la Passion, affliction	Ps 117:24: Jour de la résurrection, fête	Contre le jeûne dominical des Manichéens. Jeûner le jour de la Passion
12*				Si dim. 14e lune, prendre dim. suivant pour fêter la résurrection	La Passion, jour de jeûne

ANNEXE: SCHÉMA DE LA LETTRE D'AMBROISE DE MILAN AUX ÉVÊQUES D'ÉMILIE (*Epist. extra coll.* 13 (23 Maur.))

Thèmes	Images	Choix d'échéances				Pâque(s)			Jeûnes
		Païens	chrétiens			chrétiennes		juive	
			vétérotes-tamentaires	néotestamen-taires		vétérotes-tamentaires	néotestamen-taires		
§									
13	Le Temple, Jn 2:19						Jn 2:9: Triduum pascal. Dim. 18 avril 387, 14ᵉ lune: prendre dim. suivant		
14							Les précédents de 373 et de 377		
15						Dt 16:1: le mois est lunaire et le 1ᵉʳ mois commence à l'équinoxe (de printemps)			Pendant le 12ᵉ mois
16			*Consuetudino peritorum*: l'équinoxe au 21 mars; le 1ᵉʳ mois du 21 mars au 21 avril			Le concile de Nicée et l'ennéadékaétéride			
17							Pâque 380: 21 avril		
18*				1 Cor 5:7; Lev 23:5 et Num 28:16: mois et lunaison					

19			Ex 12:5–8 et 11–14: Passion et résurrection; agneau immolé et fuite d'Egypte	Ex 12:11, 29, 31, 33 et 39: jeûner le jour de la Passion
20			1 Jn 2:18: l'heure ultime	
21		La mer & la nuée, 1 Cor 10:2–4 Le pain & la boisson, Ex 16 et 17:1–7	Baptême: 1 Cor 10:2–4 Le précédent de 360. Pâque 387	
22		Porte, Col 4:6 & lèvres, Ps 140:3. Parfum, 2 Cor 2:15; Azyme, 1 Cor 5:8	[Col 4:3], Ps 140:3 et 1 Cor 5:8	

La Vulgate traduit par *vetus testamentum* l'expression *hè palia diathèkè*, et par *novus testamentum* l'expression *hè kainè diathèkè* (2 Cor 3:6–17; Lk 22:20; voir Paul (2000), 689–94). Ambroise utilise le plus souvent 'l'ancienne loi' et 'la nouvelle loi'; la lettre *Extra coll.* 13, 9 et 21 parle cependant de *vetus testamentum*.

Les références scripturaires données en crochets [ ] correspondent à des allusions, non à des citations.

L'astérisque (*) placé en exposant après le numéro de certains paragraphes signale des emprunts à la table pascale de Théophile d'Alexandrie (lettre à Théodose et prologue) relevés par Michaela Zelzer dans son édition de la lettre d'Ambroise.

IMMO WARNTJES

THE *ARGUMENTA* OF DIONYSIUS EXIGUUS AND THEIR EARLY RECENSIONS

Abstract

Dionysius Exiguus composed the earliest known computistical formulary written in Latin in 525. However, this formulary has not survived in its original form. The editors of Dionysius' computistical writings, Wilhelm Jan and Bruno Krusch, published a corpus of 16 *argumenta* from a single manuscript, namely MS Oxford, Bodleian Library, Digby 63 under Dionysius' name, since this was the only manuscript known to them that preserved the original 525 dating. Some of these 16 *argumenta*, however, contain dating clauses as late as 675, which immediately cast doubt on their ascription to Dionysius. In fact, the 16 *argumenta* edited by Jan and Krusch should more precisely be defined as a computistical formulary of 675, to be termed the *Computus Digbaeanus* of 675, which includes the original Dionysiac *argumenta*. This article, then, reconstructs the original Dionysiac corpus on the basis of new manuscript evidence. Moreover, the different stages of interpolations and additions that eventually led to the composition of the *Computus Digbaeanus* are analyzed, and with this the development of computistical formularies written in Latin in the 150 years from 525 to 675.

Keywords

Dionysius Exiguus, Cassiodorus, Maximus Confessor, Bede, Hrabanus Maurus, history of computus, computistical formularies, computistical *argumenta, Computus paschalis* of 562, *Computus Digbaeanus* of 675, *Computus Cottonianus* of 688/9, *Computus Rhenanus* of 775, *Fragmentum Nanciacense*, Sirmond manuscript, Irish computistics, Anglo-Saxon computistics, Frankish computistics.

Introduction

In the early centuries of Christianity, the West and the East followed different Easter cycles, which led to conflicting dates for Easter Sunday and continuous disputes between Rome and Alexandria.[1] An attempt to reconcile the two systems had failed in the mid-fifth century, when Victorius of Aquitaine was ordered by the papal curia to establish an Easter table that would solve the differences.[2] Victorius' system was finally approved in Rome, and it became very popular in the West, especially the Frankish Empire,[3] but his decision to list two possible dates for Easter Sunday in certain years did not solve the problem, while even his unambiguous dates did not always correspond to the Alexandrian ones.[4] Then, in 526, Easter Sunday was supposed to fall on *luna* 22 according to Victorius,[5] which was regarded as an unacceptable lunar age in the Alexandrian reckoning. It was presumably for that reason that the papal curia decided in the previous year, 525, to contact Dionysius Exiguus about this question.[6] By this time, Dionysius had made a name for himself as a canonist and translator from Greek into Latin,[7] and it probably was precisely his reputation as translator that persuaded

---

[1] For the early history of the Easter controversy (technical and historical) see Krusch (1880); Schwartz (1905); Mac Carthy (1901), xxxii–clxxvii; Ideler (1825–6), ii 191–275; Schmid (1907); Ginzel (1914), 210–20, 232–45; Rühl (1897), 107–126; Gentz in *PW* 18, 1647–53 (*s.v.* Ostern); Chaîne (1925), 19–61; Jones (1943), 6–77; Strobel (1977); Wallis (1999), xxxiv–l; Declercq (2000), 49–82; van de Vyver (1957); Grumel (1960); Lejbowicz (2006), 10–63.

[2] Victorius' Easter table and prefixed prologue are edited by Krusch (1938), 4–52. For the Victorian reckoning see Ideler (1825–6), ii 275–84; Schwartz (1905), 72–80; Ginzel (1914), 245–7; Rühl (1897), 126–8; Jones (1943), 61–8; Wallis (1999), l–lii; Declercq (2000), 82–95; Declercq (2002), 181–7.

[3] The most explicit source for the adoption of the Victorian system can be found for the Frankish Empire in form of the Acts of the Council of Orleans of 541 (*Concilium Aurelianense a.* 541 §1: CCSL 148A, 132; MGH Conc. 1, 87): *Placuit itaque Deo propitio, ut sanctum pascha secundum laterculum Victori ab omnibus sacerdotibus uno tempore celebretur.* For the adoption and use of the Victorian system in the Latin West see Krusch (1884); Poole (1918a); Schmid (1907), 38–107; Jones (1934), 412–20; Jones (1943), 65–6.

[4] For the differences between the two systems see especially Schwartz (1905), 73–80; Ginzel (1914), 246; Jones (1934), 411–2.

[5] Krusch (1938), 51.

[6] Cf. Krusch (1884), 107–8; Krusch (1938), 59; Ginzel (1914), 247; Jones (1934), 414; Jones (1943), 68; Borst (1998), 177; Wallis (1999), liii; Declercq (2000), 107–9; Declercq (2002), 203–5.

[7] For Dionysius' life and work see Jülicher in *PW* 5, 998–9 (*s.v.* 'Dionysius Exiguus'); Mordek in *LM* 3, 1088–92 (*s.v.* 'Dionysius Exiguus'). For Dionysius' origin being Alexandrian / Egyptian rather than Scythian see Zeller (1991), 167–8.

the papal curia to ask for his help: In the end, the task to be accomplished was to translate the Alexandrian Easter table and instructions from Greek into Latin.[8] Dionysius composed his computistical works in two stages, and it is only the first that will concern us here:[9] 1) He translated, converted, transformed, and continued the 95-year Easter table of Cyril of Alexandria for the subsequent 95-year period, from 532 to 626, providing a manual for this table in a prologue addressed to an unidentified bishop Petronius, and adding a Latin translation of the letter of Proterius and a series of *argumenta*. 2) After Dionysius had accomplished this task, the letter of Paschasinus was found in the papal library, which made the papal curia wonder about the workings of the 19-year lunar cycle. Again, they consulted Dionysius about this problem, and his reply in form of a letter of 526 marks the second stage.

Now, the focus of the present article is on the above-mentioned *argumenta* which Dionysius attached to his Easter table. Dionysius introduced them in his prologue in the following way:[10]

Nec non et argumenta Aegyptiorum sagacitate quaesita subdidimus, quibus, si forsitan ignorentur, paschales tituli possint facile repperiri; id est, quotus annus sit ab incarnatione domini et quae sit indictio, quotus etiam lunaris circulus, decennovenalis existat ceteri aeque simili supputationis compendio requirentur.

'We have also supplied *argumenta*, which we obtained through the acuteness of the Egyptians, with whose aid the data of the columns of the Easter table can easily be found, should they perhaps not be known; namely, the year which happens to be from the incarnation of the Lord, and which indiction this year happens to have, and also which year in the *cyclus lunaris* and *decemnovennalis* it happens to be; and the data of the other columns should likewise be sought by similar computational methods.'

It is very unfortunate that Dionysius mentions only four *argumenta* explicitly in this paragraph and does not supply the reader with a comprehensive list. Such a list of all *argumenta* compiled by Dionysius in 525 would have saved modern scholars much trouble,

---

[8] Cf. Neugebauer (1979), 101–5; Neugebauer (1982).

[9] Dionysius' computistical writings were edited by Jan (1718), republished in PL 67, 453–520, and by Krusch (1938), 59–86.

[10] Krusch (1938), 67. The translation is mine. A different translation of this passage can be found in Teres (1984), 181.

since the manuscript transmission of these *argumenta* is not straightforward. Only two modern editions of them exist, one published by Wilhelm Jan in 1718, the other by Bruno Krusch in 1938;[11] both scholars edited these *argumenta* on the basis of a single manuscript, namely MS Oxford, Bodleian library, Digby 63, because this was the only manuscript known to them that preserved the 525 dating. Yet, this corpus of 16 *argumenta* published by Jan and Krusch contains only nine with dating clauses for 525, whereas the remaining seven have either no or later dating clauses, the latest for 675. Still, Jan, and Krusch following him, decided to publish the entire corpus among the computistical works of Dionysius Exiguus. This corpus of 16 *argumenta*, preserved uniquely in MS Oxford, Bodleian Library, Digby 63, 72v–79r, will be referred to as the *Computus Digbaeanus* of 675 in the following.[12]

The *Computus Digbaeanus* is one of the earliest known corpora of computistical formulae and / or short texts (which I will term a computistical formulary in the following) written in Latin. The earliest is, in fact, the original corpus of the Dionysiac *argumenta*. Unfortunately, the Dionysiac corpus has not survived in its original form. The only other known formulary from the sixth century is a text commonly known as *Computus paschalis* of 562, controversially attributed to Cassidorus.[13] This formulary seems to be nothing but original Dionysiac *argumenta* with the examples being accommodated to match the chronological data of 562. Consequently, the *Computus paschalis* of 562 transmits valuable information about the extent of the original Dionysiac corpus. From the entire seventh century, the *Computus Digbaeanus* of 675 and the so-called *Computus Cottonianus* of 688/9 are the only known formularies.[14] Interestingly

---

[11] See n 9.

[12] For this text see pp. 45–53, 64 below.

[13] For this text see pp. 67–8 below.

[14] For the *Computus Cottonianus* see pp. 73–5 below. Some of the *argumenta* of the *Computus Cottonianus* with seventh-century dating clauses found their way into the *Computus Rhenanus* of 775 and from there into the tenth century *Canones lunarium decemnovennalium circulorum* (cf. pp. 75–6, especially n 111 below). Other than that, I am presently aware of only one further manuscript that contains *argumenta* with seventh-century dating clauses, namely MS München, Bayerische Staatsbibliothek, Clm 14725: a section on folio 14r–v consists of the Dionysiac *Argumenta I, II, V,* and *VIII* with explicit dating clauses for 695. These should be compared with the *Computus Cottonianus* and the *Computus Rhenanus*. The three *argumenta* of the *Fragmentum Nanciacense* with dating clauses for 625 were later incorporated into the *Computus Digbaeanus* and are discussed in that context below (cf. pp. 69–72, 89–91, 95).

enough, the *Computus Digbaeanus* includes all the *argumenta* of the original Dionysiac corpus, as far as can be established, and consequently also of the *Computus paschalis*; in fact, it preserves all known *argumenta* with dating clauses for 675 or earlier. Therefore, a study of the *Computus Digbaeanus* actually results in a study of the development of computistical formularies written in Latin over the first 150 years, i.e. from 525 to 675.

Additionally, the *Computus Digbaeanus* marks a watershed in the development of computistical formularies: The number of only 16 *argumenta* preserved in the *Computus Digbaeanus* is an impressive witness to the slow process of development in this field over the first 150 years. However, this process accelerated rapidly in the last quarter of the seventh century and then especially in the eighth and ninth centuries, a development in which the *Computus Digbaeanus* played a major part: It had opened up the Dionysiac corpus which was previously strictly defined. This allowed further formulae and texts to be added to that corpus, which were invented in considerable numbers once the Dionysiac reckoning was introduced into Ireland and Britain, and later into the Frankish realm. In this process the original Dionysiac *argumenta* lost their attribution and thus their authoritative connotation, which led to them being revised either by the addition of further explanations, or by generalisation; due to this revision they lost their original phrasing and shape, while the corpus itself disintegrated with selected *argumenta* being incorporated in larger computistical formularies or compendia.

It is the purpose of this article, then, to analyze the different stages in the development of computistical formularies before this significant change occurred, i.e. in the 150 years between 525 and 675, by means of examining the different layers of the *Computus Digbaeanus*. For this, it is at first necessary to introduce the 16 *argumenta* of the *Computus Digbaeanus* at some length as the basis of discussion, and at the same time the final stage of the development in question. The starting point of this development, i.e. the original corpus of Dionysiac *argumenta*, is, however, highly disputed. The different views on this first computistical formulary in Latin will be reviewed, before an attempt is made to reconstruct that corpus. From that basis it will be possible to trace the intermediary stages in the development from the original corpus of Dionysiac *argumenta* of 525 to the *Computus Digbaeanus* of 675. Concerning terminology, of the 16 *argumenta* in question, the ones that are likely to have been compiled by Dionysius himself are termed the

Dionysiac *argumenta*,¹⁵ the ones that were later added are called the pseudo-Dionysiac *argumenta*.

Description of the *Computus Digbaeanus* of 675

The *Computus Digbaeanus* of 675 comprises folios 72v–79r of the ninth-century MS Oxford, Bodleian Library, Digby 63; its provenance is highly disputed.¹⁶ The beginning of this computus is well defined, starting in the lower quarter of folio 72v, with the rest of the page being left blank. Moreover, a title is given: *Incipiunt argumenta de titulis pascalis egiptiorum investigata solercia ut praesentes indicent*. The end of this computus, however, can less easily be established. Jan and Krusch in their editions believed that the *Disputatio Morini*, starting on folio 79r, marked the end, arguing that this computus comprises 16 *argumenta*. But it should be noted that at least the inclusion of *Argumentum XVI* is debatable, since it is the only one of these 16 that has a separate heading (*De racione bissexti*), and may thus have been regarded as an independent treatise. The 16 *argumenta* in question have often been described, but the mathematical formulae have not yet been satisfactorily explained in print.¹⁷ A proper understanding of these *argumenta* is, however, vital for their assessment:

Argumentum I is divided into three paragraphs, of which only the first is printed in the main body of the editions;¹⁸ §1 gives a formula for

---

¹⁵ This article will not assess the question whether Dionysius is rightly credited with the composition of these *argumenta*, or whether he should only be credited with their translation, which certainly is a matter of debate. From the citation given above it seems, however, that Dionysius acted merely as translator. For this question see especially Neugebauer (1982); Pallarès (1994), 16; Declercq (2002), 201–2.

¹⁶ Cf. Dumville (1981), 168; Dumville (1983); Ó Cróinín (1982b), 407; Ó Cróinín (1983d), 257–9. For catalogue descriptions of this MS see Hunt and Watson (1999), 64–6; Lindsay (1915), 470; Krusch (1926), 51–4; Jones (1939), 127; Borst (2001), 161–2; Borst (2006), 263–4. See also Englisch (2002), 127.

¹⁷ For discussions and mathematical notations of these *argumenta* see Springsfeld (2002), 172–82 (which is sometimes faulty); Neugebauer (1982) (where *Argumenta I–VI, VIII–XI* are explained as part of the *Computus paschalis* of 562); Pedersen (1983), 53 (who only deals with *Argumenta II–V*). The Latin text of these *argumenta* with English and German translation respectively, as well as valuable mathematical commentary, is provided by Michael Deckers and Nikolaus A. Bär on the world wide web: http://hbar.phys.msu.su/gorm/chrono/paschata.htm and http://www.nabkal.de/dionys.html.

¹⁸ §1: Krusch (1938), 75 ll. 1–4; §2: Krusch (1938), 75 variant c ll. 1–3 (ending with *conputa*); §3: Krusch (1938), 75 variant c ll. 3–9 (starting with *item ad aliam*).

calculating the AD date from the indiction; the example given is AD 525, the general formula:

$15 \times 34 + 12 +$ indiction of the present year $=$ AD

It is composed of the following parameters: the fixed parameter 12 indicates the number of years from Christ's incarnation to the end of an indiction cycle (AD 12 = Indiction 15); from AD 12 to AD 522 another 34 indiction cycles have passed, hence 15×34; thus, the indiction of any given year between 523 and 537 added to the sum $12 + 15 \times 34$ equals the number of years from the incarnation to the year in question.

It is obvious that the number of full indiction cycles between AD 12 and the year in question increases by one every 15 years, and the formula has to be adjusted accordingly, namely by increasing the multiplicand in the product 15×34 by one every 15 years (e.g. 34 is replaced by 35 in 537). This provision is taken in §§2 and 3, which are omitted in Jan's edition, and relegated to the apparatus in Krusch's.[19] Note that in a year with indiction 15 either this number can be added as the indiction of the present year or the multiplicand with 15 can be increased by one, which leaves indiction 0 for the present year; the latter option is chosen in §3.

Argumentum II gives a formula for calculating the indiction from the AD date; the example given is AD 525, the general formula:

(AD+3)mod15 = indiction   (y mod z denotes the remainder after division of y by z)

The fixed parameter 3 is chosen because it is obviously necessary for this calculation that the count of years starts with the beginning of an indiction cycle; AD 1, however, has indiction 4, so another 3 are added to the AD date (BC 3 = indiction 1); the modulo15 calculation, then, eliminates all full indiction cycles, with the remainder being the indiction of the AD date in question.

Argumentum III is divided into two paragraphs.[20] §1 gives a formula for calculating the epact, i.e. the lunar age of 22 March, from the AD date; the example given is AD 525, the general formula:

(ADmod19×11)mod30 = epact

The logic behind this calculation is this: BC 1 is the first year of a *cyclus decemnovennalis* (i.e. the 19-year lunar cycle underlying the Dionysiac reckoning); consequently, the remainder of the AD date divided by 19

---

[19] For these two paragraphs see the previous note and the discussion on pp. 81–2 below.

[20] §1: Krusch (1938), 75 ll. 9–12; §2: Krusch (1938), 75 l. 13 – 76 l. 3.

(ADmod19) equals the number of years from the first year of the *cyclus decemnovennalis* in which this AD date occurs; since the first year of the *cyclus decemnovennalis* has epact 0, and since the epact increases by 11 per year in this 19-year cycle, ADmod19 is multiplied by 11; the modulo30 calculation then eliminates all intercalated lunar months of 30 days, with the remainder being the epact of the year in question.

§2 gives a formula for calculating the epact from the year in the *cyclus decemnovennalis*;[21] the examples given are the 10th and 11th year of the *cyclus decemnovennalis*, neither of which agrees with AD 525, since cyclic number 10 corresponds to 522+n×19; the formula is:

((year in the *cyclus decemnovennalis*−1)×11)mod30=epact

The logic is the same as for the formula in §1: the first year of the *cyclus decemnovennalis* has epact 0, hence the subtraction of 1; the epact increases by 11 per year in this 19-year cycle, hence the multiplicand 11; the modulo30 calculation then eliminates the intercalated lunar months of 30 days.

Argumentum IV is also divided into two paragraphs.[22] §1 gives a formula for calculating the *concurrentes*, i.e. the weekday of 24 March, from the AD date; the example given is AD 525, the formula:

(AD+[AD/4]+4)mod7=*concurrentes* ([y] denotes the greatest integer less than or equal to y)

The logic behind this formula is this: AD 1 has *concurrentes* 5; since the *concurrentes* increase by one every year, the number of years from AD 1 (AD−1) are added to the *concurrentes* of that year (5), resulting in AD−1+5=AD+4; hence the parameter 4; every fourth year, however, the *concurrentes* increases by 2 instead of 1 due to the bissextile day, hence the parameter [AD/4] (i.e. an extra day is added for every fourth year); since the *concurrentes* denote a weekday, and therefore never exceed 7, the modulo7 calculation is then applied, with the remainder being the *concurrentes* in question.

§2 intended to simplify the calculation by counting the years not from the incarnation, but from the year following the consulship of Tiberius Iunior Augustus.[23] It is argued that the fixed parameter had then to be changed from 4 to 1, thereby implying that the first year of the new count has *concurrentes* 2. Moreover, since no parameter is added to the number of years to be divided by 4, the first year of the

---

[21] For this paragraph see pp. 82–3 below.
[22] §1: Krusch (1938), 76 ll. 4–10; §2: Krusch (1938), 76 ll. 11–14.
[23] For this paragraph see pp. 83–4 below.

new count has to be the first year in a four-year bissextile cycle, i.e. a year immediately following a bissextile year. These chronological data match AD 581.

Argumentum V gives a formula for calculating the year in the *cyclus decemnovennalis* from the AD date; the example given is AD 525, the formula:

(AD+1)mod19=year in the *cyclus decemnovennalis*

The fixed parameter 1 is chosen, since it was obviously necessary for this calculation that the count of years starts with the first year of a *cyclus decemnovennalis*; AD 1, however, is the second year in such a cycle, and thus 1 is added to the AD date (BC 1=first year in the *cyclus decemnovennalis*); the modulo19 calculation, then, eliminates all full 19-year cycles, with the remainder being the year in the *cyclus decemnovennalis* in question.

Argumentum VI gives a formula for calculating the year in the *cyclus lunaris* from the AD date; the example given is AD 525, the formula:

(AD−2)mod19=year in the *cyclus lunaris*

Again, the count of years obviously has to start with the first year of a *cyclus lunaris*; hence the fixed parameter −2, since a *cyclus lunaris* does not start in AD 1, but two years later, in AD 3; the modulo19 calculation, then, eliminates all full 19-year cycles, with the remainder being the year in the *cyclus lunaris* in question.

Argumentum VII lists the seven years of the *cyclus decemnovennalis* in which the Easter full moon falls in March; in the remaining twelve, the Easter full moon falls in April.

Argumentum VIII gives a formula for establishing whether or not a year is bissextile, calculated from its AD date; the example given is AD 525, the formula:

ADmod4=y; if y=0, then AD is bissextile

The logic behind this formula obviously is that every AD date divisible by 4 is bissextile.

Argumentum IX can be divided into two paragraphs.[24] §1 gives a formula for calculating the lunar age of Easter Sunday from the epact and the Julian calendar date of Easter Sunday; the two examples given are: 1) a year with indiction 3, epact 12, Easter Sunday on 30 March, *luna* 20, and 2) a year with indiction 4, epact 23, Easter Sunday on 19

---

[24] §1: Krusch (1938), 77 ll. 5–15; §2: Krusch (1938), 77 ll. 16–19.

April, *luna* 21; these examples agree with AD 525 and 526 respectively; the general formula is:

a) Easter Sunday falling on y March: $(8+\text{epact}+y)\bmod 30 = $ lunar age of Easter Sunday
b) Easter Sunday falling on z April: $(9+\text{epact}+z)\bmod 30 = $ lunar age of Easter Sunday

These two formulae can be explained in the following way:

a) If Easter Sunday falls on y March, then the Julian calendar day difference between y March and 22 March, the date of the epact, equals the lunar day difference between the lunar age on Easter Sunday and the epact; in mathematical terms: $y - 22 = $ lunar age of Easter Sunday $-$ epact; this is equivalent to $y - 22 + \text{epact} = $ lunar age of Easter Sunday; since the lunar age of Easter Sunday ranges between 1 and 30, this is equivalent to $(-22 + \text{epact} + y)\bmod 30 = $ lunar age of Easter Sunday; since $(-22)\bmod 30 \equiv 8 \bmod 30$, the equation is equivalent to $(8 + \text{epact} + y)\bmod 30 = $ lunar age of Easter Sunday.
b) If Easter Sunday falls on z April, then the Julian calendar day difference between z April and 22 March, the date of the epact, is congruent modulo 30 to the lunar day difference between the lunar age on Easter Sunday and the epact (the modulo 30 congruence is necessary here, since the epact and the lunar age of Easter Sunday may occur in two successive lunations, with the lunation ending between them consisting of 30 lunar days); in mathematical terms: $(z+9)\bmod 30 \equiv $ (lunar age of Easter Sunday $-$ epact)$\bmod 30$; this is equivalent to $(z+9+\text{epact})\bmod 30 \equiv $ (lunar age of Easter Sunday)$\bmod 30$; since the lunar age of Easter Sunday naturally ranges between 1 and 30, this leads to the equation $(z+9+\text{epact})\bmod 30 = $ lunar age of Easter Sunday.

§2 then seems to supply a confused explanation of the parameters 8 and 9 fixed for March and April respectively.[25]

Argumentum X can also be divided into two paragraphs.[26] §1 gives a formula for calculating the weekday of any given Julian calendar date

---

[25] Cf. pp. 85–7 below.
[26] §1: Krusch (1938), 77 ll. 20–25; §2: Krusch (1938), 77 ll. 25–27.

from the *concurrentes* of that year; the example given is a year with indiction 3, *concurrentes* 2, and Easter Sunday on 30 April; these data agree with AD 525; the general formula is:

Let y be the inclusive number of days from 1 January to the Julian calendar date in question; then $(y+1+concurrentes) \bmod 7 =$ weekday of that Julian calendar date.

This equation can be explained in the following way: Let z be the weekday of 1 January in the year in question; $y-1$ denotes the exclusive number of days from 1 January to the Julian calendar date in question; then $(z+y-1) \bmod 7 =$ weekday of the Julian calendar date in question; the exclusive number of days from 1 January to the place of the *concurrentes*, 24 March, is 82, and since $82 \bmod 7 = 5$, $(z+5) \bmod 7 = concurrentes$; this is equivalent to $z = (concurrentes - 5) \bmod 7$, which is equivalent to $z = (concurrentes + 2) \bmod 7$ (since $-5 \bmod 7 \equiv 2 \bmod 7$); now, substituting z with $(concurrentes + 2) \bmod 7$ in the equation above leads to $(concurrentes + y + 1) \bmod 7 =$ weekday of the Julian calendar date in question.

§2 then explicitly states that this formula can be applied for any given Julian calendar date of a year.

Argumentum XI gives a formula for calculating the epact from the AD date; the example given is AD 675, the formula:

$(AD \bmod 19 \times 11) \bmod 30 =$ epact

The formula is exactly the same as in *Argumentum III* above.

Argumentum XII gives a formula for calculating the weekday of 1 January from the AD date; the example given is 675, the formula:

$(AD - 1 + [(AD - 1)/4]) \bmod 7 =$ weekday on 1 January

The method of this *argumentum* is basically the same as in *Argumentum IV*, with some important variations: The weekday of 1 January increases by one every year; hence the number of years from AD 1 ($AD - 1$) are added to the weekday of 1 January of AD 1, which is *feria* 7; in every bissextile year the weekday increases by 2 instead of 1; the bissextile day occurs in every AD divisible by 4; but since the bissextile day occurs after 1 January, the two day increase on that date happens only in the following year; hence the parameter $[(AD-1)/4]$, which provides for an extra day being added for the years following a bissextile year; since weekdays never exceed 7, the modulo 7 calculation is applied, which leads to the equation: $(AD - 1 + 7 + [(AD-1)/4]) \bmod 7 =$ weekday on 1 January; since $7 \bmod 7 \equiv 0 \bmod 7$, this is equivalent to $(AD - 1 + [(AD-1)/4]) \bmod 7 =$ weekday on 1 January.

Argumentum XIII gives a formula for calculating the lunar age of 1 January from the year in the *cyclus lunaris*; the two examples given are the 15th and 17th year of the *cyclus lunaris*, having *luna* 16 and 9 on 1 January respectively; these data do not agree with AD 675, but with AD 625 and 627, as will be seen further below; the general formula is:

(1+year in the *cyclus lunaris*×5+year in the *cyclus lunaris*×6) mod30 = *luna* of 1 January

In years 17 to 19 of the *cyclus lunaris* the fixed parameter has to be increased by 1 from 1 to 2.

The logic is this: The 19th and final year of the *cyclus lunaris* has *luna* 1 on 1 January, hence the parameter 1; the lunar age increases by 11 per year, here divided into the multiplicands 5 and 6; the modulo30 calculation is then applied to eliminate all intercalated lunar months of 30 days; the *saltus lunae* (i.e. the reduction of one lunar day) occurs at the end of the 16th year of the *cyclus lunaris*, which results in an increase of 12 instead of 11 lunar days between the 16th and the 17th year; for that reason the fixed parameter is increased by one for the three years following the *saltus lunae*.

Moreover, the period of time between the beginning of the first hour of a day and moonrise is given in points for both dates (i.e. 1 January of the 15th and of the 17th year of the *cyclus lunaris*); this is discussed in detail in Appendix II.

Argumentum XIV gives a formula for calculating the Julian calendar date and weekday of the Easter full moon, as well as the Julian calendar date and lunar age of Easter Sunday, from the epact and *concurrentes* of a given year; the three examples outlined are the first three years of the *cyclus decemnovennalis* (having epacts 0, 11, and 22); the *concurrentes* assigned to these years are 4, 5, and 6 respectively, with the lunar and solar data combined agreeing with AD 532 to 534. The general formula is:

a) Easter full moon in March:

36−epact=y March for Easter full moon

(y+*concurrentes*+4)mod7 = weekday of Easter full moon

b) Easter full moon in April:

(35−epact)mod30 = z April for Easter full moon

(z+*concurrentes*+7)mod7 = weekday of Easter full moon

With this information it was simple to calculate the Julian calendar date and lunar age of Easter Sunday, which occurred on the Sunday following the Easter full moon.

The two formulae can be explained in the following way:

a) If the Easter full moon occurs on y March, then the Julian calendar day difference between y March and 22 March, the date of the epact, equals the lunar day difference between the Easter full moon (*luna* 14) and the epact; in mathematical terms: $y-22=14-\text{epact}$; this is equivalent to $y=36-\text{epact}$, the first part of the formula; then, the Julian calendar difference between the Easter full moon (y March) and the place of the *concurrentes* (24 March) is congruent modulo 7 to the weekday difference between the weekday of the Easter full moon and the *concurrentes* (the modulo 7 congruence is necessary here, since the weekday of the Easter full moon and the *concurrentes* may occur in different weeks); in mathematical terms: $(y-24)\bmod 7\equiv$ (weekday of Easter full moon $-$ *concurrentes*)$\bmod 7$; since $-24\bmod 7\equiv 4\bmod 7$ this is equivalent to $(y+\textit{concurrentes}+4)\bmod 7\equiv$ (weekday of Easter full moon)$\bmod 7$; since the weekday of the Easter full moon naturally ranges between 1 and 7, this leads to the equation $(y+\textit{concurrentes}+4)\bmod 7=$ weekday of Easter full moon, the second part of the formula.

b) If the Easter full moon occurs on z April, then the Julian calendar day difference between z April and 22 March, the day of the epact, is congruent modulo 30 to the lunar day difference between the Easter full moon (*luna* 14) and the epact (the modulo 30 congruence is necessary here, since the epact and the Easter full moon may occur in two successive lunations, with the lunation ending between them consisting of 30 lunar days); in mathematical terms: $(z+9)\bmod 30\equiv(14-\text{epact})\bmod 30$; this is equivalent to $z\bmod 30\equiv(5-\text{epact})\bmod 30$; since z naturally does not exceed 30 (it ranges between 1 and 18) and $5\bmod 30\equiv 35\bmod 30$, this leads to the equation $z=(35-\text{epact})\bmod 30$, the first part of the formula; then, the Julian calendar day difference between the Easter full moon (z April) and the place of the *concurrentes* (24 March) is congruent modulo 7 to the weekday difference between the Easter full moon and the *concurrentes* (again, the modulo 7 congruence is necessary here, since the weekday of the Easter full moon and the *concurrentes* do certainly occur in different weeks); in mathematical terms: $(z+7)\bmod 7\equiv$ (weekday of Easter full moon $-$ *concurrentes*) $\bmod 7$; this is equivalent to $(y+\textit{concurrentes}+7)\bmod 7\equiv$ (weekday of Easter full moon)$\bmod 7$; since the weekday of the Easter full moon naturally ranges between 1 and 7, this leads to the equation $(x + concurrentes + 7) \mod 7 =$ weekday of Easter full moon, the second part of the formula.

Argumentum XV describes the correspondence of the Roman equinoxes and solstices with the conception and birth of Christ and of John the Baptist, as well as with the passion of Christ. Moreover, it gives a chronology of Christ's life.

Argumentum XVI explains the bissextile increase of one point (defined as a quarter of an hour) per month, amounting to 12 hours in four years (i.e. the 12 hours of daytime). Moreover, it gives a curious mathematical explanation for the alleged 3 hours bissextile increase per year by dividing the 8760 hours of a year by 7, leaving a remainder of just these 3 hours.[27]

Previous Views on the Original Corpus of the Dionysiac *Argumenta*

Which of these *argumenta*, then, can be classified as Dionysiac, i.e. as belonging to the original corpus of *argumenta* compiled by Dionysius Exiguus himself? Scholarly opinion on this important question varies greatly:

The discussion about the Dionysiac *argumenta* really started with the *editio princeps* of the *Computus Digbaeanus* by Wilhelm Jan in 1718, which is a model of scholarship for its time, and still the best edition of Dionysiac computistica available. He justified the ascription of these *argumenta* to Dionysius in detail,[28] but at the same time emphasized that some of the 16 *argumenta* published by him on the basis of the *Computus Digbaeanus* of 675 (which he checked with the *Computus Cottonianus* of 688/9) appear to be later additions, while others appear to contain later interpolations.[29] About this pseudo-Dionysiac section he was then more specific in the notes to each *argumentum*. Jan generally applied four criteria to distinguish Dionysiac from pseudo-Dionysiac *argumenta*:[30] 1) Agreement with the columns of the Dionysiac

---

[27] For this calculation see especially Springsfeld (2002), 204; Springsfeld (2003), 226. Cf. also pp. 92–4 below.

[28] Jan (1718), 79–80.

[29] Jan (1718), 48–9.

[30] Jan (1718), 54, 81–8.

Easter table; 2) the reception of the *argumenta* by early medieval computists, primarily Bede; 3) the transmission of the *argumenta*; 4) internal textual evidence, including dating clauses given in the examples. The strongest of these criteria obviously is the fourth: nine of the first ten *argumenta* include examples that refer either explicitly to the year 525 (which is also the year mentioned in Dionysius' prologue), or implicitly, with the chronological data corresponding to that year. The only *argumentum* among the first ten that does not include a dating clause is *Argumentum VII*. Interestingly enough, Jan did not comment on this particular *argumentum*, and it must be presumed that he regarded it as Dionysiac. Consequently, the first ten *argumenta* are Dionysiac according to Jan. Three of these, however, contained later interpolations:[31] 1) Jan did not publish §§2 and 3 of *Argumentum I*, presumably because he regarded them as later additions; unfortunately, he did not comment on this omission, which appears strange considering the fact that these paragraphs also appear in the *Computus Cottonianus* (albeit in accommodated and altered form), which Jan consulted for comparison; 2) since §2 of *Argumentum III* does not occur in the *Computus Cottonianus*, and because it was introduced by the phrase *Item aliud computum*, Jan argued that it was not part of the original *argumentum*; 3) the same applies to §2 of *Argumentum IV*, which is introduced by the phrase *Item nuper inuentum, melius iudicaui, si breuiter patefiat*, and which again does not appear in the *Computus Cottonianus*; moreover, Jan thought that this paragraph was not transmitted in full in his *Leithandschrift*; the mentioning of the consulate of Tiberius Iunior Augustus combined with the chronological data given led Jan to date this interpolation to *c.*581.[32] In addition to these three later additions, Jan was suspicious about §2 of *Argumentum IX*, stating that it appears to be out of place here, but he left that problem for others to solve.[33]

Concerning the remaining six *argumenta*, Jan restricted himself to a general remark in a note on *Argumentum XI* rather than to comment on each of these individually, apparently because he found them difficult to understand:[34] After outlining that this and the following *argumentum* transmit dating clauses for 675 in the *Computus Digbaeanus*, for

---

[31] Jan (1718), 81, 83–4.

[32] It seems quite probable that Jan did not establish this date himself, but simply adopted the date written in a late hand in the margin to this paragraph in the manuscript (f 74r).

[33] Jan (1718), 87.

[34] Jan (1718), 88.

688 or 689 in the *Computus Cottonianus* (i.e. one and a half centuries later than the dating clauses found in the previous *argumenta*), he concluded very cautiously that these *argumenta* were either added or at least emended by later computists.

The most detailed discussion about the authorship of these *argumenta* was written just 26 years later by one of the most neglected scholars of late antique and early medieval computistics, Johannes van der Hagen.[35] Having no further manuscript evidence at hand, van der Hagen based his study on Jan's edition, which meant that he was not aware of §§2 and 3 of *Argumentum I*. His approach is more textual and mathematical, with the main new textual criterion introduced by him being the question of the general format of an *argumentum*. It seems that in his opinion an *argumentum* had to contain a mathematical formula. Consequently, he was very suspicious of every *argumentum* among the 16 in question that does not do so. Presumably on this ground, he regarded *Argumenta XV* and *XVI* as pseudo-Dionysiac. Concerning the only other *argumentum* of that sort, *Argumentum VII*, he did not commit himself to an explicit statement of authorship. But it appears that he regarded it as pseudo-Dionysiac, because it had no formula, and because it appears out of place among *argumenta* that otherwise show very similar patterns. On this basis of structure he divided the nine *argumenta* remaining of the ten classified by Jan as Dionysiac into two groups: The first group consists of *Argumenta I–VI* and *VIII*, for the reason that *Argumentum I* provides an algorithm for the calculation of the AD date of a year, which is then used as the sole precondition for the formulae of all of remaingn six *argumenta* of this group; in van der Hagen's opinion this group is to be classified with certainty as Dionysiac, since the examples in all of these explicitly refer to 525, i.e. the year in which Dionysius composed his prologue; concerning §2 of *Argumenta III* and *IV*, he agreed with Jan that they are later interpolations, adding that §2 of *Argumentum III* does not fit into the structure of the *argumenta* of this group, since the precondition in the formula of this paragraph is the year of the *cyclus decemnovennalis*, not the AD date; moreover, he dates §2 of *Argumentum IV* to the year 703. The second group, then, consists of *Argumenta IX* and *X*. Since these also have dating clauses for 525 (though only implicit ones), van der Hagen argues that *videntur etiam esse Dionysii*. His hesitancy in attributing them to Dionysius stems from the fact that these two

---

[35] Van der Hagen (1734), 205–8.

argumenta do not share the structure of the previous ones, i.e. that their precondition is not the AD date.

In contrast to Jan, van der Hagen discussed in detail the six *argumenta* that Jan had classified as pseudo-Dionysiac. First he treated *Argumenta XI* and *XII* together, accepting Jan's classification as pseudo-Dionysiac on the basis of the examples explicitly referring to the year 675, while adding that *Argumentum XI* deals with a question that had already been discussed in *Argumentum III*, namely the calculation of the epact of a given year. He then connected *Argumentum XIII* with Maximus Confessors' *Computus ecclesiasticus* rather than Dionyisus' works, a very interesting observation that has escaped the notice of later scholars.[36] Turning to *Argumentum XIV*, he argued that it might have been composed by Dionysius himself, because the chronological data given agree with the years 532–534; but since these are the first three years of the Dionysiac Easter table, van der Hagen thought that these examples may as likely have been composed by a later computist, who found these three years suitable to illustrate his calculations. Finally, he regarded *Argumenta XV* and *XVI* as pseudo-Dionysiac, because they simply did not constitute *argumenta* in his sense of the word.

In 1901, Bartholomew Mac Carthy supported Jan's and van der Hagen's view that *Argumenta XI–XVI* must be regarded as pseudo-Dionysiac by introducing new manuscript evidence. He drew attention to the MS Vatican, Biblioteca Apostolica, Vat. Lat. 5755, which preserves a fragment of the Dionysiac *argumenta*, ending with the explicit *Finiunt argumenta paschalium titulorum* after *Argumentum X*.[37] Unfortunately, he did not compare this fragment with Jan's edition, and thus did not comment further on its implications on the corpus of Dionysiac *argumenta*.

Despite this new manuscript evidence, Bruno Krusch decided to republish the entire *Computus Digbaeanus* among Dionysius' computistical works in 1938. His edition, the latest and standard edition to this day, must be regarded as inferior to Jan's for various reasons, among them the fact that he did not compare the text with the *Computus Cottonianus* as Jan had done.[38] Krusch did not discuss the authorship of these *argumenta*, which leaves the reader with the impression that the ascription to Dionysius is valid for all 16,

---

[36] For this connection see pp. 90, 98–100, 102–3 below.

[37] Mac Carthy (1901), lvi. For the manuscript see pp. 66–7 below.

[38] Cf. also the verdicts on these editions by Jones (1943), 68–9; Pallarès (1994), 17–8; Wallis (1999), liii.

save for the passages and examples specified as interpolated or accommodated, namely §2 of *Argumentum IV* (with Krusch arguing for a date *c.*582), and the examples given in *Argumenta XI* and *XII*.[39] All this is very surprising considering the fact that Krusch was the first to draw attention to the correspondence of the *Computus paschalis* of 562 attributed to Cassiodorus with the first ten Dionysiac *argumenta*, which only led him to an analysis of Cassiodorus' authorship of the *Computus paschalis*, not of Dionysius' authorship of the *argumenta*.[40] Moreover, Lehmann's article on the *Computus paschalis*, published in 1912 and including an edition of that text,[41] must have been known to him and would have had provided him with previously unknown manuscript witnesses.

Manuscript studies of computistical texts were then generally brought to a new level by Charles W. Jones in the late 1930s and early 1940s. Concerning the Dionysiac *argumenta*, he drew special attention to two manuscripts from Mainz (now in the Vatican library: MSS Vatican, Biblioteca Apostolica, Pal. Lat. 1447 and Pal. Lat. 1448), the Vatican manuscript introduced into the discussion by Mac Carthy, and the so-called Sirmond manuscript, i.e. MS Oxford, Bodleian Library, Bodley 309.[42] Unfortunately, he never discussed the transmission of the Dionysiac *argumenta* in these manuscripts in detail, nor did he test his own views on them with van der Hagen's thorough study, which he knew.[43] He simply and persistently argued that only the first nine *argumenta* can be regarded as Dionysiac,[44] which is fairly surprising, since the first ten rather than nine appear as one body in every manuscript he had drawn attention to.[45] An explanation for this opinion is left wanting, and he never commented on possible later interpolations within these nine *argumenta*. Nevertheless, his view was later accepted

---

[39] Krusch (1938), 62 (where he mistakenly places the mentioning of the consulship of Tiberius Iunior Augustus in *Argumentum VIII* rather than *Argumentum IV*), 76–8.

[40] Krusch (1884), 113–4. For the *Computus paschalis* of 562 see the following note and pp. 67–8 below.

[41] Lehmann (1912), reprinted in Lehmann (1959), with the section on the *Computus paschalis* on pp. 47–55.

[42] Jones (1937), 207, 214; Jones (1943), 69–70. Jones (1943), 69 gives an extensive list of MSS containing Dionysius' computistical writings, stating at the end that the Palatine MSS 'provide valuable evidence about the *argumenta*'. Yet, of the three Palatine MSS listed by him only the two mentioned above contain the *argumenta*. For these MSS see pp. 65–6 below.

[43] Jones (1943), 69.

[44] Jones (1943), 70, 358, 368.

[45] Cf. Jones (1937), 214; Jones (1943), 70, 107 and pp. 64–7 below.

by Wesley Stevens, Peter S. Baker and Michael Lapidge, Dan Mc Carthy, and, as will be seen below, also by Joan Gómez Pallarès.[46] Few statements can be gathered about some of the *argumenta* Jones regarded as pseudo-Dionysiac: For *Argumentum XI* he gave the same reasoning as van der Hagen, arguing that it includes a dating clause for 675, and that the problem outlined in this *argumentum* (the calculation of the epact) had already been dealt with in *Argumentum III*, since 'Dionysius, who is always most concise, would hardly have given two *argumenta* for the same thing; had he done so, he would certainly have given them together';[47] concerning *Argumentum XIII* he argues that it is a 'complex and unsatisfactory' tract originating in the seventh century.[48] Consequently, Jones' criteria for dismissing certain *argumenta* as pseudo-Dionysiac primarily are the late dating clauses on the one hand, the belief in Donysius' competence, clarity and precision on the other. The lack of the latter convinced Jones that *Argumentum XVI* must be pseudo-Dionysiac;[49] concerning *Argumentum XIV* he merely stated that it is probably not Dionysiac, but 'older than Bede'.[50]

In 1958, Alfred Cordoliani compared the *argumenta* attributed to Dionysius with the *Computus paschalis* and what he regarded as the *Computus Cottonianus*.[51] But since the focus of his study was on the *Computus Cottonianus*, he did not comment on the authorship of the other two texts.

Undoubtedly the mathematically most skilled scholar of computistics, Otto Neugebauer, was the first to use the *Computus paschalis* of 562 as evidence for the Dionysiac *argumenta*, in an article published in 1982.[52] In his opinion, the *Computus paschalis* is nothing but a faithful reproduction ('a word-by-word copy') of the original corpus of Dionysiac *argumenta*, in which only the examples were accommodated to 562. Since the *Computus paschalis* comprises *Argumenta I–VI* and *VIII–X* of the 16 *argumenta* in question (and not all of them completely), Neugebauer regarded these as the original corpus, to which he seems to have added *Argumentum XI*. Considering that

---

[46] Stevens (1981), 90, 107; Baker and Lapidge (1995), xli; Mc Carthy (2003), 37–8; for Pallarès (1994) cf. p. 60.
[47] Jones (1943), 386.
[48] Jones (1943), 355.
[49] Jones (1943), 373.
[50] Jones (1939), 43.
[51] Cordoliani (1958).
[52] Neugebauer (1982), especially 292.

Argumentum XI does not appear in the *Computus paschalis*, that it contains a dating clause for 675, and that the problem described in it is already discussed in *Argumentum III*, the inclusion of this *argumentum* among the original Dionysiac corpus appears rather misguided. Neugebauer's argument here seems to be that *Argumentum XI* deals with 'Easter computus', while *Argumenta VII, XII–XVI* do not.[53] This argument does not convince the reader, especially since *Argumentum XIV*, e.g., deals with the calculation of the Easter full moon and Easter Sunday. His other arguments for not including *Argumenta XII–XVI* among the original Dionysiac corpus are textual (different style and different terminology compared to the previous ones).[54] It seems that Neugebauer was not aware of §§2 and 3 of *Argumentum I*, since he worked from Jan's edition of the text, but he quite rightly stated that the chronological data of §2 of *Argumentum III*, which does not appear in the *Computus paschalis*, agrees with the years 522 and 523.[55] Concerning §2 of *Argumentum IV*, which does also not appear in the *Computus paschalis* either, he argued that the mentioned consul Tiberius Iunior Augustus has to be identified with the Byzantine emperor Tiberius, who ruled from 578 to 582. Hence, according to Neugebauer the formula given in this paragraph was applicable to the years counted from 578 onwards, and was probably invented during Tiberius' reign. Yet, the altered formula does not work for years counted from 578, so that Neugebauer's interpretation must be regarded as unlikely.[56] §2 of *Argumenta IX* and *X* do not feature in Neugebauer's discussion.

In an article on Dionysius' dating-system, Gustav Teres, using Jan's edition of the *argumenta* (which he mistakenly ascribes to Benjamin Hoffmann), seems not to have studied Jan's notes in detail, since he

---

[53] Neugebauer (1982), 292, 297, 300.

[54] Neugebauer (1982), 292.

[55] Neugebauer (1982), 293–4.

[56] According to Neugebauer ((1982), 296) the modification of the parameter 4 in the formula of §1 of *Argumentum IV* (which represents the *concurrentes* of AD 1 — 1) to 1 in §2 is due to the fact that $577 \equiv 3 \bmod 7$. I do not understand the logic of Neugebauer's argument here. If he meant that the weekday difference between AD 1 and 578 was 3, then he calculated her wrongly, not taking bissextile years into consideration. Anyway, the parameter 1 in §2 of *Argumentum IV* clearly indicates that the *concurrentes* in the first year of the count were 2; moreover, since no parameter is added to the number of years to be divided by four, the first year of the count is supposed to be a year following a bissextile year; both criteria do not agree with 578, the year given by Neugebauer, since it was the second year after a bissextile year, having *concurrentes* 5. Cf. pp. 47–8 above and pp. 83–4 below.

treated the entire corpus without distinction as originally Dionysiac.[57] The only reason for this treatment appears to be the fact that these 16 *argumenta* are published among the works of Dionysius Exiguus, which illustrates the risk an editor takes when publishing dubious material among the original work of an author.

Dáibhí Ó Cróinín, in his 1988 edition of the Irish computistical textbook *De ratione conputandi*, pointed out the fact that this textbook and *Argumentum XVI* discussed similar concepts, which led him to argue that this *argumentum* was 'undoubtedly Insular', 'possibly Irish' in origin.[58]

In 1994, Joan Gómez Pallarès drew particular attention to the *Computus Cottonianus* of 688/9 as an important witness for the Dionysiac *argumenta*, as Jan had done in 1718. This computus (at least as published by Pallarès) contains the first 14 of the 16 *argumenta* in question, none of which preserving the original dating clause of 525. Pallarès edited the section of this computus in which these 14 argumenta appear, comparing their readings with Jan's and Krusch's editions in that process.[59] But this comparison of the *Computus Digbaeanus* (i.e. the *argumenta* as published by Jan and Krusch) with the *Computus Cottonianus* did not lead him to any conclusion about the authorship of the *argumenta* in question. He simply followed Jones in arguing that the first nine are authentic, because they incorporate a dating clause (525) that agrees with Dionysius' other writings; since the same, according to Pallarès, cannot be said about the remaining *argumenta*, one should be cautious with attributing these to Dionysius.[60] This argument obviously fails to convince, because the example in *Argumentum X* also refers to 525 (though implicitly), and no other reason is given why it should be excluded form the original corpus. Moreover, the differing chronological data in §2 of *Argumenta III* and *IV* are not mentioned.

Arno Borst, in his 1998 book about the Carolingian calendar reform, argued in passing that the *argumenta* incorporating dating clauses for 525 have to be attributed to Dionysius, while he located the addition of §2 of *Argumentum IV* to Italy, dating it to 582, and the addition of arguably the last six *argumenta* to Ireland, dating it to 675.[61]

---

[57] Teres (1984), especially 182–3.
[58] Walsh and Ó Cróinín (1988), 125, 161.
[59] For the edition see Pallarès (1994), 20–31. Cf. pp. 73–4 below.
[60] Pallarès (1994), 18.
[61] Borst (1998), 177.

In this view he is followed by Kerstin Springsfeld, who more specifically stated that *Argumenta I–X* are Dionysiac because of their dating clauses (save for §2 of *Argumentum IV*, which, according to her, is datable to 582, and was probably added to the corpus in Italy), the remaining six, however, pseudo-Dionysiac, presumably composed in Ireland in 675.[62] She possibly saw confirmation for this view in the manuscript evidence, since she was aware of two manuscripts that transmitted *Argumenta I* to *X* as an entity, namely the Sirmond manuscript (MS Oxford, Boldeian Library, Bodley 309) and MS Vatican, Biblioteca Apostolica, Pal. Lat. 1448.[63] Interestingly enough, Springsfeld observed that the chronological data in §2 of *Argumentum III* do not agree with the year 525, but she did not draw the conclusion that this section may not be Dionysiac.[64] Concerning *Argumentum XIV* she argued that it may have been Dionysiac, as the examples given agree with 532–534, basically repeating van der Hagen's observation. The only reason, in her opinion, that could speak against Dionysius' authorship of this *argumentum* is that certain irregularities occur at the end of it; yet, it should be noted that these irregularities only exist in Krusch's edition, not in the manuscript from which he worked.[65] The fact that she connected some of the concepts found in *Argumenta XV* and *XVI* with Irish computistica may have convinced her to ascribe the whole body of the last six *argumenta* to an Irish anonymous.[66]

As mentioned above, Faith Wallis had earlier, in 1999, drawn special attention to the Sirmond manuscript for this question, arguing

---

[62] Springsfeld (2002), 172–82, the Irish origin of *Argumentum XIV* explicitly on pp. 70, 87, 89, 99.

[63] Cf. Springsfeld (2002), 69, 84. When discussing the Dionysiac *argumenta*, she referred to Jones' table of contents of the Sirmond manuscript (Springsfeld (2002), 172).

[64] Springsfeld (2002), 173.

[65] Springsfeld (2002), 178. Note that her argument here contradicts her statement on pp. 70, 87, 89, 99 that *Argumentum XIV* is of Irish provenance. The two irregularities in *Argumentum XIV* mentioned by Springsfeld are the following: 1) It is argued that the Julian calendar date of the Easter full moon falls on a certain, previously calculated number of days *a Kalendis Januarii*; but it is obvious from the context that the calends of either March or April are meant, depending on the year in question; in fact, *Januarii* is Krusch's own (and wrong) addition, as he clearly indicates in the apparatus. 2) It is argued that if the previously calculated number of days happens to be 30, then the Easter full moon occurs *XXX die Aprilis*; this is obviously an impossible reading, since the Easter full moon falls between 21 March and 18 April; in fact, *Aprilis* is Krusch's correction, as he clearly indicates in the apparatus; the manuscript gives the correct *a Kl* (referring to the calends of March).

[66] Springsfeld (2002), 176.

that the corpus of the first ten *argumenta* found in it has to be regarded as Dionysiac, while disregarding the following *Argumentum XIV* as pseudo-Dionysiac.[67] Unfortunately, Wallis did not discuss these *argumenta* any further, and thus did not mention the chronological and textual problems within these ten *argumenta* as outlined by Jan, van der Hagen, and Neugebauer.

In recent years, the interest in the Dionysiac *argumenta* has led to the reproduction of their text on the world wide web. On one of these sites, Nikolaus Bär provided Krusch's text with a German translation and some commentary.[68] In his 'Schlußbemerkung', he argues that only the first nine *argumenta* formed the original Dionysiac corpus, and among these §2 of *Argumenta III, IV* and *IX* represented later additions, without giving further details. He was more specific only in his notes to §2 of *Argumentum IV*, repeating Krusch's argument that this addition must date from the year 582 or later. It seems that his opinion on the original corpus is a more detailed adoption of Jones' view, and thus equally unsatisfactory, especially in terms of *Argumentum X*, whose implicit dating clause for 525 is left unexplained.

The most differentiated treatment of this question since the days of van der Hagen can be found in a brief note by George Declercq in his article about the Dionysiac reckoning, published in 2002.[69] He followed Neugebauer's approach by arguing that the *Computus paschalis* of 562 represents the original corpus of Dionysiac *argumenta* (not adding *Argumentum XI*, as Neugebauer had done). He found confirmation for this view in the fact that every single one of these *argumenta*, and no other among the 16 in question, includes a dating clause for 525. Hence, in Declercq's opinion §2 of both *Argumenta III* and *IV*, *Argumentum VII*, probably §2 of *Argumentum IX*, and *Argumenta XI* to *XVI* are pseudo-Dionysiac. As every scholar before him, however, he seems not to have been aware of §§2 and 3 of *Argumentum I*, being deceived by the editions; at least, he did not include them in his discussion.

Summary

This reassessment of scholarly opinion on the original corpus of the Dionysiac *argumenta* illustrates that the basis of discussion generally was the *Computus Digbaeanus* of 675, the *Computus Cottonianus* of

---

[67] Wallis (1999), lxxiv.
[68] www.nabkal.de/dionys.html
[69] Declercq (2002), 199.

688/9, and the *Computus paschalis* of 562, while the important manuscript evidence mentioned by Jones was sometimes referred to in passing, but never analyzed in detail. Moreover, Jan's and van der Hagen's discussion of Dionysius' authorship, both published in the early eighteenth century, remain the best studies of this question to the present day, even though they are based on only a fractional amount of the evidence known to us today. The neglect of these studies during the past century or so led to widely differing comments on the extent of the Dionysiac corpus of *argumenta*, with only Neugebauer and Declercq commenting on this question in some detail. Yet, since both scholars touched this question only in passing and in the end did not agree, neither their accounts, nor the valuable, but outdated discussions of Jan and van der Hagen can be regarded as authoritative in this question. The lacuna of a proper analysis of the extent of the Dionysiac corpus of *argumenta* will be dealt with in the following, since only such an analysis can provide the basis for the present study.

Manuscript Evidence for the Original Corpus of the Dionysiac *Argumenta*

As the previous discussion has shown, only five manuscripts containing *argumenta* with the original dating clause of 525, the year in which Dionysius evidently composed his prologue, have been discovered over the past three centuries.[70] For that reason alone, these five manuscripts obviously are the primary witnesses for what may have been the original corpus of the Dionysiac *argumenta*. Since the two existing editions of the *argumenta* are based on only one of these five manuscripts, and none of the remaining four has ever been analyzed in detail concerning the extent and structure of the original Dionysiac corpus, these are discussed here first. But it has also become apparent in the preceding discussion that some of the later recensions, most notably the *Computus paschalis* of 562 and the *Computus Cottonianus* of 688/9, also preserve valuable information concerning this question. Yet, these later recensions are numerous; for that reason they will be classified

---

[70] The sole manuscript witness for the Dionysiac *argumenta* listed by Cordoliani (1943), 60 is MS Vatican, Biblioteca Apostolica, Reg. Lat. 1260, fols 118–125, which is, in fact, a later recension and does not include the 525 dating. The same holds true for MS Basel, Universitätsbibliothek, F III 15k, mistakenly mentioned by Springsfeld (2002), 73, 76, 84 to include the Dionysiac *argumenta*. Cf. p 77 below.

and analyzed according to their relevance for the transmission of the original Dionysiac corpus of *argumenta*. Finally, an overview of this manuscript evidence is provided as the basis for establishing the original structure and subsequent development of the Dionysiac *argumenta*.

MSS Preserving the Original Dating Clause of 525

Group A – MS Oxford, Bodleian Library, Digby 63, 72v–79r (*Computus Digbaeanus* of 675):
This manuscript formed the basis for Jan's and Krusch's edition of the *argumenta*, as well as the present analysis, and is discussed in detail above.[71] Concerning the dating clauses, nine of the 16 *argumenta* transmit the original date of 525, namely *Argumenta I–VI* and *VIII* explicitly, *Argumenta IX* and *X* implicitly. §2 of *Argumentum IV*, *Argumentum XIII*, and *Argumentum XIV* incorporate implicit dating clauses for 581, 625, and 532 to 534 respectively, as is discussed elsewhere.[72] The data given in §2 of *Argumentum III* mentions the 10th and 11th year of the *cyclus decemnovennalis* (corresponding to 522+n×19), while *Argumentum XI* and *XII* explicitly refer to 675. As mentioned above, this computus has an incipit, but no clearly defined end.[73] The incipit reads: *Incipiunt argumenta de titulis pascalis egiptiorum investigata solercia ut praesentes indicent.*

Group B – MSS Oxford, Bodleian Library, Bodley 309, 81r–82v (Sirmond Manuscript); Vatican, Biblioteca Apostolica, Pal. Lat. 1447, 6v–8v; Vatican, Biblioteca Apostolica, Pal. Lat. 1448, 13r–17v; Vatican, Biblioteca Apostolica, Vat. Lat. 5755, 3–4:
Characteristically, the manuscripts of this group transmit *Argumenta I* to *X* identical in structure and content to the *Computus Digbaeanus* (including §§2 and 3 of *Argumentum I*, which are not included in the main body of the text in the editions), but contrary to that computus these ten *argumenta* are here strictly defined by incipits and explicits as a single corpus; to this corpus *Argumentum XIV* is attached, in the same version as in the *Computus Digbaeanus*.[74] Hence, the dating

---

[71] See pp. 45–53 above.

[72] See pp. 47–8, 51, 83–4, 89–91.

[73] Cf. p 45 above.

[74] The facsimile of MS Vatican, Biblioteca Apostolica, Pal. Lat. 1447, 6v–8v in Appendix III (*Plate 1*) illustrates the strictly defined corpus of *Argumenta I–X* with *Argumentum XIV* attached that is the characteristic feature of Group B manuscripts.

clauses in these manuscripts are the exact same as in the *Computus Digbaeanus*.

Jones drew attention to the three manuscripts that contain the full text, while Mac Carthy had earlier referred to a fragment that included only the second half of it.[75] The most important of these manuscripts probably is MS Oxford, Bodleian Library, Bodley 309, which Jones had identified as once in the possession of Jacques Sirmond and studied by the famous early modern scholars of chronology, Petavius and Bucherius, in the seventeenth century, but which was subsequently regarded as lost in the eighteenth and nineteenth centuries.[76] The manuscript itself was written in Vendôme in the eleventh century, but Jones argues that a large part of this manuscript represents the corpus of computistical texts used by Bede for his composition of *De temporum ratione*, among them the *argumenta* in question.[77] Hence, at least in Jones opinion, 'it is definitely possible that O is a copy of a manuscript written before A.D. 725'.[78] If this view is accepted, then this manuscript would contain the oldest known witness of the corpus of *argumenta* attributed to Dionysius. The first ten *argumenta* are clearly defined as one body in this manuscript by the incipt *Incipiunt argumenta Grecorum de titulis paschalibus inuestigata solertia* and the explicit *Expliciunt argumenta paschalium titulorum*. *Argumentum XIV* occurs as a separate item after that, headed *Incipit calculatio quomodo repperiri posit, quota feria .i. singulis annis .xiiii. luna paschalis, id est circuli decennouenalis* and concludes with the phrase *Haec argumenta hic finiuntur*.

Six years later, in his introduction to Bede's computistical works, Jones then listed manuscripts that contain Dionysius' computistical texts complementary to the manuscripts used by Jan and Krusch. At the end of that list he drew special attention to the Palatine manuscripts in the Vatican as primary witnesses for the study of the Dionysiac *argumenta*.[79] Yet, of the three Palatine codices mentioned in his list only two

---

[75] Cf. pp. 56–7 above.

[76] Jones (1937).

[77] The Dionysiac *argumenta* are item 14 in Jones' table of contents of this manuscript; in his opinion items 13–45 certainly constitute Bede's computus. Jones (1937), especially 214; Jones (1943), 105–10. Cf. Wallis (1999), lxxiv; Springsfeld (2002), 69–70, 76; Ó Cróinín (2003a), 202. For other descriptions of this manuscript see Jones (1939), 126–7; Borst (2001), 159–60; Borst (2006), 263; Mc Carthy and Breen (2003), 27–8; Peden (2003), xl–xli.

[78] Jones (1937), 210. Cf. Ó Cróinín (2003a), 201.

[79] Jones (1943), 69. Cf. also p 57 above.

actually contain the *argumenta*, both from Mainz or its vicinity, written in the early ninth century. Of these two, MS Pal. Lat. 1447 is regarded as a copy of MS Pal. Lat. 1448.[80] This is confirmed by the text of the *argumenta*, in which these two manuscripts differ only in form, not in content or phrasing: The *argumenta* frame an extensive Easter table from 798 to 854 in MS Pal. Lat. 1448, while they are separately copied in MS Pal. Lat. 1447. *Argumenta I* to *X* are clearly defined as one corpus in both codices by the incipit *Incipiunt argumenta Graecorum de titulis paschalibus Aegyptiorum investigata solertia* and the explicit *Expliciunt argumenta paschalium titulorum*. This corpus is then followed by *Argumentum XIV*, headed *Incipit calculatio*.[81]

Earlier, at the turn of the century, Bartholomew Mac Carthy was the first to realize the importance of a fragment that survived in the Vatican MS Vat. Lat. 5755 for the extent of the corpus of Dionysiac *argumenta*.[82] This fragment is famous among scholars of Old Irish, for it contains a considerable number of Old Irish glosses, but it has never been systematically studied by any modern computist.[83] Old Irish scholars tend to argue that the language of the glosses is datable to the eighth century, and that it does not seem likely that these glosses were copied.[84] Therefore, if an eighth century date is assigned to this fragment, it is the oldest surviving manuscript witness of the *argumenta*. The fragment starts on page 3 of the manuscript, which preserves the second half of *Argumentum VIII*, as well as *Argumenta IX* and *X* complete in the same form as in the *Computus Digbaeanus*,

---

[80] For descriptions of MS Vatican, Biblioteca Apostolica, Pal. Lat. 1447 see Schuba (1992), 257–61 (arguing for Mainz, early saec. IX); Mittler (1986), 128; Lindsay (1915), 481 (Mainz, before 813); Jones (1939), 135 (Mainz, c.813); Borst (2001), 73–4; Borst (2006), 298–9 (in both accounts Borst argues for Mainz, between 808 and 813). For descriptions of MS Vatican, Biblioteca Apostolica, Pal. Lat. 1448 see Schuba (1992), 261–5 (Trier and Mainz, saec. IX¹); Lindsay (1915), 481 (Trier, 810); Jones (1939), 135 (Trier and Mainz, saec. IX¹); Springsfeld (2002), 84–5 (first part, including the Dionysiac *argumenta*, Trier, 810); Borst (2001), 60–2; Borst (2006), 299–300 (in both accounts Borst argues that the part including the Dionysiac *argumenta* was written in Trier in 810). For MS Pal. Lat. 1447 being a copy of MS Pal. Lat. 1448 see Mittler (1986), 128; Borst (2001), 73.

[81] Cf. Appendix III (*Plate 1*).

[82] Cf. p 56 above.

[83] For descriptions of this manuscript see Zimmer (1881), xxx; Stokes and Strachan (1901–3), ii xii; Kenney (1966), 671–2; *CLA* I, 11 (No. 32). Charles W. Jones, CCSL 123A, xiii lists this manuscript among computistical texts that 'may have lain within Bede's range', but 'as yet not satisfactorily studied'. This codex also contains a fragment of the *calculus* of Victorius of Aquitaine, cf. Peden (2003), xl.

[84] Stokes and Strachan (1901–3), ii xii; Kenney (1966), 672.

including the implicit dating clauses for 525. *Argumentum X* is then followed by the explicit *Finiunt argumenta paschalium titulorum*, which also marks the end of this page. Page 4 then contains *Argumentum XIV* with the implicit dating clause for 532–534, breaking off after three quarters of the text.[85] According to these details it must be presumed that the manuscipt from which this fragment originates included the first ten Dionysiac *argumenta* as a strictly defined corpus, with *Argumentum XIV* attached, as it appears in the other three codices of this group.

Early Recensions of the *Argumenta*

Recension A – The *Computus paschalis* of 562:

According to its editor, Paul Lehmann, this corpus of *argumenta* survives in five manuscripts.[86] Since it is incorporated in the second book of Cassiodor's *Institutiones* in every single one of these codices, it is assumed that Cassiodor himself may have been the author.[87] Now, by comparing the *Computus paschalis* with the Dionysiac *argumenta* in their published form (i.e. with the *Computus Digbaeanus*), Otto Neugebauer has pointed out that the *Computus paschalis* is nothing but a copy of these *argumenta*, with only two major differences: *Argumenta VII, XI–XVI* are missing in the *Computus paschalis* (as they represent *argumenta* that were later added to the original Dionysiac corpus), and the examples in the remaining *Argumenta I–VI, VIII–X* were modified to match the year 562 instead of 525.[88] Yet, if the edition

---

[85] The full text of p 3 and half of p 4, including all Old Irish glosses, are edited in Zimmer (1881), 259–61; Stokes and Strachan (1901–3), ii 39–41.

[86] Lehmann (1959), 48; Cordoliani (1943), 56–7. The five manuscripts are: MS Karlsruhe, Badische Landesbibliothek, Aug. CLXXI, 49v–50v; MS Milano, Biblioteca Ambrosiana, D 17 inf., 52r–v; MS Paris, Bibliothèque Nationale, Lat. 2200, 70v–72v; MS Würzburg, Universitätsbibliothek, M. p. misc. F. 5a, 30v–31v. For the mistaken addition to this list of MS Padua, Biblioteca Antoniana, I 27, 14v–15r in CCSL, Clavis patristica 3A, 269, which stems from this codex being the only one mention under Cassidorus' Computus in Thorndike and Kibre (1963), 1455, see n 112 below.

[87] For the debate concerning Cassidorus' authorship and the incorporation of this corpus of *argumenta* in the second book of his *Institutiones* see Krusch (1884), 113–4; Poole (1918b), 210–1; Lehmann (1959), 47–52; van de Vyver (1931), 289–91; Neugebauer (1982), 292, 301.

[88] For Neugebauer's problematic view that *Argumentum XI* was also part of the original Dionysiac corpus, even though it does not feature in the *Computus paschalis*, see pp. 58–9 above.

of the *Computus paschalis* of 562 is compared with Group A–B texts, further significant textual differences become apparent (some of which are noted by Neugebauer himself[89]): the *Computus paschalis* conflates §§2 and 3 of *Argumentum I*, while §2 of *Argumenta III*, *IV*, *IX*, and *X* do not feature in this text. As regards the incipits and explicits of the *Computus paschalis*, these vary greatly among the five manuscripts, none of which agreeing with the ones in Group A–B texts.[90]

Recension B – MS Paris, Bibliothèque Nationale, Nouvelle acquisition latine 1615, 154r–155r (The *Computus Parisinus* of 819/20):

One important witness for the Dionysiac *argumenta* has escaped the attention of modern scholars. In the ninth century MS Paris, Bibliothèque Nationale, Nouvelle acquisition latine 1615, 154r–155r from Fleury or Auxerre the Dionysiac *argumenta* are transmitted in a recension of 820.[91] This recension comprises a clearly defined corpus of *argumenta* identical to the corpus of Group B, with *Argumenta VII* and *VIII* being switched. This switch is noteworthy in so far as it demonstrates that the early ninth-century editor of this corpus apparently regarded *Argumenta I–VI* and *VIII* as somehow connected, presumably because of the fact that only these seven include explicit dating clauses.[92] The dating clauses have been accommodated to the year 820 or 819, with few exceptions: No explicit dating clause is given in *Argumentum I*, while the number of indiction cycles that have passed since AD 12 (namely 53) dates this recension to the period between 807 and 821 inclusively; §2 of *Argumentum III* gives the 17th year of the *cyclus decemnovennalis* as example, which agrees with 814; §2 of *Argumentum IV* still preserves the count of years from the year following the consulship of Tiberius Iunior Augustus, which refers to 581;[93] *Argumentum IX* gives data that only matches 813/814 (813: indiction 6 – 7 in MS –, epact 15, Easter Sunday on 27 March, *luna* 20; 814: indiction 7, epact 26, Easter Sunday on 16 April, *luna* 21); *Argumentum X*

---

[89] Neugebauer ((1982), 295–6) notes that §2 of *Argumenta III* and *IV* are not part of the *Computus paschalis*. Cf. also pp. 58–9 above.

[90] For the incipits and explicits of the *Computus paschalis* see Lehmann (1959), 52, 55.

[91] For this manuscript see Krusch (1926), 53 (St.-Benoît-sur-Loire in Fleury, saec. IX); Jones (1939), 130 (Fleury, saec. IX); Borst (2001), 143–5 (Fleury or Auxerre, the latter more likely, c.830); Borst (2006), 282–3 (Fleury or Auxerre, c.830).

[92] This structure of the *Computus Parisinus* strengthens van der Hagen's view (van der Hagen (1734), 207) that *Argumenta I–VI* and *VIII* are to be regarded as an entity. For this question see pp. 55–6 above and pp. 78–9 below.

[93] For this dating see pp. 47–8, 83–4.

still preserves the chronological data of 525 (indiction 3, *concurrentes* 2, Easter Sunday on 30 March). This preservation of the chronological data for 525 proves that the original Dionysiac *argumenta* were the exemplar, and not, e.g., the *Computus paschalis* of 562, while the unaltered inclusion of §2 of *Argumentum IV* (and in fact the entire structure of this *Computus Parisinus*) demonstrates that the examplar was, in fact, a Group B text. These ten *argumenta* are clearly defined as one corpus in the manuscript by a blank line before the first *argumentum* (probably to be later filled by a heading) and by the explicit *Explicit argumenta paschalium titulorum*.[94] In contrast to Group B, however, this corpus is not immediately followed by *Argumentum XIV* here, which appears in a different part of the codex (fols 187r–v).

Recension C – Binding fragment of MS Nancy, Bibliothèque Municipale, 317 (356) (The *Fragmentum Nanciacense* of 625):

An important manuscript witness for the pseudo-Dionysiac *argumenta*, discovered 140 years ago, has hitherto been overlooked by scholars of early medieval computistics. In 1866, the French palaeographer Henri D'Arbois de Jubainville discovered a binding fragment in the Bibliothèque municipale de Nancy written in an Irish hand of the late eighth or early ninth century;[95] later palaeographers have argued for Bobbio as the provenance of the codex to which this flyleaf is attached.[96] It contains computistical formulae, sporadically glossed in Old Irish. These glosses have made this fragment famous among scholars of Old Irish,[97] but it has never received any attention by modern computists. The recto side of this fragment consists of four passages:[98] 1) The first deals with the length of moonlight per lunar day. 2) The second passage explains how to calculate the lunar age of

---

[94] Cf. the facsimile in Appendix III (*Plate 2*).

[95] D'Arbois de Jubainville (1866).

[96] The shelf-mark of this codex was 59 when this binding fragment was discovered by D'Arbois de Jubainville (and this number is repeated in all subsequent discussion of the Old Irish glosses found on this flyleaf), but it was changed soon thereafter, in 1873, to 317 (356) (cf. Favier (1886), 123). For a description of this codex see Favier (1886), 176–7; Lindsay (1915), 469; Bischoff (2004), 307; Zimmer (1881), xxx–xxxi; Stokes and Strachan (1901–3), ii xii; Kenney (1966), 672.

[97] The Old Irish glosses are edited in D'Arbois de Jubainville (1886); Zimmer (1881), 262; edited and translated in Stokes and Strachan (1901–3), ii 41; edited, translated and discussed in Gaidoz (1867), 70–1 and D'Arbois de Jubainville (1867). For the Irish context of this fragment see also Zimmer (1881), xxx–xxxi; Stokes and Strachan (1901–3), ii xii; Kenney (1966), 672.

[98] A facsimile of this fragment is reproduced in Appendix III (*Plate 3*) below.

the calends of each month for any given year from fixed regulars for each month[99] and the epact of the year in question; the epact of the example given is 4, i.e. the 15th year of the *cyclus decemnovennalis*. 3) The third passage demonstrates how to calculate the epact from the AD date, which ultimately derives from *Argumentum III*; the example given is AD 701;[100] this calculation is followed by a list of weekday regulars for the calculation of the weekday of the calends of each month for any given year, starting with October,[101] and another list of lunar regulars for the calculation of the lunar age of the calends of each month, starting with September, which is identical to the list given in the previous passage.[102] This passage is followed by the explicit *Finit argumenta igitur*, which is then followed by the incipit *Alia argumenta nuper inuenta incipiunt amen*. 4) The fourth passage gives formulae for calculating the day of the month and weekday of *luna* 2 of the March lunation from the lunar age of 1 February and the *concurrentes* respectively; this *luna* 2 was significant, because it had the exact same relation to the *initium quadragesimae* (the beginning

---

[99] These lunar regulars are listed from September to August, and they are identical in order and numerical value to the same list given in the Bobbio Computus 3, 22 (PL 129, 1282, 1289); *Dial. Neustr.* 21 (Borst (2006), 403); *Lect. comp.* I 4 (Borst (2006), 547–8); *Lib. comp.* II 15 (Borst (2006), 1165); Computus of Pacificus of Verona §14 (Meersseman and Adda (1966), 77). Bede lists these regulars from January to December in *De temporum ratione* 20 (Jones (1943), 220–1), as do some manuscripts of *Lect. comp.* and Hrabanus Maurus, *De computo* 70 (CCCM 44, 285), copying Bede. The *Computus Rhenanus* of 775 has both lists (MS Köln, Diözesan- und Dombibliothek, 103, 189v–190r; MS Wolfenbüttel, Herzog-August-Bibliothek, Weißenburg 91, 173r; for this text see pp. 75–6 below).

[100] The AD date is not explicitly mentioned, but for calculation purposes implicitly given as 19×30+19×6+17 (=701); the epact is then correctly calculated as 7.

[101] Usually this list begins with March in computistical texts: So in the Bobbio Computus 3, 21 (PL 129, 1282, 1288); *Dial. Neustr.* 8 (Borst (2006), 389); *Lect. comp.* I 3 (Borst (2006), 546–7); *Lib. comp.* II 15 (Borst (2006), 1165); Computus of Pacificus of Verona §6 (Meersseman and Adda (1966), 75). Bede lists these weekday regulars from January in *De temporum ratione* 21 (Jones (1943), 222), as do some manuscripts of *Lect. comp.* and Hrabanus Maurus, *De computo* 73 (CCCM 44, 289), copying Bede. The *Computus Rhenanus* of 775 has both lists (MS Köln, Diözesan- und Dombibliothek, 103, 189v; MS Wolfenbüttel, Herzog-August-Bibliothek, Weißenburg 91, 173r; for this text see pp. 75–6 below). Interestingly enough, the only computistical text to my knowledge that has the same list of weekday regulars as this fragment, i.e. beginning with October, is the *Computus Cottonianus* of 688/9 (f 75v, cf. n 107 below), while it also appears as a gloss to Willibrord's Easter table (MS Paris, Bibliothèque Nationale, Lat. 10837, 41r) and in the Irish influenced MS München, Bayerische Staatsbibliothek, Clm 14456, 66v.

[102] Cf. n 99.

of the Lenten fast) as the Easter full moon had to Easter Sunday; in fact, this *argumentum* seems to be an adaption of *Argumentum XIV* to the *initium* calculation; the first example given (for this *luna* 2 occuring in February) is the first year of the *cyclus decemnovennalis*, having *concurrentes* 7 (and no provision is made for this year being bissextile), which agrees with AD 703, 798, and 893; since palaeographers argue that the script of this fragment can be dated to the late eighth or early ninth century, 798 seems to be the most likely date for this *argumentum nuper inventum*, and thus the whole fragment may be dated to c.798; yet, I would not rule out the possibility of dating this fragment to c.703, especially since the previous passage contains a dating clause for 701; of the second example (for this *luna* 2 falling in March) only the first two lines survive.

The verso side of this fragment, then, is the one that concerns us here. It contains only part of the last sentence of *Argumentum XI*, as well as *Argumenta XII* and *XIII* complete. A finit-clause (*Finit amen finit*) appears after *Argumentum XIII*, clearly marking the end of a corpus of *argumenta*. The date given in the examples of *Argumenta XII* and *XIII* is 625, explicitly for the former, implicitly for the latter (the lunar data matches 625/7: 15th and 17th year of the *cyclus lunaris*, having epact 16 and 8 respectively).

It is obvious from this analysis that the original manuscript contained at least the end of the second example of the last *argumentum* of the recto side as well as *Argumentum XI* in the lacuna between the recto and the verso side. Furthermore, it seems plausible to suggest that the incipit before the last *argumentum* on the recto side does not correspond to the explicit on the verso side: The incipit refers to 'recently invented *argumenta*', followed by an *argumentum* with a dating clause for AD 703, 798, or 893, while the *argumenta* on the verso side have dating clauses for AD 625. Therefore I would suggest that another explicit and incipit occurred in the now lost part immediately before *Argumentum XI*. This would imply that *Argumentum XI* to *XIII* were regarded as an entity in this fragment.

The primary importance of the verso side of this fragment is twofold: 1) It preserves the earliest known dating clause for *Argumenta XII* and *XIII*, namely 625; previously, on the basis of the *Computus Digbaeanus*, 675 was regarded as the earliest date at least for *Argumentum XII*, while the few scholars who studied *Argumentum XIII* found it impossible to date it with any certainty; in fact, *Argumentum XIII* of the *Computus Digbaeanus* preserves the implicit dating clause for 625 given in the *Fragmentum Nanciacense*, which suggests that the author of the *Computus Digbaeanus* worked from an exemplar of *Argumenta XI* to *XIII* of 625 as transmitted in this fragment, accommodating the explicit examples (i.e. the ones explicitly mentioning AD 625 as the precondition), but not *Argumentum XIII* with its implicit data; consequently, it can be safely assumed that *Argumenta XI* to *XIII* are at least as old as 625. 2) Some details of *Argumentum XIII*, especially the mentioning of certain numbers of points, have caused modern scholars some trouble, and they have never been explained satisfactorily; the *Fragmentum Nanciacense* provides a different perspective on these difficulties, which are discussed in detail in Appendix II.

Since this important piece of evidence for the pseudo-Dionysiac *argumenta* has never been transcribed in full, I provide such a transcription in Appendix I.

Recension D:

The *Computus Parisinus* of 819/20 is a rare exception of a late recension of the Group B corpus. Generally, the *argumenta* had a different fate from the last quarter of the seventh century onwards: the corpus as preserved in Group B disintegrated, and the *argumenta* found their way disconnectedly into textbooks, formularies or compendia. Recension D, then, includes every text of whatever form that transmits a later recension of one or more Dionysiac *argumenta* without preserving the Group B corpus.

Since the innumerable texts of this recension obviously do not provide any information about the original corpus of Dionysiac *argumenta*, I will discuss them only briefly here. Such a discussion is nevertheless necessary, since quite a few texts and manuscripts that fall into this category have been connected to the original corpus of the Dionysiac *argumenta* by modern scholars, so that their relation to this corpus needs to be clarified. Moreover, the following analysis is designed to illustrate how the *argumenta* were transmitted once the process of disintegration of the corpus had started.

In fact, the *Computus Digbaeanus* marks the first stage in the disintegration of the Dionysiac *argumenta*: Even though the Group B corpus is preserved, the addition of further *argumenta* without clearly marking off the original corpus by means of explicits, incipits or simple headings left the extent of this corpus undefined.

Once the corpus had opened up, it was quite natural that more newly-invented formulae and texts were added, not only to the end of the original corpus, but also into it; moreover, the examples were then usually accommodated to the *annus praesens* of the later compiler. This process, which seems to have started in the last quarter of the seventh century, led to the development of ever growing compilations of computistical texts, in which the original Dionysiac *argumenta* only played a minor role. Yet, at the early stage of this development, the Dionysiac *argumenta* that were included in these compilations usually lost their original dating clause, but remained unaltered otherwise. But once they were clearly disconnected from the Dionysiac corpus, they lost their authoritative attribution, and were then more liable to alteration. Consequently, from the beginning of the eighth-century compilers and authors of computistical textbooks, formularies, and compendia started to rephrase the Dionysiac formulae, usually by adding more details and explanations to the calculations, or by generalising them.

The earliest witness to this process of progressive disintegration of the Group B corpus and eventual alteration of the original *argumenta* is the famous *Computus Cottonianus* of 688/9, preserved in one of the few eighth-century computistical codices, namely MS London, British Library, Cotton Caligula A XV, whose provenance appears to be northern France.[103] Pallarès has published the section of this text that includes *Argumenta I to XIV*, and compared it with Jan's and Krusch's editions. Yet, the section published by him does not comprise the entire text, since Pallarès was only interested in the transmission of the *argumenta* ascribed to Dionysius, and thus only transcribed as far as the middle of folio 77r;[104] his choice for

---

[103] Cf. Lindsay (1915), 461 (arguing that this manuscript was written in France, in 741); *CLA* 2, 19 (No. 183; Lowe argues that it was 'probably written in North-east France, in a centre with Insular connexions; copied from an exemplar written A.D. 743, the year mentioned on fol. 107'; earlier in this passage he describes the script as 'pre-Carolingian French minuscule, saec. VIII²'); Jones (1939), 120 (northern France, saec. VIII); Pallarès (1994), 20–1. It is the oldest manuscript preserving the Dionysiac *argumenta* in the original or as a recension (a recension in this case); cf. Borst (2006), 398–9.

[104] Pallarès (1994), 20–31. He also defines the *Computus Cottonianus* as comprising fols 73r–77r of this manuscript in two articles of 1987 and 1989, now reprinted in Pallarès (1999), where the relevant references can be found on pp. 26 and 57 respectively (in both cases he mistakenly gives f 73v as the beginning).

breaking off at this point was possibly also influenced by the fact that the following item, i.e. the *Suggestio Bonifati primiceri*, is the only item in this computus that is published separately;[105] yet, this short item, having neither incipit nor explicit that would distinguish it from the rest, seems rather to be an integral part of this computus. In my opinion, folios 73r–80r should be identified as a cohesive formulary, to be termed the *Computus Cottonianus* of 688/9: the beginning of this formulary is clearly defined by the fact that folio 72v is left blank, and thus a new text starts on folio 73r; a later hand adds the heading *Cassiodorus de computo paschali*, an obviously mistaken attribution; the first page after folio 73r, on which the end of an *argumentum* agrees with the end of a page, is folio 80r; no other feature can be found before this page that would mark the end of the text; at the bottom of folio 80r, then, a different hand adds a list of 'canonical' and 'uncanonical' lunar ages, i.e. the 19 epacts on 1 January of the Victorian reckoning and the 11 lunar ages that do not occur on that date in the Victorian 19-year lunar cycle respectively; this later addition filling a previously blank space marks the end of this formulary; on folio 80v, then, starts a recension of the so-called Acts of the Council of Ceasarea.[106]

Now, the *Computus Cottonianus* contains every *argumentum* of the Group B corpus, and at the same time *Argumenta I–XIV* as well as §2 of *Argumentum XVI* of the 16 *argumenta* of the *Computus Digbaeanus*, most of which are here accommodated to 688/9 and preserved in a structure disrupted by additional *argumenta*: The computus begins with *Argumenta I* to *VIII* (accommodated to 688 and §2 of *Argumenta III* and *IV* omitted), followed by *Argumentum XIV* (including the chronological data for 532 to 534); this is followed by a text on the Julian calendar limits for the Easter new moon, the Easter full moon, and Easter Sunday; then, *Argumenta XI* and *XII* are given (accommodated to 689); attached to *Argumentum XII* (which deals with the calculation of the weekday of 1 January) is a formula to calculate the weekday of the calends of any given month by adding the *concurrentes* of a year to given regulars for each month;[107]

---

[105] This text is edited from this codex and five other manuscripts by Krusch (1926), 55–7.

[106] For the Acts of the Council of Caesarea see Krusch (1880), 303–10; Jones (1939), 44–5; Wilmart (1933), 19–27; Strobel (1984), 80–95.

[107] This list of regulars, starting with October, is uncommon and has a parallel in the *Fragmentum Nanciacense;* cf. n 101 above. It is transcribed and discussed in Pallarès (1999), 8–9, who places it in a rather unlikely Arabic context.

this is followed by *Argumenta IX* (with chronological data for 688 and 689), *X* (preserving the chronological data for 525), and *XIII* (preserving the chronological data for 625 and the number of *puncti* as given in the *Computus Digbaeanus*[108]); the remaining part of this computus consists of arguments and texts previously unconnected with the Dionysiac *argumenta*, with the exceptions of §2 of *Argumenta XVI* and *III*, the latter preserving the original chronological data of Group A–B texts.

This description illustrates the degree of disintegration achieved by the *Computus Cottonianus*: the Dionysiac *argumenta* were embedded among numerous other formulae and texts without distinction, while most of the examples were accommodated to the *annus praesens* of the compiler. Consequently, some late seventh- and early eighth-century computists got to know these *argumenta* through a formulary like the *Computus Cottonianus* rather than through the Group B corpus. In fact, the *Computus Cottonianus* seems to have been quite influential in transmitting Insular computistical knowledge in general, and the Dionysiac algorithms in particular.[109] Its role as one of the earliest Insular sources for Frankish computistics and the extent of its influence have yet to be established, since it plays no part in Arno Borst's source analysis of eighth-century Frankish computistical texts. Direct dependency on the *Computus Cottonianus* can be found in a computistical formulary which is preserved in two late eighth- and early ninth-century manuscripts from Cologne and probably Worms, i.e. MS Köln, Diözesan- und Dombibliothek, 103, 184v–190v and MS Wolfenbüttel, Herzog-August-Bibliothek, Weißenburg 91, 169r–173v respectively.[110]

---

[108] For this problem and the possibility that the *Computus Digbaeanus* and the *Computus Cottonianus* are products from the same computistical school, see Appendix II.

[109] Cordoliani had, quite mistakenly, connected the *Computus Cottonianus* with Spanish computistics: Cordoliani (1942); Cordoliani (1958). This view was corrected by Pallarès (1999), 29–32, 60–2, 99.

[110] For descriptions of MS Köln, Diözesan- und Dombibliothek, 103 see Jaffé and Wattenbach (1874), 40–2; Weber Jones (1932), 32–3; van Euw (1998a) (arguing for Cologne, c.795); Bischoff (1998), 397 (probably Cologne, saec. VIII./XI.); Jones (1939), 116 (Cologne, c.800); Springsfeld (2002), 82 (Cologne, 810–18); Borst (2001), 63–5; Borst (2006), 238 (Borst argues in both accounts for Cologne, shortly after 810); facsimiles are available on the world wide web: www.ceec.uni-koeln.de. For descriptions of MS Wolfenbüttel, Herzog-August-Bibliothek, Weißenburg 91 see Butzmann (1964), 257–68 (Weißenburg, saec. IX^1); Borst (2006), 316–7 (possibly Worms, early saec. IX).

The latest dating clause in this formulary is 775, and thus we may term it the *Computus Rhenanus* of 775.[111] This formulary is an excellent example for the fate of the Dionysiac *argumenta* in computistical compendia of the eighth and later centuries: It copies Bede's version of *Argumentum I*, then gives a version of *Argumentum II* for 775, of *Argumentum VI* for 764, of §2 of *Argumentum III* implicitly for 764 (5th year of the *cyclus decemnovennalis*), of §1 of *Argumentum IV* for 764 and of *Argumentum V* for 764; this is followed by a generalized version of *Argumentum II*, of §1 of *Argumentum III* and *IV*, and of *Argumenta V*, *VI*, and *VIII*; then, among other passages copied from the *Computus Cottonianus*, it transmits the latter's recension (though for 689 rather than 688) of *Argumenta VIII*, *I*, and (obviously without dating clause) *VII*; finally, the first part of *Argumentum XIV* is given. Hence, even though most of the Dionysiac *argumenta* are included in this formulary, none appears in its original form, since the compiler worked from various later recensions of them.

---

[111] This formulary has not yet been identified as such, and is consequently unpublished. In the descriptions of the Cologne codex, it is denoted as *Canones lunarium decennovalium circulorum* in Jaffé and Wattenbach (1874), 41, and divided into two parts (fols 184v–187v and fols 187v–190v) in van Euw (1998a), 134, with reference to the pseudo-Bedan *Canones lunarium decennovennalium circulorum* (PL 90, 877–82) for the first part. Butzmann (1964), 267 describes this formulary of MS Weißenburg 91 more precisely as consisting to a large degree of the pseudo-Bedan *Canones lunarium decennovennalium circulorum*, with various insertions. In fact, the entire *Canones lunarium decennovennalium circulorum* are part of the *Computus Rhenanus*, in which it constitutes *c*. half of the text, and consequently only half of the *argumenta* of the *Computus Cottonianus* that are transmitted in the *Computus Rhenanus* can also be found in the *Canones lunarium decennovennalium circulorum*. Yet, the relationship between these two texts is rather the other way round, with the *Canones* being a partial copy of the *Computus Rhenanus*: The *Canones* survive in only one manuscript to my knowledge, namely the 10th century MS Köln, Diözesan- und Dombibliothek, 102, 94r–97r (for facsimiles see www.ceec.uni-koeln.de). This codex shows direct dependency on MS Köln, Diözesan- und Dombibliothek, 103, and consequently the *Canones* are copied from the Cologne version of the *Computus Rhenanus*. Jones, in his discussion of the *Canones* in (1939), 82–3, neither refers to the two Cologne codices, nor to the Wolfenbüttel MS. Borst, in his monumental edition of Frankish computistical texts (2006), published *argumenta* of the *Computus Rhenanus* in three different texts, using excerpts from the Cologne version of this computus in his edition of the *Lectiones sive regula conputi* (*Lect. comp.*), *Annalis libellus* (*Lib. ann.*), and *Libri computi* (*Lib. comp.*), and excerpts from the Wolfenbüttel version only in his edition of *Lect. comp.* Accordingly, all four of these texts share common material, but every single one of them deserves a separate treatment, the *Computus Rhenanus* of 775 included.

In general, the textual closeness to the original *argumenta* that is still preserved in the *Computus Cottonianus* was lost in the eighth and early ninth centuries, because the authors and compilers of computistical texts and formulae felt that the *argumenta* either needed more explanations, or should rather be more generalized. Hence, many of the published computistical text from that period include Dionyisac algorithms, but the wording usually varies considerably from the original. Moreover, as has been illustrated in the case of the *Computus Rhenanus* of 775, the formulae were usually copied or altered from an intermediary exemplar, not from the original. Through these intermediary texts the Dionysiac and pseudo-Dionyisac algorithms remained very popular. Some of these can be found in the major computistical works of the eighth and early ninth centuries: Bede's *De temporibus* and *De temporum ratione*, the *Bobbio Computus*, the Frankish texts *Lectiones sive regula conputi* (*Lect. comp.*), *Annalis libellus* (*Lib. ann.*), *Libri conputi* (*Lib. comp.*), *Liber calculationis* (*Lib. calc.*), the Computus of Pacificus of Verona, and Hrabanus Maurus' *De computo*.[112] They also appear in considerable variation in MS Basel, Universitätsbibliothek, F III 15k (with examples for 789), which Springsfeld described misleadingly as containing the original Dionysiac *argumenta*,[113] MS Vatican, Biblioteca Apostolica, Reg. Lat. 1260 (with examples for 788), which is erroneously the only manuscript listed by Cordoliani for the original Dionysiac *argumenta*,[114] and MS Padua, Biblioteca Antoniana, I 27, which is mistakenly listed as a manuscript witness of the *Computus*

---

[112] Bede's *De temporibus* 14 (Jones (1943), 304–5), *De temporum ratione* 47, 49, 52, 54, 57, 58 (Jones (1943), 266, 269, 273–4, 278); Bobbio Computus 153–155 (PL 129, 1364–6); *Lect. comp.* IIII 1–7, V 1, VI 6 (Borst (2006), 591–603, 624–6); *Lib. ann.* 2, 3, 6, 8, 15, 20, 24, 25 (Borst (2006), 683–8, 698, 712–3, 715–6); *Lib. comp.* II 7–12b, III 13 (Borst (2006), 1150–6, 1194); *Lib. calc.* 17–23 (Borst (2006), 1397–9); Computus of Pacificus of Verona §§208–9, 213, 216, 219, 249, 250, 309 (Meersseman and Adda (1966), 109–12, 116, 129); Hrabanus Maurus' *De computo* 62, 67, 69, 72, 78, 90 (CCCM 44, 278, 283–5, 288, 295, 312–3).

[113] Springsfeld (2002), 73, 76. Her statement seems ultimately to derive from Jones (1937), 214, who states more precisely: 'The same rubrics with altered formulae in Ba, fo. 37v, from an exemplar written A.D. 789.'

[114] Cordoliani (1943), 60. As so often, Cordoliani trusted the incipit more than the actual content. At least from the dating clauses he should have realized that this is, in fact, a later recension. The incipit reads (MS Vatican, Biblioteca Apostolica, Reg. Lat. 1260, 118r): *Incipiunt argumenta de titulis paschalibus Aegiptiorum inuestigata sollertia, quae Dionisius conposuit utraque lingua Grece uidelicet et Latine eruditus.* This text is somehow related to the *argumenta* in the Basel MS mentioned in the previous note (I have not yet worked out the relationship), which have a very similar incipit.

paschalis of 562.[115] Moreover, the computistical excerpts from MS Paris, Bibliohèque Nationale, Nouvelle acquisition latine 2169 (including a recension of *Argumentum I* from 817) and MS Léon, Biblioteca de la Catedral, N. 8 (including a recension of *Argumentum I* from 806) published by Pallarès fall into this category.[116]

In summary, the texts of this recension do provide valuable information about the later transmission of the Dionysiac algorithms, but only limited insight into what may have constituted the original corpus of Dionysiac *argumenta*. Yet, especially the earliest texts of this recension, which are the most likely to have had the original Dionysiac corpus as their exemplar, show an interesting feature by grouping *Argumenta I–VIII* without their later additions, sometimes excluding *Argumentum VII*: In the *Computus Cottonianus Argumenta I–VIII*, without the later additions to *Argumenta III* and *IV*, appear as one body at the beginning of the text, with all examples being accommodated to 688 (*Argumentum VII* obviously has no dating clause). Bede's *De temporibus* of 703 gives a generalized version of *Argumenta I–VI* and *VIII*, in which only the first paragraph appears in the case of multi-paragraphed *argumenta*.[117] The same holds true for the *Lectiones sive regula conputi* (*Lect. comp.*) of 760, even though §2 of *Argumentum III* occurs slightly later in the text, as it does in the *Computus Cottonianus*.[118] This fact obviously raises the question, whether some seventh and eighth century computists regarded only *Argumenta I–VI* and *VIII* as originally Dionyisac.[119] If they did, the question about the reason and source behind this view remains, since no manuscript is known

---

[115] CCSL, Clavis patristica III A, 269 lists MS Padua, Biblioteca Antoniana, I 27, 14v–15r as a witness for the *Computus paschalis* of 562, referring to Thorndike and Kibre (1963), 1455. There, the incipit *Si nosse vis quotus annus est ab* is attributed to the *Computus paschalis*. Yet, this incipit cannot be found on the mentioned folios of the Padua MS, which, in fact, are part of Hrabanus Maurus' *De computo*. It may well be that Thorndike and Kibre actually meant MS Padua, Biblioteca Antoniana, I 27, 86v–87r, with this incipit occurring on f 86v. Yet, the *argumentum* introduced by this incipit continues to calculate 881 as the *annus praesens*; this and the following *argumenta* have no connection to the *Computus paschalis* other than that both texts include later recensions of the Dionysiac *argumenta*.

[116] Pallarès (1999), 67–91.

[117] Bede, *De temporibus* 14 (Jones (1943), 301–2).

[118] *Lect. comp.* IIII 1–7, V 1 (Borst (2006), 591–603).

[119] It should be remembered that van der Hagen (1734), 207, regarded only this group of *argumenta* as certainly belonging to the original Dionysiac corpus. Cf. pp. 55–6 above.

which transmits only these 7 *argumenta* with the original dating clause of 525.

Overview

After this discussion of all known manuscripts and texts that contain information about the original corpus of the Dionysiac *argumenta*, it seems appropriate to facilitate the reader with an overview of the details of this analysis, which is done in table form here (*Figure 1*). At the same time, *Figure 1* will serve as the basis for the following discussion of the extent of the original corpus and the different layers of additions that are part of the *Computus Digbaeanus* of 675.

Defining, Dating, and Placing Dionysiac and Pseudo-Dionysiac *Argumenta*

From the manuscript evidence summarized in *Figure 1* above the following conclusions can be drawn: Since *Argumenta XI–XIII* only appear in the *Computus Digbaeanus* (Group A) and the *Fragmentum Nanciacense* (Recension C) and incorporate seventh-century dating clauses, they can safely be regarded as pseudo-Dionysiac; the same is valid for *Argumenta XV* and *XVI*, which have no dating clauses, but appear only in the *Computus Digbaeanus* of the texts listed above. Consequently, *Argumentum XIV* is embedded among pseudo-Dionysiac *argumenta* in that formulary; moreover, it appears separated from the main corpus of *argumenta* by an explicit and incipit in Group B, and does not appear in any of the recensions.[120] For these reasons it seems that it should also be classified as pseudo-Dionysiac, but the implicit dating clause for 532 to 534 suggests that it needs more analysis. Of the first ten *argumenta*, only the parts that appear in all groups and recensions and include a dating clause for 525, the year in which Dionysius evidently wrote his prologue, can, at this stage, be safely considered as Dionysiac, namely §1 of *Argumentum I, Argumentum II*, §1 of *Argumenta III* and *IV, Argumenta V, VI, VIII*, and §1 of *Argumenta IX* and *X*. Yet, doubt remains concerning the last mentioned, since they are not included in the recensions of these *argumenta* in

---

[120] Note that in the Paris MS, *Argumentum XIV* is physically disconnected from Recension B (*Computus Parisinus* of 819/20). Cf. pp. 68–9 above.

Argumentum	I			II	III		IV		V	VI	VII	VIII	IX		X	
	1	2	3		1	2	1	2					1	2	1	2
Group A	525	x	x	525	525	10, 11 of cd (579)	525	(581)	525	525	x	525	(525/6)	x	(525)	x
Group B	525	x	x	525	525	10, 11 of cd (579)	525	(581)	525	525	x	525	(525/6)	x	(525)	x
Recension A	562	y		562	562		562		562	562	VIII	562	(562)		(562)	
Recension B	post 806 pre 822	z	z	820	820	17 of cd (814)	819	(581)	819	819 819	x	VII	(813/4)	x	(525)	x
Recension C																

Argumentum	XI	XII	XIII	XIV	XV	XVI
Group A	675	675	(625/7)	(532–4)		
Group B				(532–4)		x
Recension A						
Recension B					x	
Recension C	partially	625	(625/7)			

Figure 1 The extent and chronological details of the Dionysiac and pseudo-Dionysiac *argumenta* according to their primary witnesses. AD dates explicitly mentioned in the *argumenta* stand alone, while implicit dates are given in brackets. If an *argumentum* contains no dating clause, x stands for its inclusion in the respective corpus of *argumenta*, y and z for its inclusion in variation, while a blank space indicates that it is not included. '10, 11 of cd' denotes the 10th and 11th year of the *cyclus decemnovennalis*, which represents the chronological data given in that *argumentum*. Roman numerals indicate that the *argumentum* with that number appears in that place in the respective corpus.

Bede's *De temporibus* and the *Lectiones sive regula conputi* (*Lect. comp.*), and they appear disconnected from the first eight *argumenta* in the *Computus Cottonianus*. Thus, all the paragraphs and *argumenta* which cannot readily be attributed to Dionysius as well as the evidently pseudo-Dionysiac *argumenta* need to be analyzed in detail concerning their most likely date and place of composition and addition to the Dionysiac corpus. It is apparent from the manuscript evidence that whichever parts of the first 10 *argumenta* were later added to the original corpus, these additions happened before the corpus as defined in Group B and Recension B (*Computus Parisinus* of 819/20) received a wider distribution, and hence before further *argumenta* were attached to this strictly defined corpus. Therefore, the controversial passages among the first ten *argumenta* are analyzed and summarized first, followed then by *Argumenta XI* to *XVI*.

§§2 and 3 of *Argumentum I*: The first editor of the Dionysiac *argumenta*, Wilhelm Jan, did not include these two paragraphs in his edition, even though they appear in his sole manuscript witness, while the second editor, Bruno Krusch, relegated them to the apparatus.[121] Thus, both editors did not regard them as genuinely Dionysiac. This is an odd treatment of these paragraphs considering that both editors had no scruples in publishing *argumenta* with seventh-century dating clauses in the same corpus. Anyway, the reason for discarding these paragraphs seems to have been their content: as outlined above,[122] the formula given in §1 needed to be modified every 15 years. The need for this modification is explained in general terms in §2, including the example of the next modification, i.e. the increase of the multiplicand by 15 from 34 to 35, which would become necessary in 537. §3, then, expands this example to the subsequent modification, i.e. the increase of the multiplicand by 15 from 35 to 36, which would become necessary in 552. Moreover, §3 gives more mathematical details about the application of the algorithm in years of such a modification, in this case the year 552: instead of adding indiction 15 in this year, the multiplicand by 15 is increased by one; in subsequent years, however, the indiction will then have to be added again until indiction 15 is reached; at this point the multiplicand by 15 will again have to be increased by one, from 36 to 37, which would become necessary in 567. Consequently, the argument for discarding these paragraphs as pseudo-Dionysiac seems to be that they deal with problems that would not

---

[121] Jan (1718), 81; Krusch (1938), 77.
[122] Cf. pp. 45–6 above.

need to have concerned Dionysius at his time of writing. Additionally, these detailed further explanations seem out of place considering the short and concise descriptions of the following formulae.

Yet the formula given in §1 could only be fully understood and applied in subsequent years with at least the additional explanations outlined in §2. Moreover, the parameter calculated in *Argumentum I*, i.e. the AD date, is the sole precondition of the following six *argumenta* (*II–VI* and *VIII*); consequently, the correct calculation of this parameter was vital, and the algorithm was not complete without the explanation of the modifications that become necessary every 15 years. This need for further details alone explains the different structure of *Argumentum I* compared to the plain formulae of the subsequent ones. If §§2 and 3 are, however, considered together, the discussion of the modifications to the algorithm seems indeed unnecessarily extensive compared to the very brief and precise style of the subsequent *argumenta*. Moreover, every detail concerning the modifications to the algorithm was readily and precisely explained in §2; §3 only supplies further nuances, which are illustrated by the example of the year 552. On these textual grounds it seems to me that only §2 was part of the original corpus of Dionysiac *argumenta*, while §3 was added in or just before the year 552. This view is supported by the manuscripts evidence: while both paragraphs occur in Groups A and B and in adapted form in Recension B, §3 does not appear in Recension A, i.e. in the *Computus paschalis* of 562, which is a crucial recension for this analysis of the shape of the original Dionysiac corpus. In consequence, this implies that a paragraph was added to the original corpus as early as *c*.552, while the author of the recension of 562 worked from the original, pre-*c*.552 corpus; but it seems also possible that this third paragraph was added at a later stage, with its author regarding an extension of the example of §2 as the best way of providing more details for the modification of the formula given in §1.

§2 of *Argumentum III*: There are various conclusive reasons for dismissing this paragraph as pseudo-Dionisiac: it does not appear in Recension A; it explicitly states that it was invented later (*Item alium computum nuper inuentum*); the calculation differs from the general structure of calculating from the AD date; the chronological data refers to the 10th and 11th year of the *cyclus decemnovennalis*, which is incompatible with 525, the date given in §1 of this *argumentum*. The data corresponds, in fact, with $522 + n \times 19$.[123] Since this paragraph appears in Groups A and

---

[123] Cf. p 47 above.

B, as well as Recension B, it seems plausible to suggest that it was inserted at this place before the *argumenta* received a wider circulation, i.e. before the end of first quarter of the 7th century. The 10th year of the *cyclus decemnovennalis* occurred in the following years in the period between 525 and 625: 541, 560, 579, 598, and 617. Of these, none is really close to the date suggested for the addition of §3 of *Argumentum I*, i.e. c.552. In fact, the similarity in style of the first sentence of this paragraph (cited above) with the first sentence of §2 of *Argumentum IV*, which can be dated to 581 or shortly thereafter, seems to suggest that the paragraph in question here was composed close to this date. Hence, I regard 579 or one or two years later (applying examples of the years just passed) as the most likely date for its invention, and that it was then added to the Dionysiac corpus in or slightly later than 581.

§2 of *Argumentum IV*: That this paragraph is a later addition is explicitly expressed in its first sentence: It was invented later for the purpose of simplifying the calculation of the preceding paragraph (*Item nuper inventa melius iudicavi, si brevius patefiat*). This simplification was achieved by counting the years not from Christ's incarnation, but from the first year after (*post*) the consulship of Tiberius Iunior Augustus; because of this change in the precondition, the fixed parameter given in the algorithm had to be modified accordingly from 4 to 1. Since the number of years was counted inclusively, this fixed parameter denotes the *concurrentes* of the first year of the count minus one. Moreover, the count of years in the formula of §1 obviously starts with the first year of a four-year bissextile cycle, i.e. a year following a bissextile year (namely AD 1), since no parameter is added to the number of years to be divided by four, i.e. every fourth year of the count was bissextile; as no further modifications in this respect are mentioned in §2, the same must be valid for the first year of the new count.[124] Hence, the first year after the consulship of Tiberius was supposed to have *concurrentes* 2 and to follow a bissextile year. Various suggestions have been made concerning the AD date of the year in question, but these suggestions were almost never accompanied by any further explanation.[125] There

---

[124] For the details of the algorithm cf. pp. 47–8 above.

[125] A late hand in the margin of MS Oxford, Bodleian Library, Digby 63, 74r notes 'AD 581' as comment to this paragraph, which was adopted by Jan (1718), 84; the present analysis proves that this is the correct date. Van der Hagen (1734), 207 regards 703 as the date in question. Krusch (1938), 62, 76 argues for 582, and is followed in this view by Borst (1998), 177; Springsfeld (2002), 172; Bär, http://www.nabkal.de/dionys.html. Neugebauer (1982), 296, makes a case for 577. Cf. pp. 53–62 above.

can be hardly any doubt that the Tiberius referred to in this passage is the Byzantine Emperor Tiberius Constantinus. According to the generally accepted chronology, Tiberius ruled from 578 to 582.[126] Yet, of these years only 581 agrees with the above mentioned chronological data, and thus this must have been the year regarded as the first year after Tiberius' reign by the author of this paragraph. Consequently, this paragraph seems to have been composed in 581 or shortly thereafter, presumably in a place in or with close connections to the Byzantine Empire; since it was apparently added before this corpus got a wider distribution, some place in Italy with the just mentioned attribution seems to be most likely.[127]

Argumentum VII: It should be remembered that Dionysius compiled the *argumenta* to provide algorithms for calculating the data of the columns of his Easter table.[128] From that perspective, *Argumentum VII* seems well placed here, since it deals with the Easter full moons, which are listed in the column following the years of the *cyclus lunaris*, the topic of *Argumentum VI*. Yet, van der Hagen doubted that this *argumentum* had been part of the original Dionysiac corpus for the reason that it has a structure different from the other *argumenta*, as it provides no formula, but only a list.[129] More conclusive than this argument is the manuscript evidence: *Argumentum VII* does not occur in Recension A and in a different place in Recension B. Consequently, this *argumentum* appears to be a later addition, which was presumably inserted into the corpus in or after 552, i.e. the possible date for the earliest addition (§3 of *Argumentum I*), but before it got a wider circulation in the early seventh century. It seems rather unlikely that it was added at the same time as §2 of *Argumenta III* and *IV*, since the phrasing is different, as is the quality of the *argumentum* (formula vs list). As will be discussed in the following, it may have been added at the same time as §2 of *Argumentum IX*.

Argumentum IX: As mentioned earlier, two aspects suggest that this and the following *argumentum* may not be Dionysiac: 1) In some later transmissions of the Dionysiac algorithms *Argumenta I–VIII*

---

[126] For Tiberius and the period of his reign see Schreiner in *LM* 8, 760–1 (*s.v.* Tiberios).

[127] Borst (1998), 177; Springsfeld (2002), 172 also argue for Italy. Cf. pp. 60–1 above.

[128] See the citation given on p 42 above.

[129] Van der Hagen (1734), 207.

appear as a well-defined group, while these two *argumenta* either do not appear at all, or are transmitted disconnected from this group;[130] 2) the structure of these two *argumenta* is strikingly different when compared to the previous ones, since they do not have the AD date as their precondition. On the other hand, there are many conclusive arguments that *Argumentum IX* was part of the original Dionysiac corpus: it supplies an algorithm to calculate the lunar age of Easter Sunday, which constitutes one of the columns in the Dionysiac Easter table, and which was, at the same time, the most debated question in the Easter controversy; the chronological data of the two examples given in this *argumentum* agree with 525 and 526;[131] *Argumentum IX* is included in all Groups and Recensions, especially also in Recension A (again in considerable variation);[132] most conclusive of all, *Argumentum IX* is referred to and applied in the *Suggestio* of Bonifatius, the chancellor of the papal curia, to pope John I, written in 526.[133] From this evidence there cannot be any doubt that this *argumentum* was certainly part of the original Dionysiac corpus.

Yet the question remains whether this holds true for all parts of this *argumentum*. The reason for outlining two examples instead of just one is that different fixed parameters had to be applied, depending on whether Easter Sunday fell in April or March.[134] Thus, both examples were necessary to understand the algorithm, and for that reason alone both were quite certainly part of the original *argumentum*. Serious doubts exist, however, concerning the second paragraph of this *argumentum*. Scholars have generally avoided a discussion of this paragraph, and if they commented on it, opinions on the meaning of this paragraph vary, and it is yet to be explained convincingly.[135] The key to an understanding of this paragraph may lie in an interpretation of *Argumentum IX* to be found in the *Lectiones sive regula conputi* (*Lect. comp.*): chapter VI 6 of that text consists of a

---

[130] Cf. pp. 78–9 above.

[131] Cf. pp. 48–9 above.

[132] Cf. p 80 above.

[133] Krusch (1926), 56–7.

[134] Cf. pp. 48–9 above.

[135] Comments on this paragraph can only be found on the world wide web: Deckers (http://hbar.phys.msu.su/gorm/chrono/paschata.htm) suggests that this paragraph deals with the calculation of the lunar age of any given day of January, while Bär (http://www.nabkal.de/dionys.html) rather thinks that this paragraph is supposed to point out that the algorithm of §1 of this *argumentum* can also be applied for calculating the lunar age of any given day of a year.

generalized version of §1 of *Argumentum IX*, followed by a slightly altered version of §2 of that *argumentum*, which is then followed by the interpretation in question.[136] The editor of this text, Arno Borst, lists five manuscripts for this interpretation, none of which is older than the middle of the ninth century, while the dating clause of the oldest version suggests that it was first composed in 792.[137] According to this interpretation, §2 of *Argumentum IX* explains nothing but a different composition of the fixed parameters in the algorithm of §1.[138] Only the fixed parameter for April is discussed: if 3 is added to the number of months from September to December (4 inclusively), then the sum is 7; if the last two days of December are added to this, then the sum is 9; this is the fixed parameter applied for calculating the lunar age of Easter Sunday, if the latter happens to fall in April. It is very striking that a discussion of the fixed parameter for March is left wanting in this interpretation, which is the most confusing part of §2 of *Argumentum IX*. In that part of §2 it is argued that in bissextile years 2 instead of 3 has to be added to the number of months from September to December (4 inclusively). But a bissextile year obviously makes no difference to the lunar calculation explained in §1 of this *argumentum*. If the early ninth century interpretation reproduces the original meaning of §2, then 'bissextile years' has rather to be read as 'years in which Easter Sunday falls in March'; the reading would then basically be that the March regular consist of the number of

---

[136] *Lect. comp.* VI 6 (Borst (2006), 624–6).

[137] *Lect. comp.* VI 6D (Borst (2006), 626). The five manuscripts are: MS Bern, Burgerbibliothek, 417 (Fleury, 826); MS Karlsruhe, Landesbibliothek, Augiensis CLXVII (Soissons / Laon, *c*.848); MS Köln, Diözesan- und Dombibliothek, 83² (Cologne, 805); MS London, British Library, Harley 3017 (Fleury, *c*.863); MS St. Gallen, Stiftsbibliothek, 248 (northern France, *c*.850). Borst seems also to include MS Ge, but since no variants are given for this MS, I suspect that it should rather have been listed among the manuscripts in which this passage does not occur. All manuscripts give the dating clause for 792 (epact 23, Easter Sunday on 15 April, *luna* 17) except for the Bern MS, in which the dating clause is accommodated to 825 (epact 28, Easter Sunday on 9 April, *luna* 16).

[138] This interpretation is supported by the fact that different explanations for the fixed parameters in this algorithm were composed as early as 526: Boniface, the chancellor of the papal curia, in his *Suggestio* to pope John I of 526 explained the fixed parameter 9 for April as consisting of the number of months from September to April divided by 2 (8/2=4) plus the five days that the Egyptian year of 360 days is shorter than the Julian calendar year of 365 (Krusch (1926), 56–7). Hence, §2 of *Argumentum IX* was probably just one of a number of different explanations for the fixed parameters in this algorithm, but obviously the only one that made its way into the Dionysiac corpus of *argumenta*.

months between September to December (4 inclusively) + a fixed parameter of 2 + the last two days of December, adding up to 8. The author of §2 may have regarded the first example given in §1 of this *argumentum* as a bissextile year, which led him to the assumption that the fixed parameter for March had to be applied in every bissextile year. Whatever the case may be, the mentioning of the bissextile year is totally misplaced in this context (since it makes a difference in weekday, but not in lunar calculations), and it appears that the compiler of this paragraph did not fully understand the preceding examples. Consequently, it seems very likely that §2 of *Argumentum IX* represents a later, confused addition, which is confirmed by the fact that it does not appear in Recension A. As for a possible date for the addition of this paragraph, it obviously found its way into the Dionysiac corpus before the emended corpus as represented in Group B and Recension B got a wider distribution, i.e. before the end of the first quarter of the seventh century. This paragraph cannot really be connected with §2 of *Argumenta III* and *IV*, since their precise chronological argument is not matched by the confused explanation here. Moreover, the purpose of this paragraph was not to add a new or simpler formula, but to provide additional explanation to the existing argument. Hence, if this paragraph was added at the same time as any of the other additions, I would tend to connect it with *Argumentum VII*, but not with §3 of *Argumentum I* and §2 of *Argumentum X*, which are more technical, and very precise. In the end, however, the most likely scenario seems to be that it was added separately.

Argumentum X: As has been mentioned in the discussion of *Argumentum IX*, *Argumentum X* does not appear in some later transmissions of the Dionysiac *argumenta* and its structure is strikingly different compared to the first seven *argumenta* (*I–VI, VIII*), which casts doubts on its inclusion in the original corpus. Additionally, *Argumentum X* deals with weekday calculation and thus is not immediately connected to any column in the Dionysiac Easter table. Yet, the example given here is to calculate the weekday of Easter Sunday of 525,[139] and thus this *argumentum* may have been included in the corpus as a method for double-checking the solar Easter data, and thus would have had an immediate application for the Easter table (it could obviously also be used for calculating the weekday of the Easter full moon). Moreover, the implicit date in

---

[139] Cf. pp. 49–50 above.

this *argumentum* suggests that it was part of the original corpus, and more decisively it figures in Groups and Recensions A and B discussed above. However, §2 of this *argumentum* does not feature in Recension A. In this paragraph it is argued that the weekday of any given Julian calendar date from 1 January to 31 December can be calculated with the algorithm of §1; this algorithm is here outlined again, this time not by means of an example, but in general terms. Such a generalisation of the preceding formula is not given for any of the other *argumenta*, and it does not fit into Dionysius' general scheme of explaining the parameters of his Easter table. For these reasons, I believe that this paragraph also represents a later addition, which may be connected to §3 of *Argumentum I*, as will be argued in the following; at least it was added to the Dionysiac corpus before that got a wider distribution in form of the Group B corpus.

Argumenta I–X: In summary, of the 10 *argumenta* preserved in the Group B corpus, §§1 and 2 of *Argumentum I*, *Argumentum II*, §1 of *Argumenta III* and *IV*, *Argumenta V*, *VI*, and *VIII*, and §1 of *Argumenta IX* and *X* can be regarded as Dionysiac and seem to have formed the original corpus of Dionysiac *argumenta*. On the other hand, §3 of *Argumentum I*, §2 of *Argumenta III* and *IIII*, *Argumentum VII*, and §2 of *Argumenta IX* and *X* have been identified as pseudo-Dionysiac. But can these additions be classified, i.e. were some of these added at the same time? Only §2 of *Argumenta III* and *IV* can be connected with some degree of certainty, since both have very similar phrasing in their first sentence, and they are the only additions that provide new formulae. Because of these formulae, §2 of *Argumentum III* can be dated implicitly to 579, §2 of *Argumentum IV* to 581 or shortly thereafter. Hence, it seems that these two paragraphs were added to the corpus in 581 or shortly thereafter. As for the place of this addition, the reference to the Byzantine emperor Tiberius Constantinus in §2 of *Argumentum IV* suggests a place that was either part of the Byzantine Empire, or at least in close contact with it. We need not assume that the original corpus had travelled very far at this early stage, so some place in Italy, either under Byzantine rule or with strong Byzantine connections, seems most likely. As for the remaining four additions, §3 of *Argumentum I* and §2 of *Argumentum X* have a quality distinctly different from the other two, since they provide further detailed technical explanations from the hand of a skilled computist, while *Argumentum VII* is a mere list which could have been compiled from any Easter table, and §2 of *Argumentum IX* is a confused explanation of the fixed parameters of the preceding algorithm. Thus, if these four additions were to be grouped, I would be inclined to connect §3 of *Argumentum I* with §2 of *Argumentum X*, and *Argumentum VII* with §2 of *Argumentum IX*. §3 of *Argumentum I* suggests the date 552 for its addition, but it may well have been added retrospectively; I would not rule out the possibility that §3 of *Argumentum I* and §2 of *Argumentum X* were added at the same time as §2 of *Argumenta III* and *IV*. *Argumentum VII* and §2 of *Argumentum IX*, however, seem to me to be disconnected from the other four additions.

All this must obviously remain very speculative. Nevertheless, for the outlined reasons I would at least propose that §2 of *Argumenta III* and *IV* were added to the corpus in 581 or shortly thereafter, §3 of *Argumentum I* and §2 of *Argumentum X* earlier or at the same time, while *Argumentum VII* and §2 of *Argumentum IX* were presumably added later, but before the end of the first quarter of the seventh century; all these additions seem to have been included before the corpus, as it is now preserved in Group B and Recension B, left Italy.

The complexity of the different levels of the just discussed first ten *argumenta* has been noted in passing (and thus never been fully explained) by only very few scholars, most notably Jan and van der Hagen, and, more recently, Neugebauer and Declercq. Concerning the subsequent *Argumenta XI* to *XVI* in the *Computus Digbaeanus*, recent scholars like Borst and Springsfeld believe that they were possibly invented, but certainly added to the corpus in 675, presumably in Ireland.[140] This view needs serious revision, as the following discussion will show.

Argumenta XI to *XIII*: The opinion that *Argumenta XI* to *XVI* were possibly composed and added to the corpus in 675 is based on the fact that *Argumenta XI* and *XII* in the *Computus Digbaeanus* contain dating clauses for that year, which have been the earliest dating clauses for these *argumenta* known to modern scholars of computistics. Yet, an important piece of evidence, discovered 140 years ago, has been overlooked by these scholars, namely the *Fragmentum Nanciacense* discussed above.[141] This fragment suggests that the three *Argumenta XI* to *XIII* formed an entity and that they

---

[140] Cf. pp. 53–63 above.
[141] Cf. pp. 69–72 above.

were probably invented in 625, if not earlier. 625 would have been a very appropriate date for the composition of new *argumenta*, since the Dionysiac Easter table ended in the following year, 626, and the continuation of that Easter table made re-calculations necessary.[142]

Having established 625 as the most likely date for the composition of *Argumenta XI* to *XIII*, the question of their provenance remains. This early date obviously suggests that they were composed somewhere in Italy or Spain, because these were the only Latin speaking places where the Dionysiac reckoning was accepted at that time. Rome, the Frankish kingdoms, Christian Kent and Northumbria would rather have followed the Victorian reckoning, while the 84 (14)-year Easter reckoning still prevailed in Ireland and the British kingdoms, as well as Pictland.[143] In fact, two facts speak strongly for the western Mediterranean world in general, and Italy in particular: 1) The provenance of the manuscript to which the *Fragmentum Nanciacense* is attached is believed to be Bobbio, and the same may be true for this flyleaf;[144] 2) in *Argumentum XIII* computistical methods are applied, which are also described by Maximus Confessor in his *Computus ecclesiasticus* of 641, and which may ultimately go back to fifth century north African computistics.[145] Maximus wrote his computitistical treatise in the North African part of the Byzantine Empire and therefore this *argumentum* may also have been composed in a place under Byzantine control, or at least with close connections to the Empire, somewhere between Bobbio and North Africa.

There is no evidence that this group of *Argumenta XI–XIII* was attached to the Dionysiac corpus (or rather the Group B corpus)

---

[142] For the end of the Dionysiac Easter table see Krusch (1938), 74. The necessary re-calculations are explained by Dionysius himself in his prologue (Krusch (1938), 64). Since *Argumenta XII* and *XIII* deal with the calculation of the solar and lunar data for 1 January, it seems to me that these *argumenta* were invented to add columns for 1 January to the continuation of the Dionysiac Easter table. Columns for the *feria* and *luna* of 1 January can be found in later Easter tables of the Dionysiac reckoning, e.g. in an Easter table for 798–854 transmitted in MSS Vatican, Biblioteca Apostolica, Pal. Lat. 1447, 19v–22r; Vatican, Biblioteca Apostolica, Pal. Lat. 1448, 13v–16r; Vatican, Biblioteca Apostolica, Reg. Lat. 1260, 112r–114v.

[143] Unfortunately, there is no authoritative study on the adoption of the different Easter reckonings in Western Europe. Valuable, but often outdated information on this question can be found in: Krusch (1884); Poole (1918a–b); Schmid (1904); Schmid (1907); Jones (1934).

[144] Cf. n 96 above.

[145] Cf. Appendix II below.

before it was included in the *Computus Digbaeanus* in 675; consequently it seems to me that this group of *Argumenta XI–XIII* and the Group B corpus had disconnected histories before 675.

Argumentum XIV: This *argumentum* is one of the most popular in early computistical literature. The reason for this popularity obviously is its concise method for calculating the Julian calendar date and the weekday of the Easter full moon, as well as the Julian calendar date and lunar age of Easter Sunday. Its earliest version provides an implicit dating clause for 532 to 534, and it is this version which accompanied the Dionysiac *argumenta* in Group B and which was then incorporated in the *Computus Digbaeanus* and the *Computus Cottonianus*.[146] The chronological data were later adapted to match the years 684–686, which was probably done very close to these years.[147] Another and more detailed version of this *argumentum* was composed in 776 and is commonly ascribed to Alcuin.[148] Moreover, numerous variations of his *argumentum* can be found in later computistical texts and manuscripts.

Even though the implicit dating clause of the earliest version suggests 532 as the date of composition, there are good reasons to doubt this. If it was composed in 532, one would expect it to be quite naturally included in the corpus of Dionysiac *argumenta*. Yet, it is not part of, nor does it travel with Recension A of 562, or Recension B; moreover, even though this *argumentum* is attached to the Group B corpus, it is at the same time markedly separated from the main corpus of *argumenta*. Consequently, there is no evidence that it became part of a larger corpus of *argumenta* before its inclusion in the *Computus Digbaeanus* and the *Computus Cottonianus*. Thus, from the existing evidence the incorporation of *Argumentum XIV* into a larger corpus can be dated to 675.

From the manuscript evidence it seems, then, that *Argumentum XIV* was composed and attached to the Dionysiac *argumenta* after the inclusion of all additions to its main body, i.e. after the Group B corpus was established, but before this corpus got a wider distribution, since it is

---

[146] This version is discussed in detail on pp. 51–3 above.

[147] This version can be found, e.g., in MS Paris, Bibliothèque Nationale, Lat. 4860, 151r–v (cf. Springsfeld (2002), 181); MS München, Bayerische Staatsbibliothek, Clm 14456, 46v–47v. The characteristic of this version is that the *concurrentes* given in the three examples are 5, 6, and 7 respectively.

[148] For this version see especially Jones (1939), 43–4, 104–6; Springsfeld (2002), 183–5, 322–8.

attached to the Group B corpus in every manuscript of that group. Hence, I would suggest the late sixth or early seventh century for its composition, before the Group B corpus with *Argumentum XIV* attached left Italy. This is confirmed by the fact that the algorithm described in *Argumentum XIV* was at least partially known to Maximus Confessor in 641;[149] it is possible that Maximus worked directly from *Argumentum XIV*; on the other hand, it is at least as likely that the algorithm itself was an old invention of the Greek speaking East (with Alexandria as its computistical centre), but that it became known to the Latin speaking West only by the late sixth or early seventh century through *Argumentum XIV*. Either way, this would again suggest that *Argumentum XIV* was composed (in Latin) in a centre somewhere in Italy that was part of, or at least in contact with the Byzantine Empire. The examples for 532 to 534 in this *argumentum* were then obviously chosen because they represent the first three years in the Dionysiac Easter table, which were regarded as appropriate, neutral examples.

Argumenta XV and *XVI*: It has been suggested so far that none of the previous *argumenta* was actually composed in 675. Consequently, only *Argumentum XV* and *XVI* may have been original contribution by the author of the *Computus Digbaeanus*, and therefore only these two *argumenta* may provide a clue about the provenance of this computus. They are not *argumenta* in van der Hagen's strict sense, since they do not supply mathematical algorithms; for that reason they can also not be dated with certainty. From the evidence of the *Computus Digbaeanus*, it seems, however, quite certain that these *argumenta* were composed either in 675 or earlier. *Argumentum XV* deals with theological explanations of the equinoxes and solstices, as well as the chronology of Christ's life. *Argumentum XVI* discusses the bissextile day, including a curious division of the 8760 hours of a year by 7 to explain the annual increase of the bissextile day. Now, the topic of *Argumentum XV* was prominent in Bede's *De temporum ratione*, as well as in Irish computistical textbooks of the late seventh and early eighth century, and from these sources it found its way into Frankish computistics.[150] The division given in *Argumentum*

---

[149] Cf. Maximus Confessor, *Computus ecclesiasticus* I 19 (PG 19, 1235–8).

[150] Cf. Bede, *De temporum ratione* 30 (Jones (1943), 236); Munich Computus (MS München, Bayerische Staatsbibliothek, Clm 14456, 8r–46r; an edition is forthcoming) f 19r; *Computus Einsidlensis* (MS Einsiedeln, Stiftsbibliothek, 321 (647), 82–125) pp. 98–9; Bobbio Computus 29, 45 (PL 129, 1291, 1297); *Lib. comp.* I 3b (Borst (2006), 1104); *Lib. calc.* 82 (Borst (2006), 1433). For the recently discovered *Computus Einsidlensis* see Warntjes (2005) and now also Bisagni and Warntjes (2008). For the theological explanations of the equinoxes and solstices discussed in this *argumentum* see also Strobel (1977), 290–303.

XVI, however, curious as it is, cannot be found in Bede's computistical writings, but was particularly popular in Irish computistical textbooks contemporary to the *Computus Digbaeanus*, and then later in Frankish computistics, drawing on these Irish sources.[151] From this perspective it seems that one has to consider Ireland rather than Anglo-Saxon England or the continent for the provenance of the *Computus Digbaeanus*.

On the other hand, it is a very curious fact that only the dating clauses of *Argumenta XI* and *XII* were accommodated to the *annus praesens* of 675 in the *Computus Digbaeanus*. The reason for this may have been that the compiler knew about the significance of the 525 date and did not want to undermine Dionysius' authority by changing the dating clause. Only *Argumenta XI* and *XII*, then, explicitly referred to a year other than 525, namely 625, and the compiler saw no reason why he should not accommodate this to his *annus praesens*. Yet, he seems not to have been computistically skilled enough to change the implicit dating clause of *Argumentum XIII*, which still preserves the 625 dating. Anyway, the fact that the compiler of the *Computus Digbaeanus* actually worked on two of these *argumenta* is interesting in itself. None of the Irish computistical textbooks (*Computus Einsidlensis*, the Munich Computus, and *De ratione conputandi*) mentions any of the algorithms given in the first 14 *argumenta*; the calculations in these textbooks are executed by strikingly different and simpler methods. In fact, the term *annus domini* itself is not mentioned in any of them, (save for a brief reference to it as an element of the Dionysiac Easter table in *De ratione conputandi* 103), and neither is the term *concurrentes*, both being absolute essentials for an understanding of the Dionysiac formulae. From the evidence of these textbooks it seems that Irish computists applied methods of calculation to the Dionysiac reckoning that they had already used for the *latercus* (the 84 (14)-year reckoning) and the Victorian reckoning; these had proven their value over decades, if not centuries, and Irish computists felt no need to adopt formulae based on new chronological concepts they were unfamiliar with, like *annus domini*. For these reasons I suspect that the *Computus Digbaeanus* was compiled in an Anglo-Saxon centre rather than an Irish one, considering especially that Bede used the Dionysiac *argumenta* as early as 703.[152] Yet, as mentioned above, the division given in *Argumentum XVI*

---

[151] Munich Computus fols 22v–23r; *Computus Einsidlensis* p 108 (for this passage cf. Bisagni and Warntjes (2008), 89–90); Bobbio Computus 39, 40 (PL 129, 1295–6); *Dial. Burg.* 14 (Borst (2006), 366); *De Bissexto* (PL 101, 994–5; for this text see especially Springsfeld (2002), 203–14); *Lib. comp.* I 3f (Borst (2006), 1106).

[152] Bede, *De temporibus* 14 (Jones (1943), 301–2). For further arguments for an Anglo-Saxon authorship of the *Computus Digbaeanus* see Appendix II.

does not appear in Anglo-Saxon, but is prominent in Irish computistica. Therefore, my best guess is that the *Computus Digbaeanus* was compiled in an Anglo-Saxon centre in Ireland, such as Rath Melsigi, but monastic centres in Anglo-Saxon England certainly cannot be ruled out.[153]

Conclusion

This article has demonstrated that the text published as the *argumenta* attributed to Dionysius, i.e. the *Computus Digbaeanus* of 675, comprises not only the original Dionysiac corpus, but also different layers of additions, which illustrate (if not represent) the development of computistical formularies written in Latin from 525 to 675. The different stages in this development are the following:

1) The original Dionysiac corpus of 525 is the first known computistical formulary written in Latin. It has not survived in its original extent in any manuscript, but can be reconstructed, primarily on the basis of Recension A (the *Computus paschalis* of 562), as consisting of the following *argumenta*: §§1 and 2 of *Argumentum I*, *Argumentum II*, §1 of *Argumenta III* and *IV*, *Argumenta V*, *VI*, and *VIII*, §1 of *Argumenta IX* and *X*.
2) In the course of the sixth and possibly early seventh century, §3 of *Argumentum I*, §2 of *Argumenta III* and *IV*, *Argumentum VII*,

---

[153] The provenance of the manuscript (which is highly disputed; cf. n 16 above) is irrelevant for the question of the provenance of the *Computus Digbaeanus*, since a text could have travelled anywhere in the 200 years between the composition of the text and that of the codex. For Anglo-Saxon centres in Ireland in general, and Rath Melsigi in particular, see especially Ó Cróinín (1984), reprinted in Ó Cróinín (2003), 145–65. It should be noted that the hypothesis of the *Computus Digbaeanus* being composed in an Anglo-Saxon centre in Ireland depends heavily on the evidence of *Argumentum XVI*, which, in the end, may not even have been part of *Computus Digbaeanus* proper (cf. p 45 above); additionally, even though the curious mathematical explanation of the annual bissextile increase outlined in *Argumentum XVI* cannot be found in Bedan computistica, it is, interestingly enough, attributed to a certain Theodore in the *Computus Einsidlensis* p 108; if the identification of this Theodore with Theodore of Tharsus, archbishop of Canterbury (as proposed by Bisagni and Warntjes (2008), 89–90) is accepted, then the *Computus Digbaeanus* may as likely have been composed in Anglo-Saxon England in general, in Theodore's Canterbury school in particular.

§2 of *Argumenta IX* and *X* were added to the original corpus. Only the addition of §2 of *Argumenta III* and *IV* can be dated with some certainty to 581 or slightly later; it has been suggested that §3 of *Argumentum I* and §2 of *Argumentum X* were presumably added at the same time or earlier (possibly 552), while *Argumentum VII* and §2 of *Argumentum IX* may have been added later. All these additions seem to have been composed and integrated in the original corpus before that corpus got a wider distribution outside of Italy; §2 of *Argumenta III* and *IV* probably originated in a centre under Byzantine control, or at least with some connection to the Empire. This body of *argumenta* is transmitted as a strictly defined corpus in Group B manuscripts and Recension B (the *Computus Parisinus* of 819/20).

3) *Argumentum XIV* was then attached to this Group B corpus, presumably in the late sixth or early seventh century, in all likelihood also in Italy, possibly in a Byzantine centre, or a centre with contact to the Empire. Yet, this *argumentum* was kept strictly separated from the main corpus, as witnessed in Group B.

4) *Argumenta XI* to *XIII* were in all likelihood composed as a group in 625 (as illustrated in Recension C, i.e. the *Fragmentum Nanciacense*), in Bobbio or some other Italian centre with Byzantine connections. There is no evidence that this group was connected with or even attached to the Group B corpus before its inclusion in the *Computus Digbaeanus*, but it is not unlikely that it left Italy in a manuscript which also included the Group B corpus.

5) Finally, the group of *Argumenta XI* to *XIII* was inserted between the Group B corpus and *Argumentum XIV*, and *Argumenta XV* and *XVI* were added to this enlarged corpus; additionally, the marked separation of these groups was abandoned. This newly defined corpus of 16 *argumenta* survived uniquely as the *Computus Digbaeanus* of 675, which was probably compiled in an Anglo-Saxon centre, possibly rather in Ireland (such as Rath Melsigi) than in Anglo-Saxon England proper, though the latter option can by no means be ruled out.

The abandonment of the strict definition of the Group B corpus in 675 basically opened this corpus up to further additions, which ultimately led to its complete disintegration. The earliest witness to this disintegration is the *Computus Cottonianus* of 688/9. Even though the phrasing of the *argumenta* remained essentially unaltered in this text, almost all of the original dating clauses were already accommodated to the *annus praesens* of the compiler. Not long after that, in 703, Bede began the process of generalizing these *argumenta*, with only the algorithms remaining unchanged. In the eighth and ninth centuries, then, the Dionysiac formulae were usually only known through later recensions.[154]

APPENDICES

Appendix I: *Argumenta XI–XIII* in the *Fragmentum Nanciacense* (Nancy, Bibliothèque Municipale, 317 (356))

<Argumentum XI>
... <rem>anent VII: septima luna in XI Kalendas Aprilis ...

<Argumentum XII>
Si uis nosse diem Kalendarum Ianuarium per singulos annos quota sit feria, sume annos ab incarnatione Domini nostri Ihesu Christi, ut put<a> annos DCXXV; deduc assem, remanent DCXXIIII. Hos per quartam partem partire, et quartam, quam partitus es, adiecies super DC<XXIIII>, fiunt DCCLXXX. Hos partiris per septem, remanent III; tertia feri<a> Kalendarum die Ianuarii. Si IIII, quarta feria; si V, quinta feria; si VI, sexta <fe>ria; si asse, dominicus; si nihil, sabbatum.

<Argumentum XIII>
Si uis scire, quota sit luna Kalendis Ianuariis, scito quotus lunaris ciclus lunaris XV est. Tene tibi unum, id est, ipsam diem Kalendas Ia<nu>arias, et deduces quinques deciis, V quinques, faciunt LXXV; quos <ad>iecias super unum, et fiunt LXXVI. Item ducis sexies deciis, V sexies, faciunt LXL; quos adiecias super LXXVI, et fit numerorum summa CLX<VI>, in quibus partiris tricissima, remanent XVI; sexdecima luna est in Kalendis Ianuariis, et puncta XLII. Isto modo per X et VIIII ciclos lunares conputa<bis> semper in Kalendis Ianuariis, ut quota sit luna absque errore re<pperi>es.
Dum autem ueneris ad XVII ciclum lunarem et duxeris quinques deciis <se>ptus super Kalendas Ianuarias, qui faciunt dies LXXXV.[155]

---

[154] For a brief sketch of the fate of the Dionysiac formulae from the late seventh to the early ninth century see pp. 72–9 above.

[155] Here follows mistakenly: Deinde d<u>cis sexies, octuagenta v.

Si partiris sexagissima et adiecias ipsum as<sem>, fiunt LXXXVI. Deinde duces sexies deciis septus, fiunt CII; quos adiecias <super> LXXXVI, et fiunt CLXXXVIII. Partiris ibi tricissima, remanent VIII; octaua luna est in Kalendis Ianuariis et puncta XXIIII. Sic et in XVIII et XVIIII ciclo facies. A primo uero ciclo lunari usque in XVII non partiris sexaginsimam partem ne in errorem incedas. Finit Amen fini<t>.

Appendix II: Explanation of *Argumentum XIII*

Argumentum XIII has puzzled modern computists over the past three hundred years and has never been satisfactorily explained in every detail. Jan admitted that he did not entirely understand *Argumenta XI* to *XVI*,[156] and this was arguably especially true for *Argumentum XIII*. It seems that the number of points mentioned in this *argmentum* irritated him more than anything else; since he did not know how to interpret these numbers, he referred to Bede (*De temporum ratione* 24), who described the increase and decrease of moonlight by four points per day (with five points defining an hour), and to *Argumentum XVI*, where an hour is defined as consisting of four points; he then left it to the reader to establish, which of these two theories was to be applied to make sense of the points mentioned in this *argumentum*, if any sense could be made of them. Additionally, Jan notes various variants (especially in numbers) from the *Computus Cottonianus*, illustrating that even medieval copyists were confused by this text. Van der Hagen, then, drew attention to the fact that this *argumentum* has parallels with the *Computus ecclesiasticus* of Maximus Confessor. At the same time, he pointed out that the algorithm of *Argumentum XIII* can be found in *De temporum ratione* 57, but that Bede does not mention any superfluous points.[157] It is through this chapter in *De temporum ratione* that the formula given in *Argumentum XIII* is generally well explained,[158] and I have described the general method in detail above.[159] This

---

[156] Jan (1718), 88.

[157] Van der Hagen (1734), 208.

[158] Cf. Jones (1943), 278, 388–9; Wallis (1999), 141 (both argue somewhat misleadingly that the lunar age of 1 January is 1 in the first year of the *cyclus lunaris*; correct is rather that the fixed parameter 1 in the formula refers to the lunar age of 1 January of the preceding, i.e. the 19th and final year (cf. p 51 above); they are confused by the fact that Bede started the *cyclus lunaris* a year early in his description of it in *De temporum ratione* 56; for this inconsistency in Bede's account see also Pillonel-Wyrsch (2004), 344–5).

[159] Cf. p 51 above.

formula became quite popular in eighth- and early ninth-century computistics, since it can also be found in the *Bobbio Computus, Lectiones sive regula conputi (Lect. comp.), Annalis libellus (Lib. ann.), Libri computi (Lib. comp.), Liber calculationis (Lib. calc.)*, and Hrabanus Maurus' *De Computo*.[160]

Nevertheless, the formula is not as clearly and plausibly explained in *Argumentum XIII* as it is in these other computistical texts (with the exception of *Lect. comp.*), which is primarily due to the fact that different computistical techniques are used in *Argumentum XIII*; moreover, some additional details are mentioned in *Argumentum XIII*, like the superfluous points, that cannot be explained by any of the other versions of the formula. How confusing the account in *Argumentum XIII* appears to the modern computist is illustrated by the fact that Springsfeld could not make sense of *Argumentum XIII*, even though she explains the same algorithm in *Lib. ann.* perfectly.[161] Still, good mathematical discussions of this *argumentum* by Nikolaus A. Bär and Michael Deckers can now be found on the world wide web.[162] Yet, Bär does not discuss the problem of the meaning of the points, while Deckers cautiously suggests that they may refer to the annual increase of the *saltus lunae*.

In general, there are three aspects that make this *argumentum* difficult to understand:

1) The first is the partition of the multiplicand 11 into 5 and 6 (i.e. 11=5+6). This may slightly confuse the reader, but it obviously makes no difference in the calculation. Such a partition is very uncommon in early medieval computistics, since it does not really simplify the algorithm. Van der Hagen had drawn attention to the fact that Maximus Confessor referred to people who followed the same practice;[163] yet, for their calculations this

---

[160] Bobbio Computus prefix (PL 129, 1281); *Lect. comp.* V 5 (Borst (2006), 607; Borst argues that this chapter deals with the calculation of the lunar age of any given Julian calendar day; yet, the algorithm given rather refers to the calculation of the lunar age of 1 January; interestingly enough, the multiplicand 11 is here also divided into 5 and 6, as in *Argumentum XIII*); *Lib. ann.* 42 (Borst (2006), 738); *Lib. comp.* IIII 3 (Borst (2006), 1201); *Lib. calc.* 31 (Borst (2006), 1402); Hrabanus Maurus, *De Computo* 79 (CCCM 44, 296).

[161] Springsfeld (2002), 175–6, 356–7.

[162] Nikolaus A. Bär: http://www.nabkal.de/dionys.html; Michael Deckers: http://hbar.phys.msu.su/gorm/chrono/paschata.htm.

[163] Maximus Confessor, *Computus ecclesiasticus* I 11, 12, II 1, 5 (PG 19, 1227–30, 1251–4, 1261–2).

partition of 11 into 5 and 6 was a necessity.[164] Since the formulae described by Maximus Confessor are different from the one in *Argumentum XIII*, but similar in their method of calculation, it may be inferred here that the author of *Argumentum XIII* came from a similar, if not the same computistical school as the computists referred to by Maximus Confessor. Since Maximus Confessor wrote his treatise in North Africa, and since the computistical methods described by him ultimately go back to fifth-century Northern African computistics,[165] the author of *Argumentum XIII* possibly also came from that region; at least it is hardly likely that this *argumentum* was composed outside the western Mediterranean world.[166]

2) More problematic is the second example given in *Argumentum XIII*. After explaining the algorithm by means of the 15th year of the *cyclus lunaris*, which undoubtedly refers to the *annus praesens* of the composition of this *argumentum*, i.e. AD 625,[167] the author does not choose the following year as his second example, but rather the 17th of the *cyclus lunaris*. The reason for this is that for the 17th to the 19th year an alteration of the formula became necessary, which the author wanted to illustrate here: the *saltus lunae* is applied in the 16th year, which led to an increase of 12 instead of 11 lunar days between the 16th and 17th year; consequently, 1 had to be added to the existing formula.[168] Now, this addition of 1 was executed in the most unlikely manner in this *argumentum*, namely by arguing that the product of the year in the *cyclus lunaris* multiplied by 5 has to be divided by 60 and the integer from that division has then to be added to the product. This division by 60 was to be applied only for years 17 to 19 of the *cyclus lunaris*. Now, since the products of 17, 18 and 19 multiplied by 5 range between 60 and 120 ($17\times5=85$, $18\times5=90$, $19\times5=95$), the integer when divided by 60 is 1 in all three cases. This odd division by 60 is a reminiscence of calculations described by Maximus Confessor, in which

---

[164] For these calculations see especially Schwartz (1905), 82–4. Maximus Confessor himself (*Computus ecclesiasticus* I 27 (PG 19, 1245–6)) calculated the lunar age of 1 January from the year in the *cyclus decemnovennalis* without partitioning the multiplicand 11.

[165] See especially Schwartz (1905), 70.

[166] For this question see also pp. 89–90 above.

[167] For this *annus praesens* see pp. 51, 71, 89–90 above.

[168] For this alteration of the algorithm see also p 51 above.

5/60=1/12 denoted the increase of the *saltus* per year, i.e. the 12-year *saltus* of the *Supputatio Romana*, and this again points to North African computistics.[169] Because of this association of the division by 60 with the *saltus lunae*, the addition of one which resulted from the *saltus lunae* being applied in the 16th year of the *cyclus lunaris* is described in these odd terms here.

Unfortunately, this second example is not transmitted uncorrupted in the three primary witnesses of *Argumentum XIII*, i.e. the *Fragmentum Nanciacense*, the *Computus Digbaeanus* and the *Computus Cottonianus*. The problem is that in all three accounts the fixed parameter 1 in the algorithm, which was introduced in the first example, is not explicitly mentioned in this second example: The *Fragmentum Nanciacense* calculates consistently without this fixed parameter, thus establishing the lunar age 8 for 1 January of the 17th year of the *cyclus lunaris*; in the *Computus Digbaeanus* and the *Computus Cottonianus* this parameter is silently added: in the *Computus Digbaeanus* 102+86=189, in the *Computus Cottonianus* 85+1=87, eventually resulting in *luna* 9 on 1 January of the 17th year of the *cyclus lunaris*.

How, then, did this corruption occur? There cannot be any doubt that the original *Argumentum XIII* mentioned this fixed parameter 1, since the second example was designed to demonstrate the impact of the *saltus lunae*, i.e. the extra addition of 1, on this calculation; the formula would *de facto* not have been altered, if this fixed parameter 1 was omitted. Consequently, *Argumentum XIII* must have been tampered with later, and for a specific reason: In the original *Argumentum XIII* the *saltus lunae* was applied before 1 January of the 17th year of the *cyclus lunaris*, presumably at the end of the previous November lunation; yet, a different tradition also existed, namely to apply the *saltus* at the end of the following March lunation, i.e. after 1 January of the 17th year of the *cyclus lunaris*. Hence, I suspect that followers of this March place of the *saltus* altered this example for calculating the lunar age of the 17th year of the *cylus lunaris* by omitting the fixed parameter 1, and thus arrived at *luna* 8, which they regarded

---

[169] See especially Maximus Confessor, *Computus ecclesiasticus* I 12, II 5 (PG 19, 1227–30, 1261–2) and Schwartz (1905), 82–3.

as correct, rather than *luna* 9. This altered version of the original *Argumentum XIII* is transmitted through the *Fragmentum Nanciacense*, and this alteration may well have been executed in Bobbio.[170] It was such an altered version, then, that received a wider distribution throughout Western Europe, while it was later silently corrected by followers of the November place of the *saltus*, presumably by Anglo-Saxon computists, as illustrated in the *Computus Digbaeanus* and *Computus Cottonianus*.[171]

3) Yet the aspect that puzzled modern commentators on *Argumentum XIII* most is the mentioning of the points. First of all it needs to be pointed out that the *Computus Digbaeanus* and the *Computus Cottonianus* transmit identical numbers, which disagree with the numbers of points given in the *Fragmentum Nanciacense*. As will be presently seen, this is due to the fact that different concepts are underlying these two versions of *Argumentum XIII*. I will deal with both in turn:

  a) The *Computus Digbaeanus* and the *Computus Cottonianus* give i) *luna* 16 and 16 points for 1 January of the 15th year of the *cyclus lunaris* and ii) *luna* 9 and 26 points for 1 January of the 17th year of the *cyclus lunaris*. What is meant by these points here is the exact time of the day at which moonrise occurs, i.e. the number of points between the beginning of the first hour of that day and moonrise. The algorithm is explained in Frankish computistics, namely in the *Computus Rhenanus* of 775, *Lectiones sive regula conputi (Lect. comp.)*, *Annalis libellus (Lib. ann.)*, *Libri computi (Lib. comp.)*, *Liber*

---

[170] In later computistics, the March place of the *saltus lunae* was a specifically Irish application, while the November place became standard in Anglo-Saxon computistics. This is not only evident from Irish and Bede's computistical textbooks (as well as Alcuin's letters), but it is also explicitly expressed by the Irish computist Dicuil, writing in the Frankish kingdoms in 814–816 (Dicuil, *Liber de astronomia* I 5 (Esposito (1907), reprinted in Esposito (1990), 338): *Etsi lunarem saltum in vigesimo quarto die mensis Novembris, secundum Anglos, complere volueris [. . .]. Sed si secundum Grecorum ac Latinorum regulam, quam mea gens in Hibernia in hac ratione semper custodit, praedictum saltum in vigesimo secundo die mensis sequentis, iuxta primum tempus creationis lunae rationabiliter observaveris*. The Irish may well have adopted a practice that was earlier followed in some Italian centres with Irish connections, most notably Bobbio.

[171] For the association of the place of the *saltus lunae* in November with Anglo-Saxon computistics see the previous note. For further arguments for an Anglo-Saxon authorship of the *Computus Digbaeanus* see pp. 92–4 above.

calculationis (*Lib. calc.*):[172] Count the number of days from 1 January to the day in question; add the product of the year of the *cyclus lunaris* in which this day occurs multiplied by 5; divide this sum by 60; then the remainder denotes the number of points between the beginning of the first hour and moonrise; in mathematical terms:

> (number of days from 1 January + year in *cyclus lunaris* × 5) mod 60 = points between beginning of first hour and moonrise

If this theory is applied to the examples in *Argumentum XIII*, then the results are exactly the numbers of points mentioned:

i) the number of days from 1 January to 1 January is 1 (counted inclusively); the year of the *cyclus lunaris* is 15, and 15×5=75; hence, (1+75)mod60=16, i.e. the number of points given in *Argumentum XIII*; since 5 points equal 1 hour in this algorithm, this means that on 1 January in the 15th year of the *cyclus lunaris* the moon rises in the first point of the fourth hour of the day.

ii) the number of days from 1 January to 1 January is 1 (counted inclusively); the year of the *cyclus lunaris* is 17, and 17×5=85; hence, (1+85)mod60=26, i.e. the number of points given in *Argumentum XIII*; since 5 points equal 1 hour, this means that on 1 January in the 17th year of the *cyclus lunaris* the moon rises in the first point of the sixth hour of the day.

Again, this algorithm shows parallels to the calculations described by Maximus Confessor, even though the applications are different.[173] Nevertheless, the feature of multiplying the year of the *cyclus lunaris* by 5 and dividing the product by 60 seems in both cases to be a reminiscence of the 12-year *saltus* of the *Supputation Romana*, since 5/60=1/12. The general idea of this division by 60 seems to

---

[172] *Computus Rhenanus* (MS Köln, Diözesan- und Dombibliothek, 103, 185v; MS Wolfenbüttel, Herzog-August-Bibliothek, Weißenburg 91, 170r; for this text see pp. 75–6 above); *Lect. comp.* V 9 (Borst (2006), 609); *Lib. ann.* 40 (Borst (2006), 736); *Lib. comp.* IIII 2 (Borst (2006), 1201); *Lib. calc.* 32 (Borst (2006), 1403). Borst's commentary to these chapters is misleading.

[173] Cf. Maximus Confessor, *Computus ecclesiasticus* I 12, II 5 (PG 19, 1227–30, 1261–2) and Schwartz (1905), 82–3.

have been to calculate the impact of the *saltus lunae* on the lunar calculation of any given year of the *cyclus lunaris*, even though the 12-year *saltus* had obviously no connection to a 19-year lunar cycle. The wrong application of mechanisms used for the *Supputatio Romana* to a 19-year lunar cycle, and thus the close resemblance with calculations described by Maximus Confessor (ultimately deriving from fifth-century North African computistics) appears to suggest that this algorithm either originated in North Africa in Maximus Confessor's time or earlier, or that methods introduced by North African computists later found their application in the West.

b) On the other hand, the equivalent passage in the *Fragmentum Nanciacense* reads '*luna* 16 and 42 points' and '*luna* 8 and 24 points'. These numbers of points can be explained by an uncommon system for calculating the length of moonlight per lunar day. The common theory of this question in the early middle ages is that the length of moonlight increases and decreases with the waxing and the waning of the moon by 4 points per lunar day: the length of moonlight for *luna* 1 is 4 points, for *luna* 2 8 points, etc. until it is 60 points for *luna* 15; from *luna* 15 it then decreases by 4 points, so that it is 56 points for *luna* 16, etc. until it is 4 points for *luna* 29; 5 of these points constitute an hour, so that the longest period of moonlight is 12 hours on *luna* 15. This theory can be found in the *Bobbio Computus*, *De temporum ratione*, *Annalis libellus* (*Lib. ann.*), *Libri computi* (*Lib. comp.*), the Computus of Pacificus of Verona, and Hrabanus Maurus' *De computo*.[174] The theory applied in the *Fragmentum Nanciacense* version of *Argumentum XIII* is slightly different in that it can be established that the increase and decrease of the length of moonlight is 3 points per lunar day rather than 4: for *luna* 8: $3 \times 8 = 24$ points; for *luna* 16: $15 \times 3 - 1 \times 3 = 42$ points. In this theory it seems that an hour is defined as consisting of 4 points rather than 5, so that the

---

[174] Bobbio Computus 65 (PL 129, 1305); Bede, *De temporum ratione* 24 (Jones (1943), 226–7); *Lib. ann.* 44, 45 (Borst (2006), 740–1); *Lib. comp.* IIII 5 (Borst (2006), 1202), Computus of Pacificus of Verona §§337–342 (Meersseman and Adda (1966), 136–7); Hrabanus Maurus, *De computo* 43 (CCCM 44, 255–6). For this theory cf. Jones (1943), 359; Springsfeld (2002), 358; Pillonel-Wyrsch (2004), 196–200 (including a rather meaningless comparison with modern observations).

longest period of moonlight, i.e. for *luna* 15, would be 45 points=11¼ hours, which can be compared to the 12 hours given in the common theory. Now, this interpretation of the numbers of points in the *Fragmentum Nanciacense* would have to remain speculative, were it not for the fortunate fact that the theory described is transmitted in a computistical MS, namely the often cited mid-eighth century (and therefore one of the oldest computistical codices that have survived) MS London, British Library, Cotton Caligula A 15, 106v:

> Incipit de luna quantum lucit in noctae
> Momenta XII faciunt uncia una, et uncias III faciunt puncto I, et puncti quattuor faciunt ora una.
> Luna prima lucet uncias VIIII, qui faciunt punctis III; luna secunda hora et dimidia lucet; luna tertia horas II et uncias III; luna quarta horas III lucet; luna quinta horas III et uncias VIIII; luna sexta horas IIII et uncias sex; luna septima horas V et uncias III; luna octaua horas sex; luna nona horas sex et uncias VIIII; luna X horas septem et uncias sex; luna XI horas VIII et uncias III; luna XII horas nouem; luna XIII horas VIIII et uncias VIIII; luna XIIII horas X et uncias sex; luna XV horas XI et uncias III; luna XVI horas XI; luna XVII horas X et uncias III; luna XVIII horas X et uncias VI; luna XVIIII horas VIIII et uncias VIIII; luna XX horas nouem; luna XXI horas VIII et uncias III; luna XXII horas VII et uncias VI; luna XXIII horas VI et uncias VIIII; luna XXIIII horas sex; luna XXV horas V et uncias tres; luna XXVI horas IIII et uncias VI; luna XXVII horas III et uncias VIIII; luna XXVIII horas tres; luna XXVIIII horas II et uncias III; luna XXX hora una.

Note that this passage agrees with the theory outlined above only in its first half, namely by recording a consistent increase of moonlight by 3 points per lunar day from *luna* 1 to *luna* 15. In the second half, however, this passage is inconsistent by recording a decrease of only one point from *luna* 15 to 16, an increase (!) of a point from *luna* 17 to *luna* 18, and a decrease of 5 points from *luna* 29 to *luna* 30. It may well be that the copyist was only familiar with the common theory, realized when copying the data for *luna* 15 that the theory he copied does not agree with what he was used to

(11 ¼ instead of 12 hours of moonlight for *luna* 15), and then tampered with the following data to get closer to his own data.

A few words remain to be said about the transmission of *Argumentum XIII*. As has been pointed out above, it seems that the original *argumentum* (α) has not survived, since the mentioning of the fixed parameter 1 was deliberately omitted in the second example (I denote this version, in which the fixed parameter was omitted in the second example, as β). Moreover, since two different concepts for the calculation of the mentioned points are applied in the surviving recensions of this *argumentum*, the question remains which of these concepts was applied in α. Since the methods of calculation used in the concept transmitted in the *Computus Digbaeanus* and the *Computus Cottonianus* have parallels with the general algorithm of this *argumentum*, I believe that the number of points transmitted in these two texts represent the original. They remained unaltered in β, and were only changed according to a different concept when copied in the *Fragmentum Nanciacense*. The deliberate omission of the fixed parameter 1 in β was retained in the *Fragmentum Nanciacense*, since this seems to have suited the computistical custom of the Irish copyist. In the *Computus Digbaeanus* and the *Computus Cottonianus*, however, this parameter was silently added, because their compilers, being probably trained in an Anglo-Saxon computistical milieu, followed the same practice as the author of the original *argumentum*. Since this silent addition was made in different parts of the calculation, it seems unlikely that *Argumentum XIII* of the *Computus Cottonianus* was directly copied from the *Computus Digbaeanus*. Hence, the content of these three versions suggests the following basic stemma (*Figure 2*), even though only a proper edition of *Argumentum XIII* can solve the question of their relation:

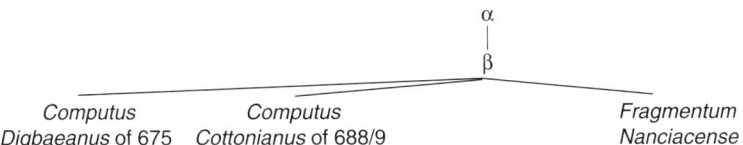

Figure 2 Tentative stemma of the early transmission of *Argumentum XIII*.

Appendix III: Facsimiles

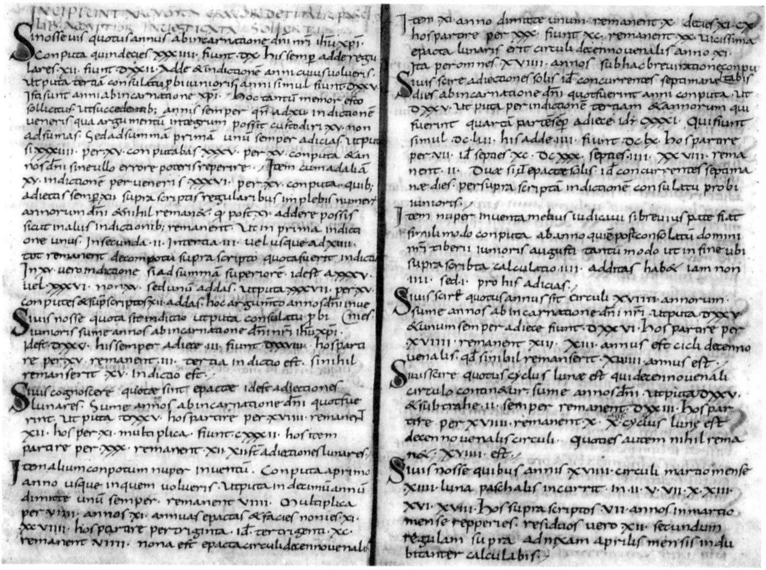

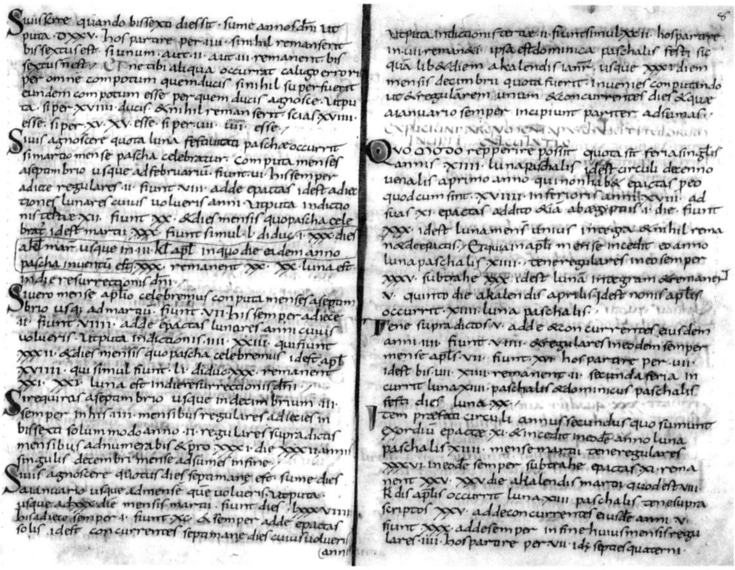

Plate 1 *Argumenta I–X* as a strictly defined corpus with *Argumentum XIV* attached as found in Group B manuscripts (here MS Vatican, Biblioteca Apostolica, Pal. Lat. 1447, 6v–8r).

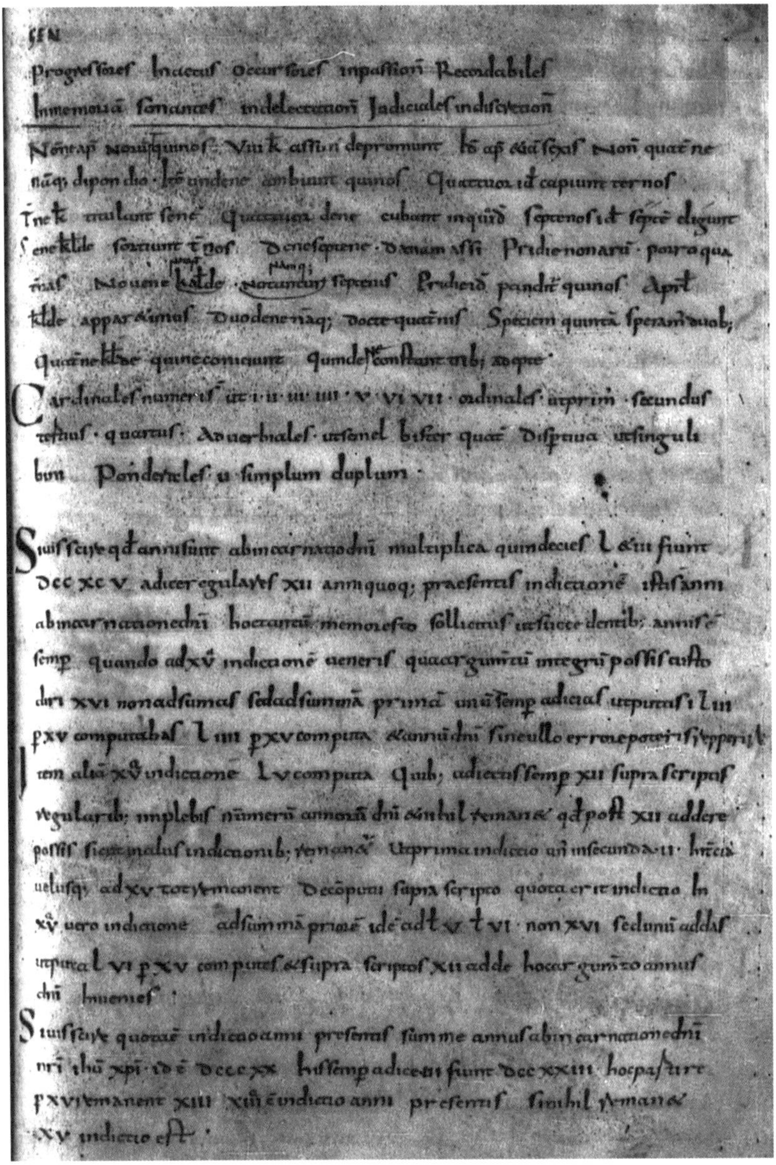

Plate 2 The *Computus Parisinus* of AD 819/20 (MS Paris, Bibliothèque Nationale, Nouvelle acquisition latine 1615, 154r–155r). *(Continued)*

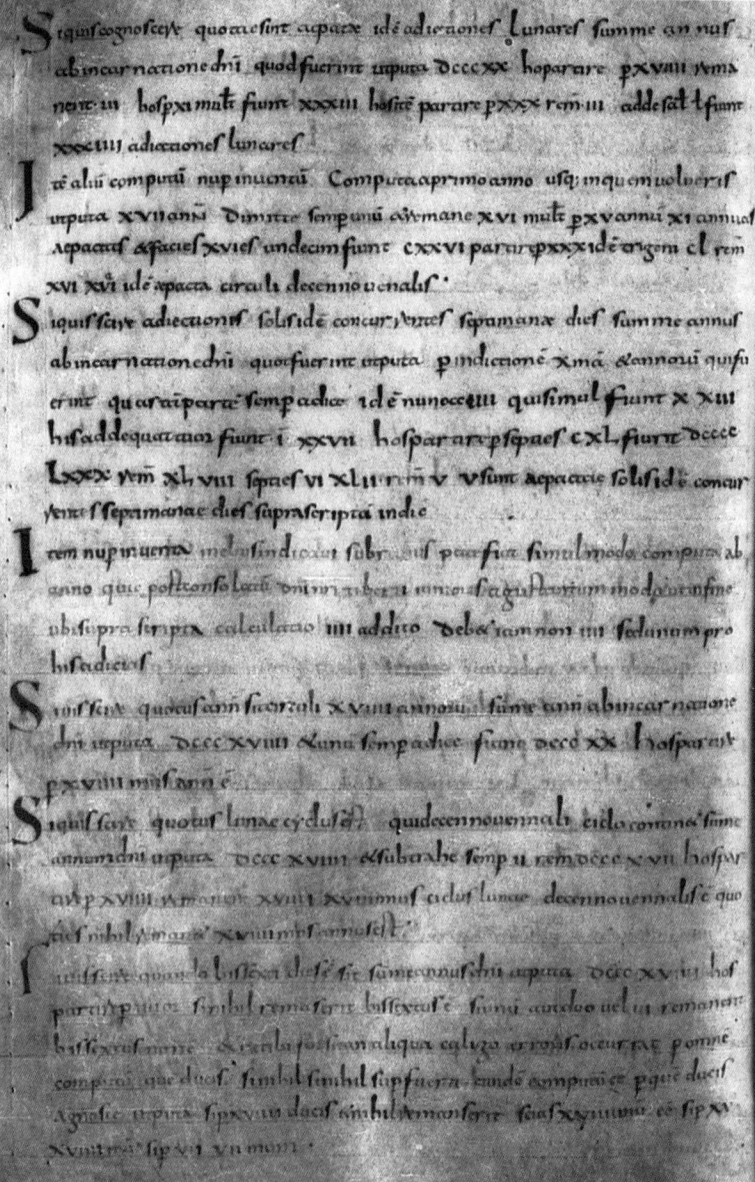

Plate 2 *(Continued)*

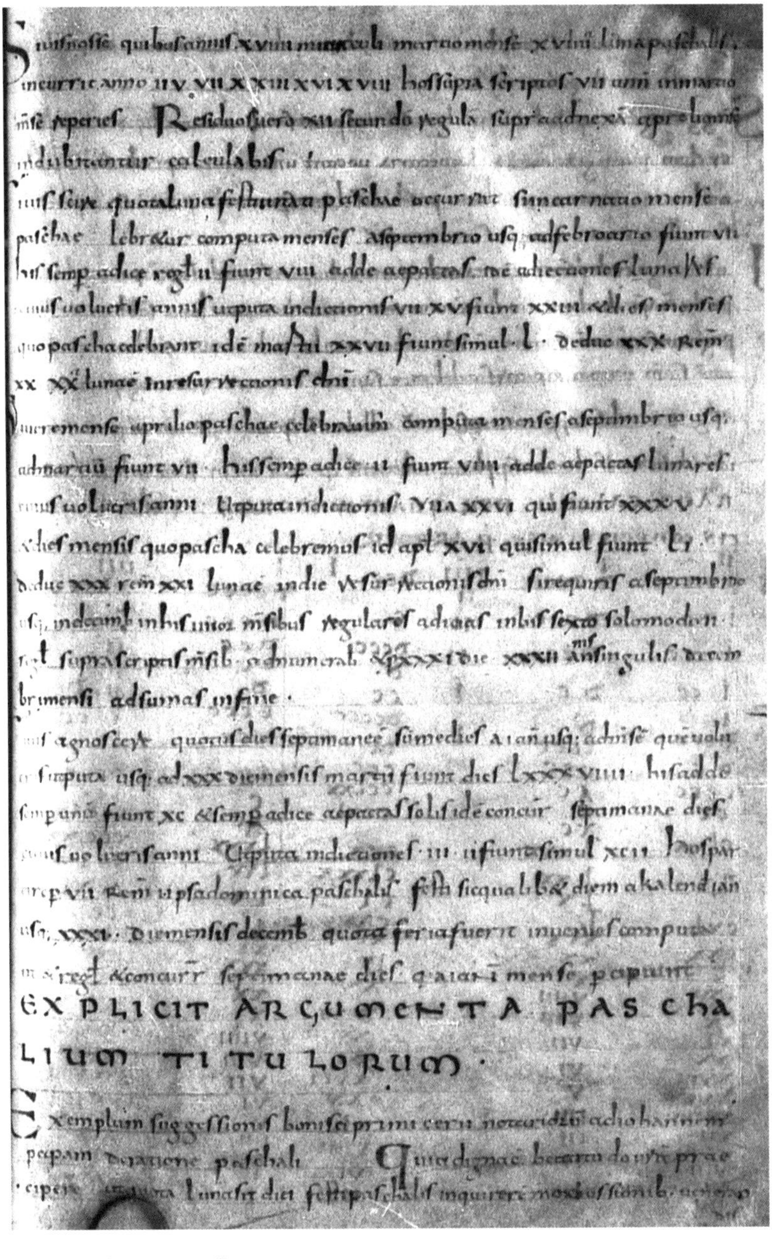
EXPLICIT ARGUMENTA PASCHALIUM TITULORUM ·

Plate 2 *(Continued)*

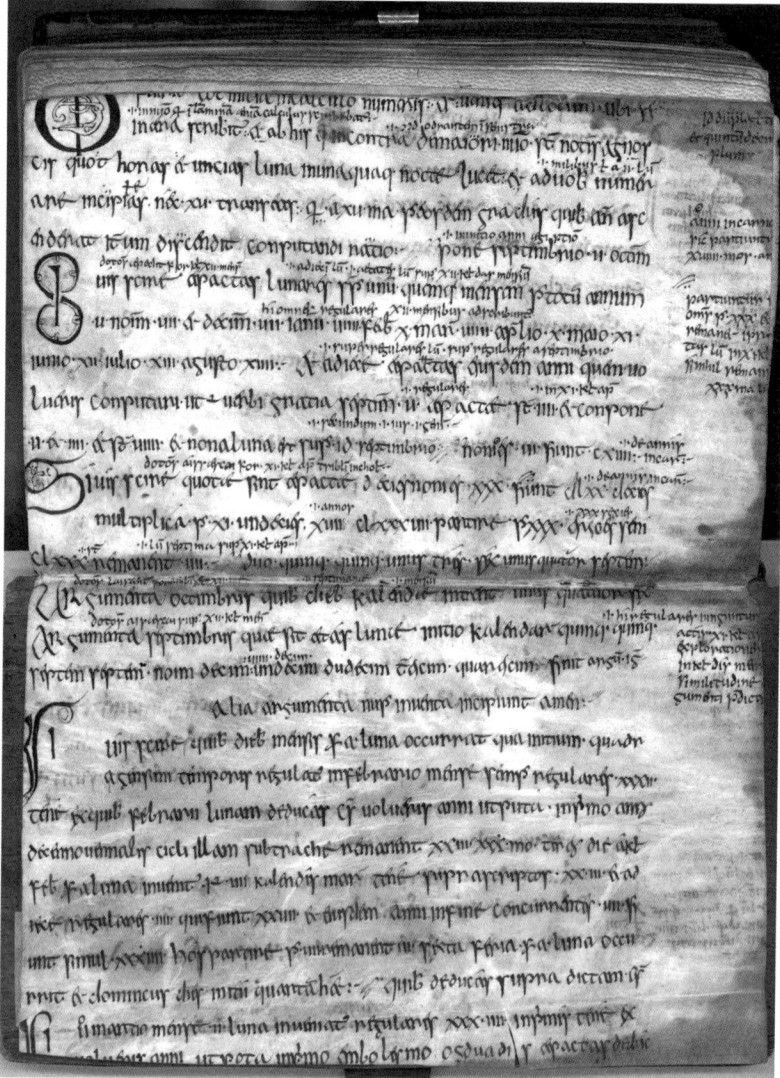

Plate 3 The *Fragmentum Nanciacense* (MS Nancy, Bibliothèque Municipale, 317 (356)). *(Continued)*

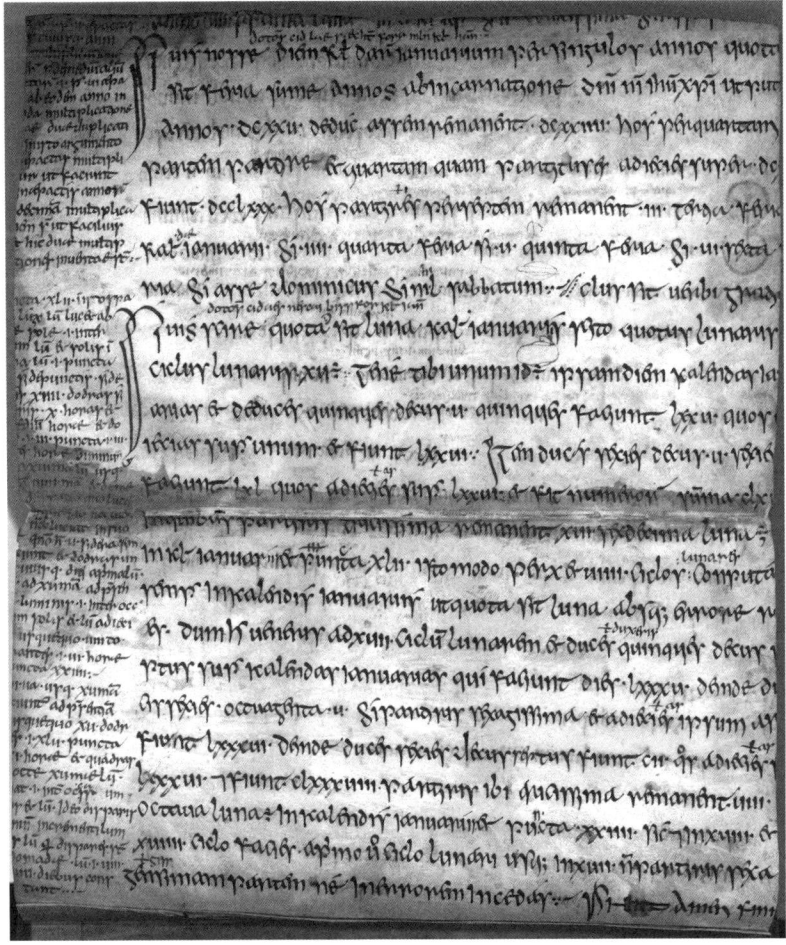

Plate 3 *(Continued)*

ERIC GRAFF

THE RECENSION OF TWO SIRMOND TEXTS: *DISPUTATIO MORINI* AND *DE DIVISIONIBUS TEMPORUM*

Abstract

This paper examines the tradition of the Sirmond computus, comparing Jones' diagrammatic representations with new information from manuscript collations of *Disputatio Morini* and *De diuisionibus temporum*. These works represent two major components of the Sirmond compilation: the paschal letters collection and the texts used for teaching computus in the schools. By tracing their individual recensions, this paper aims to refine our understanding of the origin of these works and their places in the history of early medieval computistics.

Keywords

Irish computus, time, Sirmond, Latin text editing, forgeries, Dionysiac Easter reckoning, Victorian Easter reckoning, Insular Easter, Textual transmission, Medieval Latin, Pseudo-Morinus, Pseudo-Cyril, Marinus of Arles.

Introduction

This paper compares the manuscript recension of two Sirmond texts that appear to share almost nothing in common. One is a short mystical tract on Easter chronology that has been called an Irish forgery. It is not a forgery and only a part of it is potentially Irish, but the fact remains that the *Disputatio Morini* has been misunderstood even by those who would treat it sympathetically.[1] The other text is a schoolroom introduction to the history of the calendar and to the terms and concepts necessary for learning the computus. Whereas the *Disputatio* was (as I argue in section 3 of this paper) an early text that was nearly forgotten before the end of the seventh century, *De diuisionibus temporum* was composed in the second half of the seventh century and heavily influenced the Bedan works on time that would come to dominate in European schools over the next several centuries.[2] There are two reasons, however, for bringing them together here. First, since both texts are components of the Sirmond collection firmly identified with the Irish contribution to pre-Bedan computistics, and since the Foundations of Irish Culture Project was designed to test and define the limits of this contribution, it only seems fitting to make a joint investigation. Second, and most fortunately, both of these works attracted considerable attention from Charles W. Jones, whose ideas about the transmission of pre-Bedan *computi* form the basis of this discussion. Jones' representations of the relationships between the early medieval computistical collections can be tested against the textual data that emerge from collations of *Disputatio Morini* and *De diuisionibus temporum*. Using the results of this test, it will be possible to refine our understanding of the origin of these works, and to say something new about their places within the Irish computistical tradition.

The Sirmond Computistical Collection

The phrases 'Sirmond computus' and 'Sirmond computistical collection' refer both to the codex and by extension to the grouping of works

---

[1] The full title of this brief work is *Disputatio Morini Alexandrini episcopi de ratione paschali*. It has been edited many times: by Du Cange (1688), App. 23, 480–1; by Muratori (1713), 195–6 = PL 129, 1357–8; by Pitra (1852), 14–5 (cf. xii–xiv); by Cordoliani (1945–6), 30–4; and in part by von den Brinken (1961), 218. Charles W. Jones intended to make a new edition but did not publish it. Nevertheless, his comments on the work remain the most thorough of all treatments; Jones (1943), 97 and Jones (1937), 216. More recently Strobel (1984), 116–20, and Walsh and Ó Cróinín (1988), 40–1.

[2] The work *De diuisionibus temporum* has never been printed in full. The version printed in PL 90, 653–7 is not representative of the text in its earliest and most complete form. Jones discusses the textual tradition briefly in Jones (1939), 48–51, but an adequate history of the tract remains to be written. Editions of both *Disputatio Morini* and the *De diuisionibus temporum* have been undertaken as part of the work of the Foundations of Irish Culture Project. They have been facilitated in countless ways by materials and guidance provided by Prof. Ó Cróinín of NUI Galway.

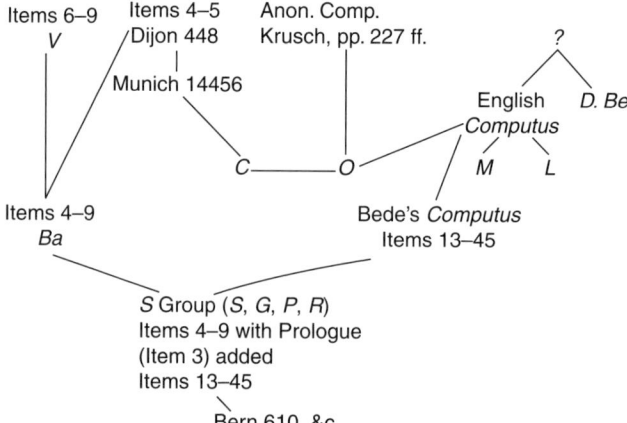

Figure 1 Jones' diagram from his 1937 publication: 'The Lost Sirmond Manuscript', 213.

For MSS abbreviations see *Figure 7* and additionally:
Ba. MS Basel, Universitätsbibliothek, F III 15k [s. ix]
Be. MS Besançon, Bibliothèque Municipale, 186 [s. ix]
C.   MS Köln, Dombibliothek, 83² [s. ix]
D.   MS Leiden, Universiteitsbibliotheek, Scaliger 28 [s. ix]
V.   MS Vatican, Biblioteca Apostolica, Vat. Lat. 642 [s. xi]

found in MS Oxford, Bodleian Library, Bodley 309. For many years scholars had puzzled over this manuscript, which had been used in the publication of the major 17th-century works on the medieval calendar, but which had since slipped into obscurity. In 1937, however, Charles W. Jones announced the rediscovery, confirmed by collation with published texts, of this precious witness to early medieval computistica.[3] Jones' subsequent studies of the materials in the Sirmond computistical collection were based on Bodley 309, but his observations frequently characterized a group of four manuscripts that he considered to be in the 'direct line' of transmission for this pre-Bedan collection of computus texts. The first of Jones' diagrams reprinted here as *Figure 1* shows these four manuscripts (the *S* group) as aggregators of disparate individual works and as agents of transmission to new manuscripts.

Two manuscripts (*S* and *G* in *Figure 1*, namely Bodley 309 and MS Genève, Bibliothèque Publique et Universitaire, 50) carry these texts

---

[3] Jones (1937), 204–8.

in a sequence that can be described as indicative of the exemplar, while the other members of the *S* group represent a more advanced stage of compilation and re-arrangement. Manuscripts *S* and *G*, therefore, were taken by Jones to be especially reliable as witnesses of the earlier history of the collection.

The content of the Sirmond computistical collection is drawn from three kinds of materials: historical and introductory works, such as parts of Bede's *De temporum ratione* (*DTR*) and all of *De diuisionibus temporum* (*DDT*), paschal letters and extracts, and the arguments, notes, and tables formerly described as 'bits', but now published in full in Dáibhí Ó Cróinín's recent article 'Bede's Irish Computus'.[4] Even this expression, 'Bede's Irish computus', though accurate from the point of view of transmission history, seems ill-suited to the materials as a whole, which have come under the name of Bede only by the accidence of their common contribution to his works on time. In other words, I would like to re-iterate that the Sirmond manuscript *S* and its twin *G* represent a computus in the broadest sense, being composed as much of canon law and exegesis as of computistical method. A titles list from the manuscript has been reproduced here for reference as Appendix I.[5]

A survey of the works included in *S* confirms the Bedan character of the collection; *DTR* heads the list, and excerpts from this work are inserted several times throughout the manuscript. Moreover, the proportion of Bedan material is even greater in *G*, where short selections from Bede's work appear frequently as glosses on the older material. These glosses once led the Swiss scientist Jean Senebier to describe the Geneva manuscript (dated to the year 804 by Jones) as Bede's own copy.[6] Unfortunately, no detailed study has ever compared the texts of *S* and *G* in order to determine precisely their relationship to each other and to the non-Bedan works in the computistical collection.

Of course, it was Jones' intention to edit the scientific works of Bede, and it was a natural, if regrettable, consequence of that intention that Bede began to seem like the inevitable arbiter of all prior computus traditions. In *Figure 1*, above, the aggregation of various

---

[4] Ó Cróinín (2003), 201–12.

[5] The titles list from Bodley 309 has been printed in Jones (1937), 213–9, Springsfeld (2002), 69–72, and in Ó Cróinín (2003), 202–3. The reader should note that the reprint of Jones' article in Jones (1994), X lacks the final pages of this capitular list.

[6] Senebier (1779), 126–41. Jones called Senebier's description 'ecstatic' but noted that it remains the most thorough treatment of the manuscript. The modern catalogue description by Bischoff gives just enough information to establish the early ninth-century date and the origin at Saint Martin's, Massay: Bischoff (1998), 284, §1351.

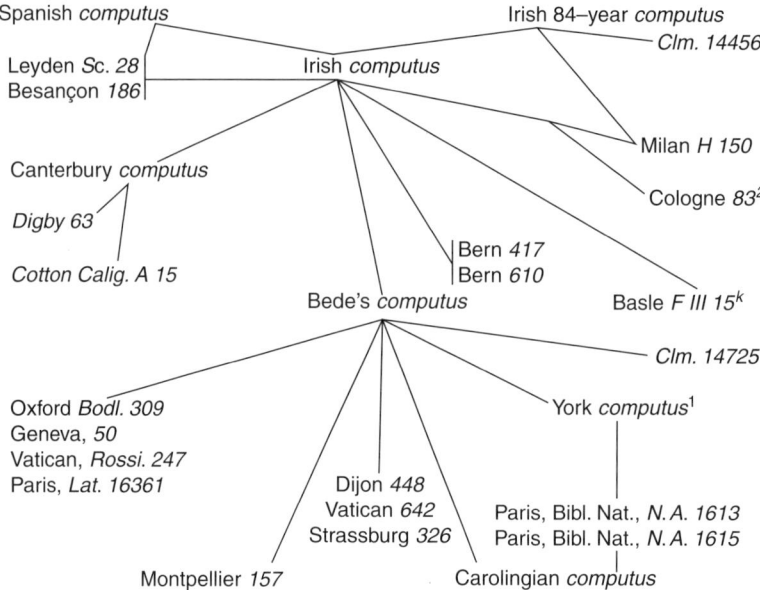

Figure 2 Large computus collections before and after Bede, according to Jones' 1943 publication: *Bedae opera de temporibus*, 112.

texts (the numbered items listed in Appendix I) appears to happen in several stages. For example, texts 13–46 are not counted together with items 3–9 as part of Bede's computistical sources. Jones would later revise his notion of these pieces and declare that all of them had been gathered into the Sirmond computus in time to be available to Bede for his work on *DTR*. This change in Jones' thinking was brought on by his realization of the extent to which *DDT* supplied the underlying structures of *DTR*. His diagram of six years later (*Figure 2* here) in *Bedae opera de temporibus* sketches the transmission of computi on a larger scale:

This sketch, it seems to me, wonderfully depicts the abiding state of confusion about these traditions. It is generally agreed that Bede's source materials derived from an Irish computus, but the individual relations of all the manuscripts listed here remain untested more than sixty years after this diagram first appeared.[7] Jones was careful to say in each case that these figures were not textual stemmata, and that there was need to verify them by checking with accurate texts.

[7] See Springsfeld (2002), 74–80, for a continuation of Jones' analysis of shared contents among the manuscripts.

The Recension of *De Diuisionibus Temporum* and *Disputatio Morini*

We may address this need in a limited way by seeking to establish the recension of two Sirmond texts (numbers 5 and 24 in the list of Appendix I) using manuscript collations. In this section, the editorial work is preliminary, and for *DDT* the collations are from a small selection of witnesses. In section three, the results of testing Jones' diagrams against the textual readings of *DDT* and *Disputatio Morini* will be applied to the larger context of the Irish synthesis of computistical materials during the seventh century.

De Diuisionibus Temporum

De diuisionibus temporum teaches that there are 14 divisions of time, beginning with the *athomus* and ending with the *mundus*. The structure presents an internal logic built around the conventions of time-reckoning and the implied symmetry of philosophical and Biblical thought, wherein time and space both agree at the maximal and the minimal limits of definition. An *athomus* is the smallest unit of time and the smallest unit of space, and the *mundus* is all of creation understood both spatially and temporally.[8] Apart from these circumscribing limits, the divisions can be either natural (as the *mensis* and the *annus*, related to observable phenomena) or artificial (as the *minutum* or the *punctum*). Throughout the work, which gives a history of the calendar as well as the quantitative

---

[8] There is nothing particularly innovative in the use of these words except, perhaps, in the symmetry of their framing here. *Athomus* – with or without the 'h' – was simply taken over from Greek with both spatial and temporal meanings intact. *Mundus* is not generally acknowledged in dictionaries as a temporal word, but its tranference from the spatial 'universe' is logical. Varro had defined *tempus* as an interval of the world's motion (*intervallum mundi motus*: Varro, *De lingua Latina* VI 2, §3 (Kent (1951), i 174–5). Since the totality of the world's motion is coterminus with the total space of the world itself, it follows that *mundus* can be used also with a temporal sense, meaning 'all of time'. This assertion is not without philosophical consequence, because it implies that time is not infinite, but rather bounded and created in the same way that space is bounded and created. The Christian utility of this idea is evident. Isidore had given an etymology of *mundus* from *motus*, but significantly he never cited the word in its temporal sense. Twice in the Etymologies he cites the definition from motion, at III 29, *mundus est appellatus, quia semper in motu est* (Lindsay (1911)), and at XIII 1, *Mundus Latine a philosophis dictus, quod in sempiterno motu sit* (Lindsay (1911)). He makes no mention of time in *De natura rerum* IX, '*De mundo*' (Fontaine (1960), 206–9). Among the medieval dictionaries, the *Catholicon* cites Isidore's derivation (Balbi (1460), *s.v.* '*mundus*'), but the 11th century grammarian Papias ascribes this meaning of *mundus* to 'the philosophers' (Papias (1496), *s.v.* '*mundus*'; for the *Elementarium* of Papias, see also now de Angelis (1977–80)).

relationships of time-reckoning, there is a remarkable sense of the mutability of time as it is understood by human beings.

The work as a whole confidently articulates the mechanics of temporal concepts even while it assimilates the often contradictory writings of patristic figures and major authorities like Augustine and Isidore. This is especially apparent in the additions to the text that ordinarily appear under the rubric *Sententiae Augustini et Isidori in laude computi* and in the final chapters of the work. In the Sirmond titles list in Appendix I, *DDT* is item number 5. Items 3 and 4 are usually grouped with *DDT* and collectively titled *De computo dialogus*. I have dealt only with *DDT* here because I believe items 3 and 4 are additions to the original dialogue which introduce issues not relevant to the primary stage of the textual history.

The history of the text is not straightforward.[9] For instance, MS Leiden, Universiteitsbibliotheek, Scaliger 28 (see *Figure 2* above), like many others, does not treat the final sections about the *aetas*, the *saeculum*, or the *mundus*. Other manuscripts add new divisions of time, or misunderstand the 14 divisions, subsequently renumbering them as 15 or even 16.[10] Typical of this confusion is the addition of the gloss *tempora* to the term *uicissitudo triformis*, both of which refer to the seasons of the year.[11] Nevertheless, selected collation from a number of manuscripts yields an adequate picture of the textual relations, given in *Figure 3*.

The stemma bears some relation to Jones' graphic of 1943, but there are discrepancies. While the pair S–G is confirmed, a new pair emerges, D–Bn, which might challenge Jones' assumption of δ's priority in a small way.[12] The independence of Basel is confirmed. The

---

[9] See the very instructive comments by Jones (1939), 48–51. Jones would identify the work *De computo dialogus* (including Sirmond items 3–5, in Appendix I) with the seventh authority on Cummian's list of AD 632 (Walsh and Ó Cróinín (1988), 84–7), which the Irishman referred to simply as *Augustinus*. Because Augustine died before the Alexandrian reckoning was introduced to the West, Cummian's citation of him as an authority supporting it can only be indirect. Jones supposed that he might have referred to a work like *De computo dialogus*, but noted that the reference would have to be to an earlier version of the work without Isidorean material.

[10] The Dijon manuscript (MS Dijon, Bibliothèque Publique, 448), in fact, adds the *cyclus* throughout the text, but not to the initial list of 14 divisions.

[11] On the definition of the term *uicissitudo triformis* and its possible source, see Walsh and Ó Cróinín (1988), 124, note to *De ratione conputandi* 13, l. 3.

[12] Jones was interested in the text of the exemplar of S and G (δ) because it represented the version closest to Bede's, but with respect to the archetype (Ω) the agreement of D and Bn (β) must be given some weight as the best representatives of α. Vu is a copy derived from the Carolingian Council of 809 and shows contamination (in the form of corrections) from the δ branch. It would take a more detailed study of Vu's sources to determine exactly which of its agreements with β can help to establish α.

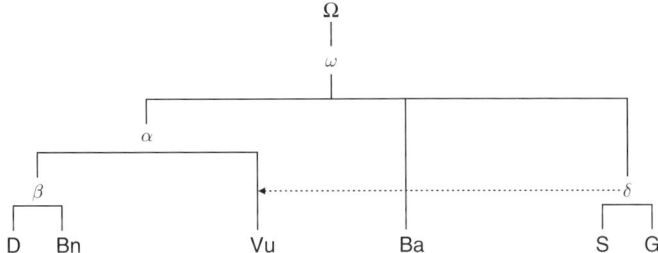

Figure 3 Textual stemma for six witnesses of *De diuisionibus temporum*.

Ba. MS Basel, Universitätsbibliothek, F III 15k, 21r–34v [s. ix]
Bn. MS Bern, Burgerbibliothek, 417, 47r–61v [s. ix]
D.   MS Dijon, Bibliothèque Publique, 448, 29–37 [s. x]
G.   MS Genève, Bibliothèque Publique et Universitaire, 50 (inv. 122), 139r–148v [s. ix^{in}]
S.   MS Oxford, Bodleian Library, Bodley 309, 62v–73v [s. xi], 'Sirmond computus'
Vu. MS Vatican, Biblioteca Apostolica, Urb. Lat. 290, 34v–41r [s. xi]

Vatican manuscript (Urb. Lat. 290) was not included in Jones' table, but shows the result of some contamination from the delta group.[13]

Most strikingly, however, the text in all its extant forms carries a major disruption of its structure in the addition of a section on the decennovenal cycle. Somehow, a work whose introductory plan does not mention any cyclic temporal period has come to devote a large proportion of its text to the history and implementation of the Alexandrian paschal table. In fact, the evidence of this early formal change still appears in several manuscripts; in S the *cyclus* is not in the list of divisions (64v), yet the discussion of the Metonic and decennovenal cycles is present as elsewhere after the chapter *De anno*. In Vu, a manuscript derived from the Carolingian Council of 809, the master in the dialogue declares that there are 14 divisions, and then proceeds to name 15, including a marginal or interlinear *cyclus*.[14] This disruption, like the coda of *Disputatio Morini* (which we will deal with in the next section) is a diagnostic feature that pegs the work as early and Irish.

---

[13] The delta group is separately determined by large omissions, by additions, by agreements in error, by agreements in the manner of stating numbers, and by agreements in the spelling of unusual words (*fetuntem*, *phatmos*, etc.).

[14] See, for example, MS Vatican, Biblioteca Apostolica, Urb. Lat. 290, 34v.

The text of *DDT* is explicit about many of its sources, especially about Isidore and Augustine, whose works are cited in nearly every section of the text. Far from simply following its sources, however, *DDT* frequently changes or supplements the earlier definitions with new material. In order to understand the great disruption of introducing the 'cycle' as a topic, it is first necessary to show the coherence and logic of the plan of 14 divisions via a comparison of Isidore's definitions.

Figure 4 illustrates the degree of independence in *DDT*'s scheme. Six of the 14 temporal units are not in the source, and the *lustrum* – a five-year period already antiquated in the seventh century – has been omitted from this list (though it is discussed in the chapter *De anno*). The rank in magnitude of *saeculum* and *aetas* has been reversed, and the very important first and last positions are entirely new. The framing symmetry of *athomus* and *mundus* has been mentioned already; the effect created by using terms with both spatial and temporal senses at the extremes of the progression here should

The divisions of time according to Isidore, *Etymologiae* V 29, §1 (Lindsay (1911))	The divisions of time according to *DDT*
....	**athomus**
momentum	momentum
....	**minutum**
punctus	punctus
hora	hora
....	**quadrans**
dies	dies
....	**ebdomada**
mensis	mensis
....	**uicissitudo triformis**
annus	annus
lustrum	***
saeculum	aetas
aetas	saeculum
....	**mundus**

Figure 4 The divisions of time in Isidore's *Etymologiae* compared to *DDT*. Bold type shows the units introduced by *DDT*, and *** indicates the absence of *lustrum* in *DDT*.

not be overlooked.[15] *DDT* does not offer a discussion of the implications, but it seems clear that the *mundus* referred to is one created by a God (and not eternal) and that the *tempus* of the title is defined by the motion of particles (and not merely by human convention). The units are mixed between natural divisions such the *annus* and artificial ones such as the *ebdomada*. Some divisions are given multiple definitions, such as *athomus*, which has five applications, and *dies*, which has both natural and artificial meanings.[16] The logic of this organization is traditional and pedagogical; it proceeds from the smallest to the largest division of time, it defines each term with reference to authorities and derivation, and it relates the history of many terms in a way suitable for students. It is also in keeping with the exegetical use of the number 14 in structures of time, such as the 14 generations from Abraham to David, and from David to Christ, the 14 lunes of a paschal moon in the Old Law, and many others as defined, for example, in Isidore's *Liber numerorum*.[17]

In light of the unity of these 14 divisions, it is a disruption of the highest magnitude to insert a new section on the paschal computus.

Figure 5 shows that the 19-year cycle of Dionysius Exiguus has been added to the text after the chapter *De anno* and before *De aetate*. There is some reason to think that the cycle is a kind of round or *annus*, and

---

[15] At the time of the composition of *De ratione conputandi* (*DRC*) – the mid-seventh century, according to Ó Cróinín (1982c), repr. in Ó Cróinín (2003), here 127 – the author mentions that some texts did not include the *athomus* in a list of 13 divisions of time (*DRC* 13 (Walsh and Ó Cróinín (1988), 124): *Sed tredecim sunt secundum alios, quia atomos non numerant*. Nevertheless, a plan of 14 divisions informs both *DRC* and *DDT* when it appears elsewhere. Lists of divisions in later versions of the text vary in their number and make-up, as for example in Hrabanus Maurus, *De computo* 10, where the *ostentus* is named after the *atomus* and the *mundus* is discarded: *D. Diuisiones temporis quot sunt? M. XIIII. D. Quae? M. Atomus, ostentum, momentum, partes, minutum, punctus, hora, quadrans, dies, mensis, vicissitudo, annus, seculum, aetas* (CCCM 44, 217). Dr Immo Warntjes supplies the details of another list that omits the *mundus* in an unpublished prologue to Bede's *De temporibus* in MS Vatican, Biblioteca Apostolica, Pal. Lat. 1448, 1v–2r; MS Vatican, Biblioteca Apostolica, Pal. Lat. 1447, 3r; MS Schaffhausen, Stadtbibliothek, Ministerialis 61, 22r.

[16] PL 90, 654: *D. Quot sunt genera atomorum? M. Quinque. D. Quae sunt? M. Atomus in corpore, atomus in sole, atomus in oratione, atomus in numero, atomus in tempore*. PL 90, 656: *D. Quibus modis dies dicitur? M. Duobus. D. Qui sunt? M. Dies naturalis et dies artificialis. D. Dies naturalis quid est? M. Dies legitimus, ab ortu solis, donec rursus oriatur; et ille dies habet viginti quatuor horas in se. D. Quot sunt divisiones diei naturalis? M. Duae, id est, dies et nox.* [. . .] *D. Quis est ergo ille dies artificialis? M. Ab ortu solis usque ad occasum, dies artificialis dicitur, id est, praesentia solis super terram.*

[17] Cf. Isidore, *Liber numerorum* 15 (PL 83, 193–4).

	The divisions of time according to *DDT*	Additions to the fourteen units of time in the *Computus Hibernicus*
1.	athomus	
2.	momentum	
3.	minutum	
4.	punctus	
5.	hora	
6.	quadrans	
7.	dies	
8.	ebdomada	
9.	mensis	
10.	uicissitudo triformis	
11.	annus	
(11a.)	....	cyclus decennovenalis (ἐννεακαιδεκαετηρίς), cyclus lunaris
12.	aetas	
13.	saeculum	
14.	mundus	

Figure 5 The 14 divisions of time in *DDT* and the insertion of the paschal cycle in manuscripts of the *Computus Hibernicus* (cf. Appendix II).

so the placement itself is not as much a problem as the extent and character of the addition.

The new section is long and complex. In the Sirmond manuscript it covers folios 71v to 73r, or nearly one fifth of the length of *DDT*. It contains a mixture of technical and historical, or pseudo-historical, information including the famous legend of the Egyptian monk Pachomius and the assertion that a true and perpetual Easter cycle was established at the Council of Nicea.[18] After noting Victorius of Aquitaine's admonition about the intercalation of a lunar month in embolismic years – namely, that Easter should come at the beginning

---

[18] For a detailed history of the conflation of the Pachomian legend with the attribution of an Easter table to the holy fathers of the Nicene Council, see Jones (1943a). Where Jones links the Pachomian legend with the 'Greek computus' received at Bangor during the abbacy of Mo Sinu Moccu Min (i.e., before AD 610), Ó Cróinín (1982a) connects the Bangor episode instead with a tract on finger-reckoning and the acceptance of the Greek list of alpha-numeric equivalencies, noting on p 46 that 'so early a date as 610 for the first attestation of the Dionysiac cycle in northern Ireland would be surprising'.

of a new year and not at the end of an old one, and so should never be celebrated within the same period of 12 lunar months, and only after 13 lunar months in embolismic years – the chapter *De cyclo* continues with an insertion from the extended Cyrillan letter about the custom of the Hebrew lunar embolism.[19] Then the student of the dialogue asks who first discovered the pattern of paschal full moons over 19 years. The master answers that these *termini paschales* were observed in the old law, but not known as such by the Christians until an angel of the Lord revealed them to Pachomius. The passage continues to narrate the acceptance of these paschal terms *semper stabiles inmotasque* by the 318 bishops at Nicea. It defines and explains the alternation of common and embolismic years, and the Greek *ogdoas* and *endecas*, all with reference to Dionysius Exiguus. Finally, the chapter *De cyclo* closes with a comparison of the decennovenal cycle with the Greek lunar cycle, explaining in numerical detail the differences between them.

Considering the length and detail of this interpolation, it is difficult to think that the intended reader of this passage is the same student for whom the original *DDT* was written. The dialogue form has been maintained, but the content of this section is far more advanced than the simple definitions that predominate elsewhere. Moreover, the change in audience and source material sometimes coincides with the omission of the remaining parts of *DDT*. Even in the Sirmond manuscript, where the final chapters are present, the capitular list of the *Computus Hibernicus* (*CH*; Appendix II) indicates that the compiler would have abandoned the plan of *DDT* after the discussion of the various cycles. Thus, in *CH* the absorption of this elementary textbook into a more sophisticated context was already underway in about the year 658.[20] It would be complete when Bede composed his *DTR* from these materials in 725.

---

[19] I.e., the Cyrillan letter in the form published by Krusch (1880), 344–9 and long considered a forgery. For discussion see Jones (1943), 93–7 and Walsh and Ó Cróinín (1988), 38–40.

[20] The dating clause to AD 658 appears in full in Ó Cróinín (1983c), repr. in Ó Cróinín (2003), here 177–8, and is discussed again in Ó Cróinín (2003a), 207. Recent discussions of the Sirmond manuscript have emphasized its representation of Carolingian computistical units (see e. g. the table in Springsfeld (2002), 78), and have sought to divide the material of the manuscript according to an assumed order of compilation. So the dating clause on fols 95v–95^(bis)r is considered by Springsfeld as appropriate only to what she terms 'Bede's Computus' (Sirmond items 13–45). Note, however, that in this she followed Jones' early estimation of the division of materials. For Jones' latest view on the matter, in which he establishes that all these materials were present in

As Jones knew, the material contained in the chapter *De cyclo* is taken from a Dionysiac letters collection that included the *Epistola Cyrilli* in its extended form.

The original letter of Cyril of Alexandria supplied the story of the inspiration of Pachomius, and the additions which match that text to the year 607 include the endorsement of the Nicene fathers' authority. These details and others from the letter of *Bonus primicerius* are added with reference to Dionysius, who is cast as a continuator of the Alexandrian tradition of Theophilus and Cyril. All of this material is supplied as a corrective to the Victorian lunar reckoning. It is important to note here that nowhere does the text denigrate Victorius even when it proposes to replace everything to do with the lunar data in his table. Instead, the text emphasizes a forward-looking unity of Christian observance, and attempts to pass over the worst of the historical strife over Easter.

When we look at the recension of *DDT*, we are able to see clearly the logic of Jones' account of the history of Cyril's letter and of the introduction of the Greek computus into Ireland. In fact, the development of *DDT* rather supports his theory about the letter, for if Pope Gregory or any of his proximate successors sent a dossier of Dionysiac documents to his English mission, it surely also circulated among the Irish, whose manuscripts are the only remaining evidence of these works.[21] That Cummian cites them all is a convenient marker for the assembly of all these materials in Ireland before 633. Tracing the incorporation of these items into the several Irish computi, we can see that there was never a standard for the treatment of the Dionysiac materials within *DDT*, but that their incorporation was subject to the needs of individual compilers.

the Sirmond archetype, see Jones (1943), 106, where he refers to his earlier opinion: 'the original statement was conservative. I then maintained that only folios 80v–140v were derived from Bede's computus, but additional study shows that all items in the first folios were also known to Bede and are therefore probably a part of that codex into which he copied *DTR*'. The Irish computus *De ratione conputandi* 13 demonstrates the existence of a version of *DDT* in the mid-seventh century outside the Sirmond tradition (see Walsh and Ó Cróinín (1988), 123–4), and the variant *complatio* of G, the twin manuscript of S, attests to the mixture of texts from several 'parts' of S in another copy. I have followed Jones and Ó Cróinín in considering the pre-Bedan materials in S as a dossier of mixed but not unrelated texts containing a datable reference to 658 and a *terminus ante quem* of 725, when extensive direct citations were included in Bede's *DTR*.

[21] See Jones (1943), 94–6.

Four major Irish computi include the text of *DDT* together with the Cyrillan version of the Pachomian legend in some form. In *De ratione conputandi* (*c.*650s), the Dionysiac material is added after the computus, forming part of a section on the lunar calendar that follows discussion of the solar calendar.[22] In the *Computus Hibernicus*, according to the titles list in Appendix II, the material appears in a section on cycles inserted into the text of *DDT* after *De anno*. In the Munich computus, *DDT* and the Dionysiac materials appear separately and have not been joined in any way.[23] The Bobbio computus (*Liber de Computo*) contains a version of *DDT* with 16 divisions (adding the redundant *tempus* (for *uicissitudo triformis*) and *cyclus* to the original list), and places the Dionysiac material together with *DDT* in an arrangement altogether more elaborate than the Sirmond version of the *CH*.[24] The recently discovered Einsiedeln computus (MS Einsiedeln, Stiftsbibliothek, 321 (647), 82–125) differs markedly from these others; although it shows an awareness of the Dionysiac letters, it has no text corresponding to *DDT*, and its arrangement of titles is unique. For this reason I have left it out of the illustration below.

In *Figure 6* the individual treatment of *DDT* indicates that the text was considered basic to computistical discourse well before the creation of the major Irish computi. The decision to place the discussion of the cycle within the text of *DDT* had far-reaching consequences, for it set a standard arrangement of topics that would be followed both by Bede and by the compilers of the Carolingian computus.

Disputatio Morini

Like the *Epistola Cyrilli* as cited in *DDT*, the second text considered here, the *Disputatio Morini Alexandrini episcopi de ratione paschali* (*DM*), has also endured the slander of forgery. In fact, there has not been a time when a modern scholar did not cite it as a deception. Even Jones, who worked toward a new edition of it, never fully cleared the work of this charge.

Fortunately, there are few enough manuscripts that a full collation of all witnesses has been possible. In the stemma in *Figure 7*, the

---

[22] The Dionysiac material begins with *DRC* 71 (Walsh and Ó Cróinín (1988), 179).

[23] An edition of the Munich Computus (MS München, Bayerische Staatsbibliothek, Clm 14456, 8r–46r) by Immo Warntjes will appear in print shortly; the *DDT* section ends on f 25r, while the Dionysiac material starts on f 35v.

[24] Bobbio Computus 89 (PL 129, 1315D–1316A). Note that the following text diverts from these divisions after the discussion of *quadrans*.

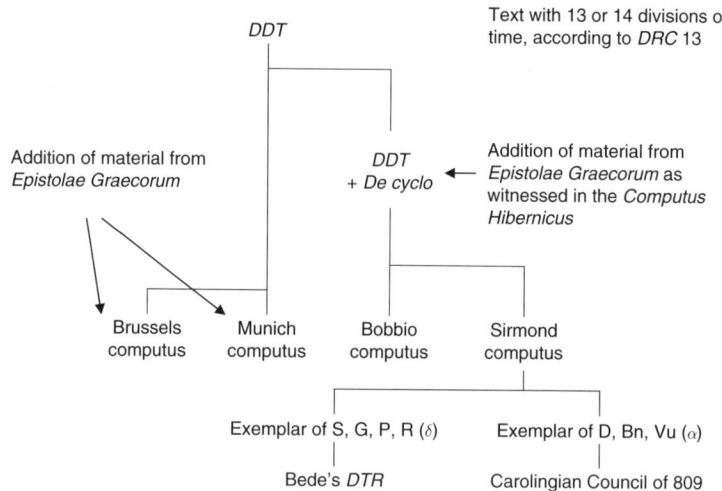

Figure 6 The descent of *DDT* through some of the Irish computi.

Brussels Computus (*DRC*) = MS Bruxelles, Bibliothèque Royale, 5413–22, 77v–107v
Munich Computus = MS München, Bayerische Staatsbibliothek, Clm 14456, 8r–46r
Bobbio Computus (*Liber de computo*) = MS Milano, Biblioteca Ambrosiana, H 150 inf
Sirmond Computus (*CH*) = MS Oxford, Bodleian Library, Bodley 309
For MSS abbreviations see *Figure 3* and additionally:
P. MS Paris, Bibliothèque Nationale, Lat. 16361 [s. xii]
R. MS Vatican, Biblioteca Apostolica, Rossianus 247 [s. xi]

manuscript of Marianus Scottus (MS Vatican, Biblioteca Apostolica, Pal. Lat 830) has been omitted because the citations from the *Disputatio* in the first part of his chronicle (on f 32r) are not sufficient to establish the manuscript's filiation.

In his seminal essay of 1937 ('The 'Lost' Sirmond Manuscript of Bede's Computus', p 216), Jones had this to say about the curious little text: 'Because of the unintelligibility of the printed editions comment has been avoided'. He was referring to the printed texts by Bucherius in 1634, by C. Du Cange in 1688, by Muratori in 1713 (reprinted in PL 129 in 1853), by Jean Baptiste Pitra in 1852, and by Cordoliani in 1945.[25] So it was not for utter lack of attention that the

---

[25] See note 1, above, for full publication details.

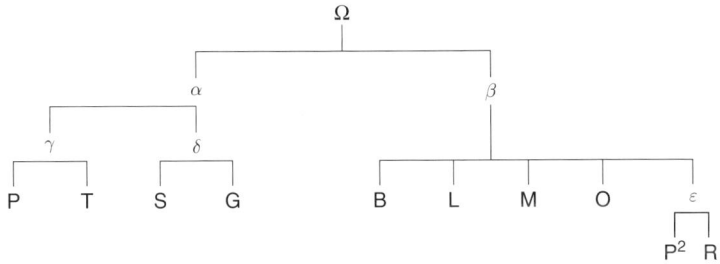

Figure 7 Textual stemma for *Disputatio Morini*.

B. MS Bruxelles, Bibliothèque Royale, 5413–22, 100v; 103r [s. ix–x]
G. MS Genève, Bibliothèque Publique et Universitaire, 50 (inv. 122), 131v–132r [s. ix^{in}]
L. MS London, British Library, Cotton Caligula A XV, 82v–83v [s. viii]
M. MS Milano, Biblioteca Ambrosiana, H 150 inf, 80r [s. ix^{in}]
O. MS Oxford, Bodleian Library, Digby 63, 79r–81r [s. ix]
P. MS Paris, Bibliothèque Nationale, Lat. 16361, 238–240 [s. xii]
P². MS Paris, Bibliothèque Nationale, Lat. 4860, 150r–150v [s. x]
R. MS Vatican, Biblioteca Apostolica, Reg. Lat. 586, 114v–115v [s. x]
S. MS Oxford, Bodlleian Library, Bodley 309, 94r–94v [s. xi]
T. MS Tours, Bibliothèque Municipale, 334, 16v–17r [s. ix^{in}]

work remained inscrutable. In fact, Jones also undertook to edit the work, and though he never published it, he did provide a series of tantalizing hints about its purpose and origin.

Appendix III contains a full text of *DM* generated from collation of all the witnesses. It should not be taken as a finished edition, but will be useful here for the purpose of citation by line number. Although a full treatment of the issues contained in *DM* will await the published edition, two items are important for the present discussion. First, a new approach to the problem of authorship will, I hope, begin to unravel the knotty tangle of exegetical and computistical matters. Second, a direct application of the advice given in the coda will show that the work belongs among the same Dionysiac materials that carried the *Epistola Cyrilli* to Ireland.

Recent years have brought considerable progress in the treatment of difficult computistica. Jones' career showed the beginnings of a recovery in these studies, which really was the emergence of a serious attitude to the details of the manuscript record. Instead of denigrating imperfect or composite texts, for example, he began to assemble a history of the tradition of computus that accepted compilation and confusion as the natural result of scientific development and manuscript survival over centuries. While this new attitude continues to push research forward on many fronts, it has especially improved the lot of the so-called forgeries. Because we are here concerned with one of these neglected texts, some mention should be made of the discovery, deciferment, and publication of *De ratione paschali* of Anatolius of Laodicea and of the Easter table discovered with it.[26] The process of restoring an accurate text of *De ratione paschali* shows that tracts on the Easter reckoning were far more subject to corruption than to deliberate alteration. It also shows the profit to be had from discarding the assumption that corrupt texts have been falsified. Had Jones seen the restored Anatolius, for example, he might not have hesitated to follow a different line of inquiry about the *Disputatio Morini Alexandrini episcopi de ratione paschali*.

The 'forgery' label had been attached to the *Disputatio Morini* long before Bartholomew Mac Carthy's time, but Mac Carthy's formulation of the charge is still the most impressive. In his introduction to the fourth volume of the *Annals of Ulster*, Mac Carthy attributes the *Disputatio* to what he believes was a vast Irish conspiracy to deceive. Consequently, Morinus was part of 'the most ludicrous tissue of fact and fiction in existence' (cxxxix–cxl). Like Cyril and Anatolius, 'Morinus is the Irish fictitious Morianus, fictitious bishop of Alexandria, whose epistle, *De ortu Paschali*, was written against the Victorian Cycle'. As it happens, Mac Carthy's bilious dismissal of this work has got almost everything wrong, from the title of the work to the author, right down to mistaking which side of the paschal controversy is supported by the *Disputatio*.

The best that can be said of the prevailing attitude of derision toward this and other works is that often the only available texts were unedited or poorly printed, and often from the worst manuscripts. So it was with Mac Carthy, whose only example of *DM* was from Muratori's edition in the *Patrologia Latina*. Muratori printed a faulty

---

[26] Where the so-called forgeries are concerned, this publication marks a watershed. It is also a remarkable example of scholarly collaboration, from the discovery of the Padua table (*latercus*) by Ó Cróinín to the final systematic reconstruction of its technical content by Mc Carthy, to the establishment of a critical text and commentary of Anatolius' tract by Mc Carthy and Breen. The textual and mathematical difficulties which had confounded earlier scholarship on the 84-year table used by the Irish and its authenticity have been largely solved. For an account of this process of publication, see Mc Carthy and Breen (2003), 10–3.

transcription of the Milan manuscript (M in *Figure 7*, above), which contains so many singular errors and contradictions that it must be considered the most unreliable witness of any now known.[27] The title *De ortu paschali* appears only in M; elsewhere the work is more conventionally described as *De ratione paschali*. M is also the only witness to describe the work explicitly as an *epistola* rather than a *disputatio*. This distinction is significant because it shows that most witnesses had an awareness of the indirect nature of the work, that it was not a simple copy of a letter but rather a series of excerpts from a larger exegetical text.[28]

The attribution of pseudonymous authorship is perhaps the strangest fact of the history of this text. Even though the *Disputatio Morini* has been included in the list of Irish forgeries from time out of mind, almost no one has suggested a candidate whom the supposed forger meant to impersonate. Since it is usually the intention of a forger to associate himself with a recognized authority, the neglect of this central question – who was Morinus? – signals the degree to which the *Disputatio* has been mistreated by researchers in the field. After he reserved comment in his essay on the Sirmond manuscript (in the passage I cited above), Jones later succeeded in identifying some new manuscripts as he prepared a working text. Dáibhí Ó Cróinín again added to the list of known witnesses and published the earliest references to the *Disputatio Morini* together with Maura Walsh in the 1988 edition of *Cummian's Letter De controversia paschali*. But the only scholar to propose a candidate for authorship is August Strobel, who in 1984 nominated Marius of Avenches, a Burgundian chronicler from the late sixth/early seventh century.

> 'One almost gets the impression our author is an exponent of the Gallic equinox-feast. At the end of the 6th century there flourished such a man in West Switzerland, Marius bishop of the church of Avenches, a chronicler. Was Morinus a misspelling of his name? The manuscript tradition

---

[27] Bobbio Computus 146 (PL 129, 1357–8). The difficulty of this witness appears from the start, where the words *Id congruum* (l 3) are changed to *Incongruum*, effectively vitiating the logic of the rest of the text. The most important feature of M is that it lacks a text of the coda (ll. 59–69), the only part of the work that Jones considered intelligible. No wonder Mac Carthy and others thought little of it!

[28] There are viable candidates for such a text, but a study of these and their intricate relations will appear together with the final edition.

offers one supporting witness (Geneva Univ. 50). However, we can get little certainty for lack of sufficient evidence on this matter.'[29]

The proposal is not without its merits, but I have not been able to see how the ascription to Marius might help to explain any of the peculiarities of our text, from the double equinox dates (cf. ll. 4–5, 15, 21–23) to the advice in the coda attached to almost all the manuscripts.

Proposing another candidate for the ascription will not settle all the details of the text, but I believe it can help to understand some of its more mystifying features. The spelling of Morinus differs in almost all manuscripts; Maurinus, Morianus, and Morinus all appear in at least one witness. Strobel's Marius followed on the more speclative comment by Jones that 'Morinus or Morianus sounds like Moran'.[30] Perhaps, then, it will not stretch belief too far to nominate a Marinus as the authority cited in the title. For there was a Marinus whose authority on Easter was known to the Irish, and whose *Disputatio* would naturally favour the Alexandrian Easter reckoning.

Marinus of Arles was bishop of that city when it was the chief seat of Gaul under the Emperor Constantine. He was also the chief judge appointed for the trial of Caecilian in Rome in October of 313. He presided over the Council of Arles in 314 and penned the letter to Pope Sylvester, reiterating the council's judgment in favour of Caecilian and condemning his Donatist detractors.[31] On account of the shape of that controversy, Marinus is more commonly associated with the canon law of baptism and the question of rebaptism among African Christians. We should remember, however, that the first canon of the Synod of Arles is also the first ever Western canon regarding Easter: 'Concerning the observance of the Lord's pasch, that <the Pope> should direct letters to everyone, as is customary,

---

[29] Strobel (1984), 120: 'Fast gewinnt man den Eindruck, unser Autor sei ein Vertreter der gallischen Äquinoktiumsfeier. Im ausgehenden 6. Jahrhundert betätigte sich in der Westschweiz *Marius, episcopus ecclesiae Aventicae* (= Avenches), chronistisch (Th. Mommsen, Chron Min. II S. 225ff.). Ob sein Name verschrieben wurde? Die Handschriftliche Überlieferung würde eine solche Verifizierung befürworten (s. den Genfer Codex Univ. 50). Indessen können wir mangels ausreichender Nachrichten hierüber keine Gewissheit erlangen (s. Bardenhewer V S. 378).'

[30] Jones (1943), 98.

[31] See Eusebius, *Historia ecclesiastica* X 5 (Schwartz and Mommsen (1903–9), 883–90) and Hefele (1907), 201–19.

so that we may observe it on one day and at one time through the whole world'.[32]

The unity of observance for Christianity's central feast has always been a chief concern for the Church. A brief look at the *DM* in Appendix III will confirm the relevance of Christian unity to the text. Moreover, the unity here is not just with regard to other current customs, but also relative to the timeframe of other major events in salvation history. The logic is centred on the words from Matthew's Gospel (Mt 26:2): *Post biduum pascha fiet*, and interrogates the chronology of the week as it pertains to creation, the exodus, the passion and resurrection, and the end times (ll. 2–13).

This is not to say that the *DM* as we have it was ever intended for this form of publication. In fact, the rapid succession of topics and hasty interpretations read like bullet-point summaries of a larger discourse. In this context, the attribution to Morinus is best understood as a reference to a famous canonist, whose interpretive logic is being sketched here for the purpose of answering a new case. We have, in fact, no other context for *DM* than in manuscripts of computus and canon law. Already in Cummian's letter Morinus was grouped with Dionysius and Cyril in the list of paschal authorities.[33] Moreover, among the several minor surprises in this work has been the discovery (or rediscovery) of the *DM* within the same manuscript Krusch used for his edition of Dionysius Exiguus (MS Vatican, Biblioteca Apostolica, Reg. Lat. 586). Krusch must have seen the text when he prepared his dissertation, but neglected to mention it where it would naturally be placed alongside the Cyrillan letter and the Cologne Prologue.[34]

Looking over the manuscripts of *DM*, and considering this recent find, it becomes clearer than ever that the text was always closely linked to the works of Dionysius. It was almost certainly a part of the collection of materials that came to be known as the *Epistolae Graecorum*.[35] Cummian's letter suggests as much when it places Morinus next to Dionysius and Cyril in the list of paschal authorities, and when he cites the first

---

[32] *Concilium Arelatense a.* 314, §1 (CCSL 148, 9): *I. Primo in loco de obseruatione Paschae dominicae: Vt uno die et uno tempore per omnem orbem a nobis obseruaretur, ut iuxta consuetudinem litteras ad omnes tu dirigas.* For more on the canons of Arles, see also Hefele (1907), 275–98.

[33] Cummian, *De controversia paschali* (Walsh and Ó Cróinín (1988), 84–7).

[34] Krusch (1880), 227–35 and 337–43.

[35] A point recognized by Jones (1926), 204.

Synod of Arles, and its first canon, explicitly.[36] To Cummian and to other expert scholars of his generation, the name Morinus was already attached to the Easter question, for which they routinely cited the Synod of Arles as the major Western council enjoining uniformity among the various traditions. It seems improbable that a connection to Marinus of Arles is not intended in the title of *DM*.

Finally, the context of this discovery can help to confirm a close reading of the meaning of the coda (ll. 59–69):

> 'After all these things I will say briefly that the pasch was never held among the Hebrews before 12 kalends of April [March 21] because the moon [that is] born on 8 Ides of March [March 8], is the full on 12 kalends of April [March 21]. Observe therefore the lunar course according to the rule of the Greeks (in the manner of the Egyptians[37]), and not according to the Epacts, that is the lunar adjections, because there it comes from the 4th lune to the 16th, but here to the 15th according to the reckoning of Eusebius, who first composed the cycle of 19 years, of Athanasius, Theophilus, Cyril and Dionysius the small, up to the time when Victorius wrote to Pope Hilary the bishop of the city of Rome. Then the disputants of Alexandria and Antioch ceased to write [the calendar] according to other cycles.'

There are many oddities here, from the attribution of the 19-year cycle to Eusebius to the use of the term 'Epacts' to the distinction 'in the manner of the Egyptians' and more. However, there is a hard core to the paragraph that clarifies its purpose and date. The paschal new moon on 8 Ides of March happens in year 13 of the *cyclus lunaris* (year 16 of the Dionysiac *cyclus decennouenalis*) when the

---

[36] Cummian, *De controversia paschali* (Walsh and Ó Cróinín (1988), 70–3, 84–7).

[37] This could mean that the author urges [a] a reckoning of the lunar year from the beginning of September (see Ó Cróinín (1993), 46 n 37, for citations about September-reckoning from Alcuin, Pseudo-Alcuin, and a Bedan commentator, and Savage (1928) for the custom among Alcuin's Irish students), [b] a method for finding the 14th lune that includes the lunar *saltus* in the 19th year of the cycle (see *DRC* 106 (Walsh and Ó Cróinín (1988), 209), [c] a reckoning of the beginning of the day from the beginning of the night, as at *DRC* 26 (Walsh and Ó Cróinín (1988), 134). Of a revision of the 19-year cycle in Alexandria, Declercq (2000), 66–7 says: 'The motive for the revision was apparently the establishment of the spring equinox on 21 March, which necessitated the adjustment of some Paschal full moons. At the same time, the commencement of the cycle was changed to a year in which a calculated new moon fell on 29 August, i.e. 1 Thoth, the first day of the Egyptian solar calendar. The first year of the Emperor Diocletian (AD 284–285), exactly one or two cycles before the revision, was such a year, and this explains probably why the official starting point of the Alexandrian cycle was fixed in this year.'

epact (as we call it) on 22 March proceeds from 4 to 15; these conditions describe exactly the situation facing a computist in the year 604, when coincidentally the Victorian table offers a discrepancy between a paschal lune 16 or lune 22. The contrast in usage between 'here' and 'there' can only refer to the difference between Dionysiac and Victorian epacts, since only the Victorian table of epacts contains a leap from 4 to 16. Evidently, the comparison between lunar data was inexact, for the *saltus* occurs in Victorius not in AD 604, but in 594 or 613 or 632, when the epacts (or lunar adjections, as the coda says) on 1 January jump from 4 to 16.[38] Nevertheless, the message is clear; adopt the lunar calendar of Dionysius and discard the lunar data as given in Victorius.[39]

The Seventh-Century Synthesis of Computistical Materials

The suggestions made in this section depend on two significant facts: first, that both the *DM* and *DDT* are integral parts of the computistical collection generally known as the Sirmond computus, after the best-known manuscript witness of the material. By this I mean that both works were present in their current manuscript form in the Irish computistical collection used by Bede in composing his *opera de temporibus*. The second fact is that within that collection, both texts make direct appeals to their readers regarding the implementation of some feature of the computus. From these facts, two further inferences can be made regarding the character of that collection and the identity of its intended readers.

---

[38] Jones reads the coda from the point of view of the Victorian data, and quite rightly notes that the year 632 coincides with Cummian's activity. He even suggests that Cummian may have written the coda himself: Jones (1943), 97–8. This is not impossible, but the motive for Cummian would have been to explain the Victorian usage, and if he is the coda's author, to correct its deficiencies. Yet the coda itself takes the Dionysiac data first, starting with the paschal new moon on March 8, and so the year in question must be year 13 of the *cyclus lunaris*. The Victorian table for AD 632 contained nothing remarkable besides the *saltus*, and agreed on an Easter Sunday of April 12 with both the 84-year table used by the Irish (*latercus*) and the Dionysiac. In 604, however, the Victorian paschal lune of 22 exceeded the Alexandrian and Irish limits.

[39] The 84-year table that was at the heart of the Irish debate does not seem to enter into consideration here. Confusion on that score probably led Mac Carthy and others to think that the tract was an attack on Victorius. The phrasing of the coda suggests instead that a Dionysiac lunar table had been grafted onto the Victorian table in an effort to improve the known faults of the Roman system.

The character of the Sirmond collection has always been the chief obstacle to an understanding of the Irish computus; the materials that comprise the collection come from so many different times and places, and cover such a variety of topics and opinions that the collection as a whole seems confused and contradictory. Even drawing the precise boundaries of the *Computus Hibernicus* is an exercise complicated by uncertainty surrounding the manuscripts. The normal reaction of scholars before the twentieth century was to see in this uncertainty the hand of a forger. Now with the development of new editions, it emerges that one of the strengths of the Irish computistical collections is the critical juxtaposition of multiple calendar traditions. The two works under consideration here exhibit this strength in the very instance of the textual interruption that once called them into question.

The list of authorities cited in the Sirmond versions of both *DDT* and *DM* is indicative of a conciliatory view toward controversy; where Columbanus had openly scorned Victorius in his letter to Pope Gregory, both *DDT* and *DM* place Victorius and Dionysius together in a single tradition. Dionysius appears as the mediator of an Alexandrian system perfected by Cyril and associated with Theophilus, Pachomius, and the Council of Nicea. Victorius is presented not as a rival to Dionysius but as the creator of a superstructure onto which the Dionysiac Easter can be added. *DDT* inserts the Pachomian legend directly after an approving citation of Victorius (S, 71v), then paraphrases the letter of Dionysius to Petronius while asserting that Theophilus and Cyril maintained the tradition of Nicea.[40] Thus, while

---

[40] The text transcribed from the Sirmond copy reads as follows (*De diuisionibus temporum* cap. *De cyclo* (MS Oxford, Bodleian Library, Bodley 309, 71v–72r); italics indicate direct quotations):

    Et iste cyclus duo nomina habet apud Grecos, hoc est ogdoas et endegas. Ogduas autem in Greco .viii. in Latino autem sermone sonat. Endegas uero .xi. interpretatur. Et sic .viii. et .xi. simul fiunt .xviiii., qui dicuntur termini pascales, hoc est *Nonae Aprilis, .viii kal. Aprilis*, et cetera.

    Δ. Quis primus nominauit istos terminos paschales scire nos oportet.

    M. Isti termini paschales fuerunt observati in ueteri lege secundum rationem communium et embolismorum, sed non fuerunt nominati illo uocabulo quomodo nominantur apud christianos. Angelus enim domini hos terminos paschales primum ad Pachomium monachum in Aegypto ex reuelatione Domini monstrauit per litteras quasdam. Vnde Cirillus episcopus dixit, *Indicabo uobis quod Pachomius monachus insignis factus apostolicae graciae egre-* [S, 72r] *gius fundator aegypti coenobiorum edidit ad monasterium quod lingua egiptia uocatur pabum litteras quas angelo dictante perceperat ut non in errorem incurrerent christiani in solemnitatis paschalis ratione, scirentque lunam primi mensis in anno communi et embolisemi.* Hunc ergo cyclum decennouenalem (hoc est terminos paschales), qui per .viiii. et .x. annos cum ratione

clearly suppressing the Victorian lunar calendar, *DDT* nevertheless includes Victorius among the paschal authorities still relevant to an understanding of the *termini paschales*.

DM is more direct in its juxtaposition of Victorius and Dionysius, and more expansive in its list of authorities. It credits Eusebius as the author of an Easter table, names Athanasius also, and finally casts the paschal disputations broadly as a contest between Alexandria and Antioch. Significantly, these disputations are said to be finished, and the same note of conciliation obtains here as in the citations of authorities in *DDT*.

The identity of the readers of these works can be inferred from the level of sophistication required to understand the crucial textual passages. While the whole of *DDT* is crafted for students, the material taken from the *epistolae Graecorum* and inserted into the text is difficult enough, both technically and historically, to warrant the view that in these places we hear the discourse of one accomplished computist to another. In *DM*, the coda centers on a specific technicality, the Victorian *saltus* and the change in epact during the 13th year of the *cyclus lunaris* (year 16 of the *cyclus decennouenalis*). This note is too precise for normal teaching. Jones was tempted to call *DM* an Irish Bull, and not without reason, for its instruction is meant as a permanent correction to the content of what will be taught about the computus.

In the two examples treated here, the Irish versions of these texts offer direct computistical instructions that were not original to the author. Now that we are no longer bound to think of this alteration as forgery, we may perhaps begin to value these precious examples of personal communication between seventh century scholars. The 'I' behind the telling of the Pachomian legend in *DDT* is the same person who altered the Cyrillan letter for a 'modern' application. Jones

communium embolismorum obseruatione discurrit, illi sancti patres qui in niceno concilio fuerunt christiana obseruatione tradiderunt obseruandum. Inde Dionysius dicit ad papam Leonem <*lege* Petronium>, *Venerabiles tricenteni .xviii. pontifices qui apud niceam ciuitatem Bythiniae contra uesaniam Arrii heretici conuenerunt. Quartas decimas lunas paschales obseruantiae per .xviiii. annorum ciclum semper stabiles inmotasque fixerunt, quae cunctis saeculis eodem quo repetuntur exordio sine varietatis excursu labuntur. Hanc autem regulam praefati circuli non tam peritia seculari quam sancti spiritus illustratione sanxerunt. Et ueluti anchoram firmam ac stabilem huic rationi lunaris dimensiones obposuisse cernuntur. Et Theophilus et Cyrillus ab hac sinodi reuerendae constitutione minime dissentiunt. Immo potius eundem decem nouenalem cyclum, qui enniakaideca dereda* <*sic*> *Greco uocabulo noncupatur, sollicite retinentes, paschalem cursum nullis diuersitatibus interpellasse monstrantur, et sancti concilii traditione ad obseruandas quartas decimas lunas paschales per omnia* tradiderunt.

thought it was a papal voice, a relic of the materials sent to the Canterbury mission. It now echoes only in the Irish manuscripts which gathered together so many diverse languages and instructions. In *DM* the coda offers advice that could only come from a seventh-century computist, and could only be meant for another contemporary scholar. Here the 'I' offers a solution to one of the outstanding problems of the Victorian computus, and urges a compromise for the sake of Christian unity. Might the coda be a contribution from the famous Cummian who wrote to Segene with the same purpose? If not, then it was at least during his lifetime and from his side of the debate about Dionysius and the authentic Paschal observance.

Conclusion

In these preliminary studies of the manuscript affiliations for two texts of the Sirmond computistical collection, Jones' and Ó Cróinín's understanding of the role of the Irish computus in Bede's work is confirmed. The Dionysiac Easter reckoning, which had reached Ireland by the early seventh century, is the source of textual confusions in the mid-seventh century texts of the Sirmond manuscript. In addition, we are now able to say that the recension history of both *DDT* and *DM* indicates that one or more computists intervened in both texts in order to instruct his counterparts in the management of a Dionysiac paschal table within a framework still credited to Victorius of Aquitaine. The source of both interventions textually was a collection of *epistolae Graecorum* received sometime before Cummian wrote his letter in 632. In *DDT*, an early textbook on time and the calendar was modified to include a new chapter on the Dionysiac Easter, treating of the paschal cycle and claiming for its authority the Council of Nicea in 325. The text of *DM*, if I am correct in identifying the attribution, likewise represents a return to the age of Constantine for its legitimation; as a series of excerpts from a fourth-century disputation, it depends on the seventh-century coda to draw its exegetical logic to a conclusion, namely, that the true Easter derives from Alexandria through Theophilus, Cyril, the Council of Nicea, and the 'joint' efforts of Victorius and Dionysius Exiguus. Undoubtedly, more information will be forthcoming as new editions of these works proceed, but already the outlines of what was meant to be a compromise on the calendar appear.

APPENDICES

Appendix I: A Short-Title Description of the Sirmond Texts in Bodley 309 (from Ó Cróinín (2003), 202–3)

1. fols. 3v–61v, Bede's *DTR*
2. fol. 61v, A list of Greek numbers
3. fol. 62r–v, Prologue and capitula to the Irish computus *De ratione temporum uel de compoto annali* (see Appendix II here)
4. fols. 62v–64v, *Sententiae Sancti Augustini et Isidori in laude compoti*
5. fols. 64v–73v, *De .xiiii. diuisionibus temporum*
6a. fols. 74r–76r, *De bissexto*
6b. fols. 76r–78r, *De saltu lunae pauca dicamus*
7. fol. 78r–v, Argumentum *De saltu lunae monstrando*
8. fol. 78v, Argumentum *De materia bissexti*
9. fols. 78v–79r, Argumentum *Si nosse desideras argumentum lunare*
10. fols. 79r–80r, Computistical poem *Annus solis continetur*
11. fol. 80r, Bede, *DTR* XXIII
12. fol. 80r–v, Bede, *DTR* XIX
13. fols. 80v–81r, Three uncanonical Dionysiac argumenta
14. fols. 81r–82r, Dionysiac argumenta I–X
15. fol. 82r–v, Dionysiac argumentum XIV, *Incipit calculatio quomodo reperiri*
16. fol. 83r–v, *Exemplum suggestionis Boni sancti primicerii, De sollemnitatibus et sabbatis*. The rubric is that of the *Exemplum Boni*, but the text is the pseudo-Columbanus letter *De sollemnitatibus*
17. fols. 84r–85r, Letter of Paschasinus to Pope Leo
18. fols. 85r–86r, Dionysius Exiguus to Boniface and Bonus
19. fols. 86r–87v, Dionysius to Petronius
20. fols. 88r–89v, Proterius of Alexandria to Pope Leo
21. fols. 89v–90v, Epistola Cyrilli *De pascha*
22. fols. 90v–93v, Anatolius of Laodicea, *De ratione paschali*
23. fols. 93v–94r, Excerpts from Eusebius, Gennadius, Jerome, etc., about Anatolius
24. fol. 94r–v, Pseudo-Morinus, *De pascha* [page 203]
25. fols. 94v–95v, *Epistola Philippi de pascha* [=Pseudo-Theophilus, *Acta Synodi*]

26. fols. 95v–95²r, 'Computistical bits' (Jones)[41]
27. fol. 96r–v, Victorian argumenta
28. fols. 96v–97r, Letter of Pope Leo to the Emperor Marcian
29. fol. 97r–v, Excerpt from *De sollemnitatibus* (No. 16 above)
30. fols. 97v–98r, Tract on finger-reckoning
31. fols. 98r–99r, Prologue to Easter table of Theophilus of Alexandria
32. fol. 99r–v, Argumenta
33. fols. 99v–101r, Prologue of (Pseudo-) Cyril of Alexandria
34. fols. 101r–105r, Excerpts from Macrobius, *Saturnalia* I 15, 5–6, known to Bede as *Disputatio Cori et Praetextati*
35. fols. 105v–106r, Excerpts from Isidore, *Etymologiae*
36. fols. 106v–107v, 'Computistical notes partially unpublished' (Jones)[42]
37. fol. 107v, Heading for a Cyrillan Alexandrian Easter table
38. fol. 108r, Rota of lunar and solar months
39. fol. 108r, Gennadius, *De uiris illustribus*, on Victorius
40. fols. 108r–110v, Letter of Archdeacon Hilarus to Victorius of Aquitaine, plus Victorius' Prologue
41. fol. 110v, (originally blank; additions by later hand)[43]
42. fols. 111r–113r, Excerpts from Chronicle of Eusebius-Jerome
43. fols. 113r–120r, Victorius' 532-year Easter table
44. fols. 120r–131v, Dionysiac Easter tables, AD 532–1042 (lacunose)
45. fols. 132r–140v, Victorius' *Calculus*
46. fols. 141r–165v, Computistical items of later date[44]

---

[41] Identified and published by Ó Cróinín (2003a), 208–10.

[42] See Ó Cróinín (2003a), 204–7.

[43] See Jones (1937), 218, for a citation of this passage from *ego Clemens romans pontifex* regarding confession.

[44] See Jones (1937), 219, for description.

Appendix II: The Capitular List of the *Computus Hibernicus* (MS Oxford, Bodleian Library, Bodley 309, 62r–v; Jones' Sirmond item 3, 'probably in 55 chapters'; printed in Jones (1943), 393–4; the asterisks mark a new chapter title)

De numero igitur fratres dilectissimi [. . .]
De athomis etiam tractandum est
De momentis *De minutis *De punctis
De feria monstranda in kal. Ian.
De feria in kal. .xii mensium
De feria in omni die querenda
De feria in termino paschali
De horis *De quadrante naturali *De quadrante artificiali *De diebus *De ebdomadibus *De mensibus lunaribus
De mensibus solaribus *De ordine mensium
De inuentione eorum *De numero mensium apud antiquos romanos
Deinde etiam interrogandum est ex quo tempore menses inuenti sunt et nuncupati
Et in quo numero dierum menses sunt in sole et luna hoc est in anno communi et embolismi et in anno solari *Et quomodo inter se dissentiunt et conueniunt illi anni *Et qui sunt qui primi menses solares inuenerunt
Et quomodo menses nominantur apud Hebreos et Aegyptios et Macedones et Grecos et Latinos, et quae sunt causae ex quibus mensis nomina acceperunt
Et postea inuestigandum est de quatuor temporibus anni
Deinde item tractandum est de annis et de generibus annorum
Et de numero annorum ab origine mundi usque ad incarnationem Christi
De numero annorum ab incarnatione Christi usque ad presens tempus *De bissexto *De saltu *De indictionibus Romanorum *De cyclo decennouennali *De cyclo lunari et quis inuenit primum cyclum decennouenalem et cyclum lunarem et in quo tempore
De luna in qua hora uel in quo puncto accenditur uel sua aetas commutatus omni die
De luna quot horis lucet in unaquaque nocte
De initio quadragesimi
De terminis quadragesimalibus
De prima luna primi mensis
De terminis paschalibus et de illorum regularibus
De pascha *De terminis rogationum *De rogationibus *De concurrentibus
De cyclo solari et lunari per digitos demonstrando *De concurrentibus demonstrandis per cyclum solarem *De bissexto monstrando per cyclum solarem
De zoziaco [sic] circulo et de .xii signis celi et eorum nominibus *De cursu solis et lunae per .xii signa
De .vii stellis errantibus

De ascensu solis et descensu hoc est quomodo crescit dies uel nox
De eclypsi solis et lunae
De numero momentorum et minutorum et punctorum et horum totius anni
De epacta monstranda et de feria querenda a presente qualibet die usque ad centum annos. De compoto Gregorum et Latinorum per litteras et digitos demonstrando
De epistolis Gregcorum [sic]
De Anatholio et Macrobio
De Victorio et Dionisio inuenti sunt *De cyclo epactarum in .xi kal. Apr.
De epactis quae currit in kal. Ian.
*De epactis in kal. .xii mensium
*De epactis uniuscuiusque diei per totum annum
De Boetio
De calculo

Appendix III: Text from Transcriptions of *Disputatio Morini*

Disputatio Morini Alexandrini episcopi de ratione paschali, eo quod senserunt alii diuerse de eo quod scriptum est 'Post biduum pascha fiet'.

Id congruum. Fuit pascha Iudaeorum in aequinoctio (hoc est VIII kal. Apr.) ut in aequinoctio renouatio mundi nouissima et redemptio per crucem Christi, quia in aequinoctio fuit mundi constructio et compositio, ne dispar sibi inuicem fieret utraque formatio. Deinde eadem longitudo facta est potentia dei tribus primis diebus et noctibus factis, sed addidit sol cursum dierum quando inluxit in mundo. Quarta die luna e contrario dempsit cursum noctium. Etenim ad uenerationem et honorem aequinoctii nouissimi et paschae Christi praesignata est et facta est prima mundi constructio in aequinoctio primo.

In aequinoctio quoque facta est solutio et redemptio Israhel de seruitute Aegyptiorum, et in eodem aequinoctio post XII kal. Aprelis praecepit filiis Israel celebrare pascha, quod est typus paschae Christi futuri per omnia in finem ex initio. Typus namque Christi agnus, qui occisus est, et manducatur quottidie agnus a septem uiris (hoc est a septem gradibus ecclesiae). Tantum autem mysterium dei scitote quod non est mirum si accederit iudicium, si conuenerit aequinoctium (quod est VIII kal. Aprelis, VI feria, luna XIIII et resurrectio luna XVI) post XII kal. Aprelis, quod est aequinoctium uernale, hoc est initium anni solis, et ante hoc initium non facitur pascha apud Hebraeos et Graecos et Latinos nisi in errore, quia dies dominica, in qua facta est lux, non fuit ante XII kal. Aprelis, quod est in nocte. Secunda feria anno crucis XI kal. Aprelis. Tertia feria et X kal. Aprelis. Quarta feria in quo dictum est, "Post biduum pascha fiet". Deinde sentiunt alii quod ita fiet pascha post biduum (id est post completionem et integritatem duorum testimentorum) et transcensus ad regnum dei conueniente in VI feriam aequinoctio luna XIIII et solemnitate resurrectionis die dominica, hoc est totius generis humani ex quo genere uirga de radice Iesse exiit. Quae uirga redemit Israhel de Aegypto, quae conuersa est in serpentem ut liberarentur Israhelitae ab Aegypto. Qui serpens persuasit homini (hoc est diabolus) mortem.

Si ergo mors a serpente, uirga in serpentem (id est Christus in mortem). Serpens exaltatur in aere ut sanaret uulnera corporum
40 respicientium in serpentem aereum. Ut soluit seruitutem Israhelitarum famulantium in Aegypto, sic homines morsi a serpentibus sanabantur a uenenis intuendo serpentem. Magnum sacramentum quid est intuendo serpentem sanari, nisi credendo in mortuum Christum sanari a morte peccatorum? Et ut liberaretur
45 populus ab Aegypto per uirgam in serpentem conuersam et agnum in pascha occisum (hoc est Christum in mortem), ita liberaretur totum humanum genus in pascha passionis Christi, qui est uerus serpens occidens serpentem diabolicum, et deuorans uirgas magorum et ministrorum diaboli.

50 Quid est autem expauit Moyses et fugit? Quando mortificatus est Christus, expauerunt et fugerunt discipuli.

Caudam adprehende posteriora adprehende, et illud: "Posteriora mea uidebis . . .". Primo factus est serpens, sed cauda retenta facta uirga; primo occisus, postea resurrexit cauda retenta: "uisus est
55 Cephae, deinde XI . . .". Est etiam in cauda serpentis finis seculi, quia mortalitas generis humani ambulat per mortem a diabolo inlatam, sed in fine seculi sicut a cauda redimus ad manum Dei, et efficimur stabilitum regnum Dei.

Post haec omnia breuiter dicam quod numquam factum est
60 pascha apud Iudaeos ante XII kal. Aprelis, quia luna nata in VIII id. Martii, XIIII est in XII kal. Aprelis. Obserua igitur cursum lunarem iuxta regulam Graecorum more Aegyptiorum et non secundum epactas (id est adiectiones lunares) quia ibi peruenitur a IIII luna ad XVI, hic autem ad XV iuxta computacionem
65 Eusebii, qui primus conscripsit circulum X et VIIII annorum, Athanasii, Theophili, Cyrilli Dionysiique exigui, usque dum scripsit Victorius Hilaro papae urbis Romae episcopo. Tunc cessauerunt disputatores alexandrini et antiocheni circulos post alios describere.

LEOFRANC HOLFORD-STREVENS

MARITAL DISCORD IN NORTHUMBRIA: LENT AND EASTER, HIS AND HERS

Abstract

The report that before the Synod of Whitby King Ōswiu of Northumbria sometimes kept Easter on a day that for Queen Eanflǣd was still Palm Sunday has traditionally been interpreted with reference to the conflicting rules about *luna XIIII* on a Sunday and dismissed as a mere occasional nuisance. In fact, this discrepancy occurred in over half the years of their marriage, but was never due to that cause; it was not the only difference in their Easter dates, and even when their Easter dates agreed, the king's Lent began later than the Queen's. The consequences not only for court life but for Northumbrian society in general are considered as the background to Ōswiu's abandonment of the *Latercus* at Whitby.

Keywords

Bede, computus, Dionysius Exiguus, Eanflǣd, Easter, *initium*, *Latercus*, Lent, lunar calendar, Northumbria, Ōswiu, Synod of Whitby, Victorius of Aquitaine, Wilfrið.

The present essay is an attempt to make sense, and consider the implications, of a famous passage in Bede's account of events leading up to the Synod of Whitby in 664, when the church in Northumbria abandoned the Insular reckoning of Easter in favour of the Dionysian computus, based on tables adapted from the Alexandrian calendar to suit the Roman:[1] Great controversy, he declares, arose when those who had

---

[1] Bede, *Historia ecclesiastica gentis Anglorum* III 25 (Colgrave and Mynors (1969), 296–8; cf. Plummer (1896), i 181–3). All quotations cited without source come from this passage.

come from Kent or Gaul asserted that the Irish were keeping the feast 'against the custom of the universal church'.² Indeed, it was said to have sometimes happened that King Ōswiu, the pupil of Irish monks, sometimes kept Easter on a day that for Queen Eanflæd, who followed the 'Catholic', i.e. Roman, practice she had learnt in Kent, was still Palm Sunday.³ This has traditionally been interpreted with reference to the Insular rule of keeping Easter on *luna XIIII*, should that date fall on a Sunday, instead of the next Sunday as under the Roman rule:

> 'It is plain that a divergence of a week would frequently be the result of this difference [in lunar limits]. For whenever the 14th of the moon fell on a Sunday, the Celts would celebrate Easter on that day, whereas the Romans would defer it to the following Sunday. This is precisely the case which Bede represents as occurring in the household of Oswy of Northumbria, where the king, who followed the Celtic use, would sometimes be celebrating Easter, while the queen, in accordance with the Roman rule, was still fasting in Holy Week.'⁴

It appears to have been taken as no more than an occasional nuisance; some writers even denied it had happened at all.⁵

Alas for doubters, and alas for believers. A manuscript in Padua, and a mnemonic verse recorded from a West Cork farmer from the reign of Queen Victoria, have enabled Dáibhí Ó Cróinín and Daniel Mc Carthy to recover and reconstruct what even Bede did not know: the detailed working of the lunar calendar, the so-called *Latercus*, by which the Insular Easter was calculated.⁶ With this knowledge, I have already shown elsewhere that Bede's discrepancy did indeed occur, not merely 'sometimes' (*nonnumquam*) but in roughly half the

---

² *His temporibus quaestio facta est frequens et magna de obseruatione paschae, confirmantibus eis, qui de Cantia uel de Galliis aduenerant, quod Scotti dominicum paschae diem contra uniuersalis ecclesiae morem celebrarent.*

³ *Obseruabat et* [sc. besides Paulinus' former deacon, cf. n 33] *regina Eanfled cum suis, iuxta quod in Cantia fieri uiderat, habens secum de Cantia presbyterum catholicae obseruationis, nomine Romanum. Vnde nonnumquam contigisse fertur illis temporibus, ut bis in anno uno pascha celebraretur, et cum rex pascha dominicum solutis ieiuniis faceret, tum regina cum suis persistens adhuc in ieiunio diem palmarum celebraret.*

⁴ Plummer (1896), ii 350.

⁵ See Wallace-Hadrill (1988), 125.

⁶ See Mc Carthy and Ó Cróinín (1987–8), repr. in Ó Cróinín (2003), 58–75, as corrected by Mc Carthy (1993); the Irish verse was recorded by Canon O'Leary (c.1920), 1, and passed on to Mc Carthy by Ó Cróinín.

years of Ōswiu and Eanflǣd's marriage, which historians date to between late 642 and 645;[7] see *Figure 1*, in which it is called Discrepancy Type 1.

However, the week's difference never arose, as Plummer assumed, because, both agreeing that a certain Sunday was *luna XIIII*, he celebrated and she postponed; indeed, as can be seen, from *Figure 1* there was no year in which *luna XIIII* was the same for both king and queen, whether on a Sunday or any other *feria*. To be sure, there were years in which *luna XIIII* fell on Sunday for one of the partners; but if Ōswiu had been prepared to postpone in deference to the queen's scruples against celebrating with the Jews, only in three years would the discrepancy have been avoided (Type 1a in *Figure 1*), and if Eanflǣd had yielded to the argument urged by Columbanus in his famous letter to Gregory the Great, that the Pasch belonged to God and not the Jews,[8] she would have avoided the clash in three years if she took her lunar ages from Dionysius, and four years if she took them from Victorius (Type 1b).

As even these data suggest, the fundamental cause of discrepancy was one that, because no theological point was thought to ride on it, was largely disregarded in polemic: that the lunar dates of the 84-year cycle diverged ever more from those of the nineteen-year cycle, whether the true Alexandrian cycle as propagated in the Latin world by Dionysius Exiguus or the bastardized jumble palmed off on the mathematically challenged by Victorius of Aquitaine. In *Figure 2* I have set out the epacts of 1 January according to the three calendars, adding in Roman figures Dionysius' own epact, properly that of the fifth epagomenal day in the preceding Alexandrian year,[9] but found in the West by Isidore's time to match that of 22 March.[10]

---

[7] Holford-Strevens (2005), 52–3.

[8] Columbanus, *Epistula* I 4 (Walker (1957), 6): *aut numquid ipsorum esse recte credendum est decimae quartae lunae Pascha, et non potius Dei ipsius instituentis Phase esse fatendum [. . .]?*

[9] The *sedes epactarum* on 28 August, the Roman equivalent of 5 Epagomenon, is considered at Maximus Confessor, *Computus ecclesiasticus* III 10 (PG 19, 1271); it is not explicitly stated by Neugebauer (1979), but can be inferred from data there presented, in particular (p 188) that it is the complement to 30 of the solar (civil) *quantième* of the lunar (pseudo-Jewish) New Year, which always fell in either the first lunar month or the embolism. Note that in the Dionysian calendar, as expounded both in *De ratione conputandi* and by Bede, owing to different distribution of full and hollow months, the lune of 28 August is always 1 higher than the next year's epact except in the *saltus* year (cycle 19).

[10] Isidore, *Etymologiae* VI 17, §31 (Lindsay (1911), i); thereafter *De ratione conputandi* 103 (Walsh and Ó Cróinín (1988), 207); Bede, *De temporibus* 13 (Jones (1943),

Year	Insular			Roman					Discrepancy Type
	Luna XIIII	Easter	Lune of Easter	Luna XIIII		Easter	Lune of Easter		
				Vict.	Dion.		Vict.	Dion.	
643	5A	6A	XV	8A	9A	13A	XVIIII	XVIII	1
644B	25M	28M	XVII	28M	29M	4A	XXI	XX	1 (1b Vict.)
645	13A	17A	XVIII	16A		17A	XV Lat.		—
				16A	17A	24A	XXII Gr.		1 (Vict. Gr.)
646	2A	2A	XIII	5A	5A*	24A	XVIII	XXI	1b (Dion.)
647	20A*	22A	XVI	25M	25M	9A	XXI	XVIII	1a
648B	9A	13A	XVIII	13A	13A	1A	XXI	XXI	2
649	29M	29M	XIII	2A	2A	20A	XXI	XXI	1b
650	16A	18A	XVI	22M	22M	5A	XVII	XVII	1a
651	7A	10A	XVII	10A	10A	28M	XX	XX	2
652B	26M	1A	XX	29M*	30M	17A	XXI	XXI	1b
653	14A	14A	XIII	17A	18A	1A	XVII	XVI	—
654	3A	6A	XVII	6A	7A	21A	XVIII	XVII	1a
655	23M	29M	XX	26M	27M	13A	XXI	XX	1 (1b Vict.)
656B	11A	17A	XX	14A	15A	29M	XVII	XVI	—
657	31M	2A	XVI	3A	4A	17A	XVII	XVI	1
658	18A	22A	XVIII	23M	24M	9A	XX	XVIIII	1
						25M	XVI	XV	3

146

659	8A	14A	XX	11A	12A	14A	XVII	XVI	—
660B	28M	29M	XV	31M	1A	5A	XVIIII	XVIII	1
661	15A*	18A	XVII	20M	21M	28M	XXII	XXI	2
662	4A	10A	XX	8A	9A	10A	XVI	XV	—
663	24M	26M	XVI	28M	29M	2A	XVIIII	XVIII	1
664B	12A	14A	XVI	16A	17A	21A	XVIIII	XVIII	1

Figure 1 King Ōswiu's Insular and Queen Eanflǣd's Roman Easter.

B = leap year F = February M = March A = April * = *Saltus lunae*

Discrepancy Type 1: Ōswiu's Easter = Eanflǣd's Palm Sunday: 643, 644, 645? (Vict. Gr., Dion.), 646, 648, 649, 651, 653, 654, 657, 660, 663, 664

Type 1a: Ōswiu's Easter is *luna XIIII* a week before Eanflǣd's: 646, 649, 653

Type 1b: Eanflǣd's Easter is *luna XXI* a week after Ōswiu's: 644? (Vict.), 645? (Dion., cf. Vict. Gr.), 648, 651, 654? (Vict.)

Discrepancy Type 2: Eanflǣd's Easter = Ōswiu's Third Sunday after Easter: 647, 650, 661

Discrepancy Type 3: Ōswiu's Easter = Eanflǣd's Fourth Sunday after Easter: 658

Years without discrepancy: 645? (Vict. Lat.), 652, 655, 656, 659, 662

Years in which discrepancy is not due to difference in lunar limits: 643, 644? (Dion.), 647, 650, 654? (Dion.), 657, 658, 660, 661, 663, 664

Years in which *Luna XIIII* falls on the same day for Ōswiu and Eanflǣd: None

	Epact of 1 January			22 Mar.		Epact of 1 January			22 Mar.
Year	Insular	Vict.	Dion.	EPACTAE	Year	Insular	Vict.	Dion.	EPACTAE
643	8	6	5	XXVI	654	10	8	7	XXVIII
644[a]	19	17	16	VII	655	21	19	18	VIIII
645	30	28	27	XVIII	656	2	30	29	XX
646	11	9	8/9*[b]	NVLLAE*	657	13	11	10	I
647	22	20	20	XI	658	24	22	21	XII
648	4*	1	1	XXII	659	5	3	2	XXIII
649	15	12	12	III	660	16	14	13	IIII
650	26	23	23	XIIII	661	27	25	24	XV
651	7	4	4	XXV	662	9*	6	5	XXVI
652	18	16*	15	VI	663	20	17	16	VII
653	29	27	26	XVII	664	1	28	27	XVIII

Figure 2 King Ōswiu's Insular and Queen Eanflǣd's Roman Epacts.

[a] This and succeeding leap years are not marked because none of the three calendars gave the *bissextus* a separate lune as Hippolytus and the Computist of 243 had done.

[b] Irish reckoning gave the Dionysian January epact of 646 as 8, with the *saltus* taken on 22 March (see Ó Cróinín on *De ratione conputandi* 73. 4–5, p 181); it is 9 in Bede, *De temporum ratione* 20, the *saltus* being taken on 25 November. Post-Bedan calendars follow suit, though some take the *saltus* on 30 July.

In fact we do not know which version of the nineteen-year cycle Eanflæd followed, nor whether it was the same throughout the period concerned; the clergy who had come from Gaul would have followed Victorius,[11] but if Abbot Wilfrið, that Romanizing bulldozer, had any influence over her as he had over her stepson Alhfrið,[12] underking of Deira, it would have been exerted in favour of the Dionysian system.[13] In Bede's history the quarrel is not as in *De temporum ratione* between Dionysius and Victorius, but between Roman truth and Irish error; as a result we are left in the dark on the relation between the former pair. In practice, they usually gave the same Easter date; during our period the only discrepancy is in 645, when if, as is probable, Eanflæd was still using Victorius, her spiritual advisers had to decide between a Latin Easter that coincided

---

300), *De temporum ratione* 50 (Jones (1943), 269); Dionysius himself offers no explanation, only a formula for finding it (*Argumentum III* (Krusch (1938), 75–6)). An Alexandrian epact of 26 Phamenoth is not only unattested (cf. Neugebauer (1979), 188 n 5) but impossible, since in years 3, 6, 8, 9, 11, 14, 17, and 19 of the cycle its lune is one lower than the annual epact. Byzantium preferred the θεμέλιος (foundation) of 31 March, whose lune matches that of 1 January, and which was calculated in different ways at different times. (It is also mentioned by Maximus Confessor, *Computus ecclesiasticus* loc. cit., cf. II 27 (PG 19, 1255), but calculated from the Alexandrian not the Byzantine cycle.)

[11] The Fourth Council of Orléans, in 541, had by its first canon required *ut sanctum pascha secundum laterculum Victorii ab omnibus sacerdotibus uno tempore celebretur* (CCSL 148A, 132; MGH Conc. 1, 87). However, the total indifference to Roman practice of the Gauls, even when Victorius' table bred dissension (Gregory of Tours, *Historiarum libri decem* V 17, X 23 (MGH SS rer. Merov. 1,1, 215, 514–5) says nothing of any request for papal guidance), is inconceivable at Canterbury.

[12] Since he was old enough to fight and to marry in the early 650s, he cannot have been Eanflæd's son; rather he and his sister Alhflæd were born to Ōswiu's first wife Rieinmellt (Rhiainfellt) of Rheged.

[13] We have no evidence to suppose that Rome, in the seventh century, used any tables other than Victorius' or Dionysius'; the first attestation of papal support for the latter comes from Pope Vitalianus (†672), if we accept Archbishop Ussher's ascription: see Jones (1943), 104. Bede, *Historia ecclesiastica* III 29 (Colgrave and Mynors (1969), 320; Plummer (1896), i 197) makes it clear that the Pope desired uniformity, which could not be achieved on any other basis once the Alexandrian reckoning had prevailed in the East; see too the addendum at Wallace-Hadrill (1988), 235. However, if as Ó Cróinín (1983b), repr. Ó Cróinín (2003), 87–98, has suggested, the letter cited by Bede at *Historia ecclesiastica* II 19 (Colgrave and Mynors (1969), 200–2; Plummer (1896), i 123–4) was provoked by the northern Irish clergy's intention of celebrating on 1 April 641, the Dionysian tables must already have been in use at Rome, for only in them was that day *luna XIIII*; as Victorius' Greek date it was *luna XV*, in the *Latercus luna XVIII*.

with the king's and a Greek Easter that did not;[14] in all likelihood they adopted the former for that very reason.[15]

In the years of this discrepancy, throughout what for the queen and her followers was Holy Week, including Good Friday and Holy Saturday, their pious fasting competed with the epicurism and rout of the king and his men rolling merry from the mead-hall: Wulfstan of Winchester's description of 'the Northumbrians inebriated in their usual way and departing full of cheer in the evening'[16] was surely no less applicable in the seventh century than in the tenth or the twenty-first. Moreover, the king might seek to pay the conjugal debt while the queen was observing Lenten abstinence.[17]

The Bedan discrepancy, Type 1, is not the only one. There were four years in which the queen had already celebrated Easter either three weeks before the king broke his fast (Type 2) or four weeks beforehand (Type 3); as *Figure 3* shows, that was because the previous Insular *luna XIIII* would have yielded an Easter date before 26 March, which was not acceptable because the *Latercus* required Easter to follow the Roman equinox on the 25th. In these years it is the king's men who would have been fasting while the queen's court was either feasting or at least living normally.

That leaves at most six years in which Ōswiu and Eanflǣd kept Easter on the same day; but even then there was disharmony, as set out in *Figure 4*. We do not know whether Eanflǣd began Lent at Quadragesima, following the practice of Gregory the Great, or on Ash Wednesday, the Wednesday before Quadragesima, if Rome had already switched to that date and Wilfriđ or another had told her.[18] But Ōswiu's Lent started on the fortieth day (reckoned inclusively)

---

[14] Both Easters are on lunes forbidden to the respective party, XV for the Latins and XXII for the Greeks.

[15] However, as Immo Warntjes pointed out at the Second International Conference on the Science of Computus at Galway in 2008, the only Victorian table recorded in an Insular source, that in the Sirmond MS, gives only the Latin date.

[16] Wulfstan of Winchester, *Vita S. Æthelwoldi* 12 (Lapidge and Winterbottom (1991), 24): *inebriatis suatim Northanimbris et uesperi cum laetitia recedentibus*. On the adverb *suatim* see Lapidge and Winterbottom (1991), 24 n 7.

[17] Continued cohabitation is proved by the birth of four childen: two sons Ecgfriđ and Ælfwini, two daughters Ælfflǣd and Ōsþrȳđ.

[18] See Böhne (1965). Irish Alexandrian computi, whether Victorian or Dionysian, place the *initium* on Quadragesima; they are followed by Hrabanus Maurus, *De computo* 83 (CCCM 44, 301–4; copied almost entirely, as Immo Warntjes kindly informs me, from the Fulda Computus of AD 789 preserved in MS Basel, Universitätsbibliothek, F III 15k, 43v–44r); but Bede ignores a date as easily found as Whitsun.

Year	I Jan.		Insular *Luna XIIII*		Hypothetical Easter
	Feria	Insular Epact	Paschal	Preceding	
647	Mon.	22	20A	Thu. 22M	25M
650	Fri.	26	16A	Thu. 18M	21M
658	Mon.	24	18A	Tue. 20M	25M
661	Fri.	27	15A	Sun. 21M	21M

Figure 3 Analysis of Discrepancies, Types 2 and 3.

Year	Insular Initium	Roman Ash W.	Quad.	Year	Insular Initium	Roman Ash W.	Quad.
643	26F	26F	2M	654	26F	26F	2M
644B	18F	18F	22F	655	18F	11F	15F
645	9M	2M	6M (Vict. Lat.)	656B	9M	2M	6M
		9M	13M (Vict. Gr., Dion.)				
646	22F	22F	26F	657	22F	22F	26F
647	14M	14F	18F	658	14M	7F	11F
648B	5M	5M	9M	659	6M	27F	3M
649	18F	18F	22F	660B	19F	19F	23F
650	10M	10F	14F	661	10M	10F	14F
651	2M	2M	6M	662	2M	23F	27F
652B	22F	15F	19F	663	15F	15F	19F
653	6M	6M	10M	664B	6M	6M	10M

Figure 4 King Ōswiu's Insular and Queen Eanflǣd's Roman Lent.

before Easter, on the Wednesday after Quadragesima, so that at least on the Monday and Tuesday of that week, and perhaps from the preceding Wednesday, the king was enjoying the last good days before the fast that his queen was already observing. On the other hand, if in the years of Discrepancy Type 1 she was still observing Pope Gregory's Lent, then he was fasting for four days before her in counterpoise to the conflict between his Easter Week and her Holy Week. And in the other four years, as shown in *Figure 5*, there were only two and a half, or in 658 only one and a half weeks when both were fasting together.

Year	Insular *initium*	Roman Holy Saturday	Days common to both fasts
647	14M	31M	18
650	10M	27M	18
658	14M	24M	11
661	10M	27M	18

Figure 5 Overlap of Insular and Roman Lent in years of Types 2 and 3.

To be sure, the two parties may have occupied separate halls, and even within one hall there was room enough to accommodate them both; but even if the discrepancies, as Henry Mayr-Harting has insisted, 'need have occasioned nothing unseemly or farcical',[19] it does not seem likely that no-one in all those years provoked an incident; and even if the king and queen regarded such brawls as mattering no more than any others, clergy are inclined to take divergent practices more seriously. After all, there had been conflict between the *Latercus* and the Victorian reckoning ever since Columban set foot in Gaul; a letter from Pope Honorius I expressing disapproval of Insular usage provoked a divided response, even after the southern Irish delegates to Rome, who expected to celebrate Easter on 21 April 631, were dismayed to learn that the rest of Christendom was keeping 24 March, a date not even permitted in the *Latercus*.[20]

According to Bede, everyone put up with the situation at King Ōswiu's court so long as Áedán was bishop because of his transparent holiness, recognizing his Easter dates to result from loyalty to his superiors at Iona. Since, with the possible exception of 645, there was a discrepancy in every year down to his death on 31 August 651, twice of Type 2, the rest of Type 1,[21] we shall in due course need to seek another

---

[19] Mayr-Harting (1991), 104, adducing the arrangements at Yeavering.

[20] See Bede, *Historia ecclesiastica* II 19 (Colgrave and Mynors (1969), 198; Plummer (1896), i 122); Cummian, *De controuersia paschali* (Walsh and Ó Cróinín (1988), 4–5, 94, ll. 281–3); Archbishop Ussher's dating (cited p 4) is confirmed by the reconstructed *Latercus*. Cummian's *Cum Greco et Hebreo, Scitha et Aegyptiaco* is plainly a literary mode of expression (cf. Act 2:8–11), but the substitution of *Scitha* (cf. Col 3:11) for the Latins of p 72, line 108 may reflect the segregation of those whose mother-tongue was not Latin. At any rate the 'Hebrews' cannot be the unbaptized Jews (*Iudei*) of line 104.

[21] *Haec autem dissonantia paschalis obseruantiae uiuente Aidano patienter ab omnibus tolerabatur, qui patenter intellexerant quia, etsi pascha contra morem eorum qui ipsum miserant facere non potuit, opera tamen fidei, pietatis, et dilectionis, iuxta morem omnibus sanctis consuetum, diligenter exequi curauit.*

cause; but in any case Bede's *omnes* must be the 'Catholic' party, for the 'Columbans' (as we ought to call them rather than 'Irish', which not all were) would have made allowances for Eanflæd, not for Áedán. Even so, and even if Bede is in practice talking only about the clergy, this is already not a matter for the king and queen alone, as if he had wanted to watch the football and she the soap.

Áedán's successor, Fínán, found himself challenged from within his own camp, by one Rónán, who had learnt 'the rule of ecclesiastical truth' (*regulam ecclesiasticae ueritatis*) in Gaul or Italy; however, the polemic of this *acerrimus ueri paschae defensor*, though it converted many or at least made them think about the matter, caused the bellicose Fínán to dig his heels in and defend the Irish practice with all his might;[22] nevertheless, his ten-year reign saw four years when king and queen celebrated Easter on the same date. Then, when Iona sent Colmán as his successor, the controversy became sharper, and encompassed other matters,[23] no doubt meaning the tonsure.[24] Friction was also increased by Alhfrið's adhesion to the Roman ways, which left the king's court isolated, and the new abbot of Ripon was doubtless not slow in offering his opinion, whether wanted or not, that those who followed Celtic practice were not proper Christians; at least, such a condemnation, asserted with all Wilfrið's usual vehemence, seems the likeliest background to Bede's statement that many people feared they were running or had run in vain.[25]

---

[22] *Erat in his acerrimus ueri paschae defensor nomine Ronan, natione quidem Scottus, sed in Galliae uel Italiae partibus regulam ecclesiasticae ueritatis edoctus. Qui cum Finano confligens, multos quidem correxit, uel ad solertiorem ueritatis inquisitionem accendit, nequaquam tamen Finanum emendare potuit; quin potius, quod esset homo ferocis animi, acerbiorem castigando et apertum ueritatis aduersarium reddidit.* Bede carefully omits to define this 'rule of ecclesiastical truth'; if acquired in Gaul, it was undoubtedly the faulty table of Victorius.

[23] *Defuncto autem Finano qui post illum [Aidanum] fuit, cum Colmanus in episcopatum succederet, et ipse missus a Scottia, grauior de obseruatione paschae, necnon et de aliis ecclesiasticae uitae disciplinis controuersia nata est.*

[24] About which Wilfrið had been instructed in Gaul: *nam et Romam prius propter doctrinam ecclesiasticam adierat, et apud Dalfinum [immo Annemundum] archiepiscopum Galliarum Lugdoni multum temporis egerat, a quo etiam tonsurae ecclesiasticae coronam susceperat.* We may speculate what we should be reading in the *Historia ecclesiastica* had Bede been as interested in hairdressing as he was in arithmetic; as it is, see James (1984).

[25] *Unde merito mouit haec quaestio sensus et corda multorum, timentium, ne forte accepto Christianitatis uocabulo, in uacuum currerent aut cucurrissent*; see Gal 2:2.

'Many people', says Bede. Of course, it is difficult to conduct a sociology of religion at this distance; not every subject of a Christian king was a Christian in either faith or works, as we are reminded by Bede's report that Ōswiu's contemporary, Eorconberht of Kent, was the first English king to destroy idols and incorporate Lenten observance into the royal law.[26] Nevertheless, it is not unreasonable to suppose that there were proportionately more Christians in Northumbria by 664 than there had been in 642; and although not all will have taken their obligations with sufficient seriousness either to abstain from eating whatever forbidden food remained from winter or to complain of two Easter celebrations in one year, those who did care about these matters will not have been helped to peace of mind when they heard different things from different clerics agreed only that the question was of vital importance, while the heathen and the atheist mocked.

Alas, Bede, monk that he is, pays less attention than we should like to the realities of everyday life; if he says less than he might have done about the discrepancies in Easter dates, that is because he did not know how the *Latercus* worked, but if he says little about those in the incidence of Lent, it is not only because he may not have known about the insular *initium*, but because, even when the king breaks his fast a week before the queen, it is the fact of difference that interests him, not its human consequence. Nor, apart from the vague sentence about running in vain, does Bede tell us about persons other than the king, the queen, and the grown-up Alhfrið. Nothing on the upbringing of the royal children, nothing on what the churls were being told by their clergy; did Easter differ from parish to parish? and what happened when husband and wife did not agree? With every year that passed, the likelihood of ill-feeling must have increased.[27]

Outside Northumbria, ill-feeling, and in Ireland fighting, between adherents of the different computi are on record,[28] and not unique.

---

[26] Bede, *Historia ecclesiastica* III 8 (Colgrave and Mynors (1969), 236; Plummer (1896), i 142): *Anno dominicae incarnationis DCXL, Eadbald rex Cantuariorum transiens ex hac uita, Earconbercto filio regni gubernacula reliquit; quae ille suscepta XXIIII annis et aliquot mensibus nobilissime tenuit. Hic primus regum Anglorum in toto regno suo idola relinqui ac destrui, simul et ieiunium XL dierum obseruari principali auctoritate praecepit. Quae ne facile a quopiam posset contemni, in transgressores dignas et competentes punitiones proposuit.* Bede's *Ear-* is a Northumbrianism: see Ström (1939), 105–6, 148.

[27] Some awareness of this is shown by Jones (1943), 81.

[28] See Plummer (1896), ii 302–3, 352–3. Confused as is the account of Easter celebrations in the *Fragmentary annals of Ireland* §166 (Radner (1978), 56), the events of 449 at Ephesus suffice to show that Plummer was right not to accept the marginal note *calumnia* added by the scribe of MS Brussels, Bibliothèque Royale, 5301-20.

Four years out of every 532, the Armenians, before they adopted the Gregorian calendar, used, together with the Syrian Orthodox ('Jacobites') and Assyrians ('Church of the East'), to celebrate on 13 April Old Style instead of the 6th like the Chalcedonian Orthodox, Copts, and Ethiopians; at Jerusalem, where all denominations are ready to quarrel about anything, these 'False Easters' led to fighting in both 1634 and 1729.[29] Nevertheless, as the bishops of Rome foresaw when imposing the solar limit of 21 April, it would be only too natural for feasters to become particularly boisterous in fasters' presence;[30] and with such characters as Fínán and Wilfrið on the loose, life cannot have been as eirenic as in Áedán's day. A wise king must sooner or later have intervened: brangling clergy may be not left alone when their disputes spill over into everyday life.

It is often difficult to tell why an anomalous situation is tolerated for a while and then provokes reform; if the Whig school of history and its disloyal child the Marxist leave the student wondering how any person not directly benefiting from the old order could endure it for a moment, the Tory school defends it so well that discontent appears to be either a mental disorder or a crime. In the case of the Northumbrian Easter, Bede offers an explanation by personalities; the personal is never to be excluded in history, but it is never a complete explanation. When Eanflæd first came to Northumbria, it was Áedán who was reviving Christianity in that kingdom after the fruits of Paulinus' mission had withered; she, the newcomer, might preserve her own ways, as might survivors from Ædwini's reign,[31] but this was no time to

---

[29] This discrepancy results from maintaining an older cycle, abandoned by the Byzantines in the 6th century, that places the *saltus* one year later than at Alexandria, so that the *luna XIIII* of 5 April Old Style in the mainstream calendar falls for them on the 6th; consequently, when (in Western terms) the Golden Number is 1 and the Sunday Letter is E (or the concurrent 2), *luna XIIII* falls for them on Sunday instead of Saturday, entailing postponement of the feast. The discrepancy recurred in 1824 and will recur in 2071. See Dulaurier (1859), 84–100; Brosset (1870), i xxxix–xlvii.

[30] In the 528 years of the Paschal cycle in which the Armenian Easter coincides with the Greek, the tenth week before Easter, which for Armenians is the very strict aṙacʻaworacʻ fast (the name denotes that it precedes Lent proper), is the Greek week of licence, in which even the Wednesday and Friday fasts have sometimes been dispensed with, called in mockery of the Armenians ἀρτζιβούριον. See Neale (1850–73), ii 742.

[31] *Obseruabat autem Iacob diaconus quondam, ut supra docuimus, uenerabilis archiepiscopi Paulini, uerum et catholicum pascha cum omnibus, quos ad correctiorem uiam erudire poterat*; but this relates to the controversy in Fínán's time.

interfere with the process of reconversion. However, as Northumbria became more integrated into Christendom, so the queen and her stepson could remind the king, no doubt oftener than he liked, that, despite the dispatch of bishops from Iona and Northumbria to newly Christianized Mercia and re-Christianized Essex[32], the faithful of southern England, southern Ireland, the entire Continent, and the Pope himself were ranged against him.

As we know, the Northumbrian clergy met at Strēanæshalh; the proceedings are engraved in the memory and etched in the heart.[33] Wilfrið, at thirty years old and not yet a bishop, dominated the proceedings, as the one man on the Roman side who knew what he was talking about; whereas his admirer Stephen of Ripon makes him appeal, honestly though as we now know mistakenly, to the authority of Nicaea, his less than admirer Bede puts in his mouth a wide range of arguments, together with insults against Colum Cille and the quite mendacious asseveration that St Peter, here not a metonym for his vicar in the Vatican but the Fisherman himself, had observed the lunar limits 15–21. Anyone who had read Victorius' preface knew that the Latin lunar limits had been 16–22,[34] but Bede's Wilfrið would not let a detail like that stand in his way, and his Colmán is too polite to point it out.

I shall not enquire how much of Wilfrið's speech goes back to the minutes of the meeting, and how much was made up by Bede himself to demonstrate that the right cause won but not with clean hands;[35] however, by making the turbulent abbot end on a resounding *Tu es Petrus* that induces Ōswiu to ask Colmán whether any such power had been granted Colum Cille, he makes the king's decision in favour of the heavenly gatekeeper arise naturally out of the debate, based on an argument that his lay mind, left reeling from ecclesiastical history

---

[32] Bede, *Historia ecclesiastica* III 21, 22, 24 (Colgrave and Mynors (1969), 280, 282, 292–4; Plummer (1896), i 170–1, 172–3, 179–80).

[33] See Bede, *Historia ecclesiastica* III 25 (Colgrave and Mynors (1969), 298–308; Plummer (1896), i 181–9); Stephen of Ripon, *Vita Wilfridi* 10 (Colgrave (1927), 20–2).

[34] Victorius of Aquitaine, *Prologus* 4 (Krusch (1938), 19).

[35] Debates in Hegesippus and the Acts of Caesarea might (as was suggested in discussion) have shown him how to compose a debate, but the classic model for a senior speaker defeated by a junior, the speeches of Caesar and Cato in Sallust's *Catilina*, is not cited in pre-Conquest England, though passages from *Jugurtha*, always transmitted with it, are used by Bede and others; see Lapidge (2006), 331.

and technical computistics, could latch on to and fully understand. By contrast, Stephen, who makes him of his own initiative, without any connection to the preceding speeches, ask with a smile, *subridens*, which of the two saints was the greater, gives the impression of a decision already taken in private and merely announced after a show of public disputation.

Perhaps it was;[36] Ōswiu could see that something had to give, and recognized that his beloved Insular usages not only differed from those of the Catholic mainstream, a disturbing thought in itself, but were coming under increased pressure from Continental or Continental-trained clergy. Conformity with Rome would enable him to outflank his disloyal son;[37] and if as Daniel Mc Carthy and Aidan Breen have persuasively argued, the Synod of Whitby took place after the eclipse of 1 May 664, total only in Northumbria, and the ensuing plague,[38] all but the staunchest of the Columban party, hitherto the established faction,[39] must have seen in the darkened sun a sign of God's disapproval for the aberrant practices maintained in the kingdom, and in particular espoused by the king: this opinion must have been widely held, for the discrepancy in Paschal observation had long been a cause of inconvenience and dissension, not only in the royal household, but wherever it was necessary for the adherent of either Easter to take note of the other.

King Ōswiu must have seen that the only way to settle the matter once for all was to concede;[40] nor need we deny he was genuinely worried about his reception by St Peter (after all, he had at least one mortal sin on his conscience, the murder of his brother Ōswini); but one should not discount the effect of domestic pressure, particularly if the dissensions were replicated amongst his subjects. And these dissensions concerned no mere abstraction, but the practical observation, not only of a

---

[36] Indeed, only on that basis could Stephen's assertion that Wilfrið spoke *humiliter* be swallowed for a second.

[37] Alhfrið, indeed, is seen as the prime mover by Mayr-Harting (1991), 107–8, who to his political motives adds devotion to Wilfrið (Mayr-Harting (1991), 8); so too Abels (1983), 1–25, who shows that his ecclesiastical triumph was a political defeat (hence his father's smile), but who perhaps overcompensates for Bede's (and Stephen's) concentration on church matters.

[38] Mc Carthy and Breen (1997), 24–30.

[39] A point well made by Mayr-Harting (1991), 108.

[40] Stancliffe (2003), 10.

feast, but of a fast; moreover, they must have been the more exasperated because hardly anyone on either side, even amongst the clergy, could give a coherent account of the arguments in favour of his or her position. The king and queen had only two options, to endure or to shout. Perhaps in the end Ōswiu had had enough of Eanflǣd's shouting.

DANIEL MC CARTHY

BEDE'S PRIMARY SOURCE FOR THE VULGATE CHRONOLOGY IN HIS CHRONICLES IN *DE TEMPORIBUS* AND *DE TEMPORUM RATIONE*

Abstract

Mommsen's 1898 assumption that Bede had compiled the Vulgate chronology of his *De temporibus* and *De temporum ratione* has been simply reiterated by scholars ever since. But critical collation of Bede's chronicles with the Irish Annals leads to the conclusion that their common features, including their Vulgate chronology, derive from a common source that originated in a chronicle compiled by Rufinus of Aquileia (†410). By the year 538, Rufinus' chronicle was being continued in Ireland, and this continuation was transferred to Iona before the end of the sixth century. Around 687, Adomnán, then abbot of Iona, presented to Aldfrith, king of Northumbria, a copy of the world history in the Iona annals extending as far as the reign of the emperor Justinian, who ruled 685–695, and also a copy of his own *De locis sanctis*. By 703, these works had reached Bede and he compiled epitomes of them both. Subsequently, in 725, he again edited this copy of the Iona annals to compile his world-chronicle in *De temporum ratione*. Thus it was Adomnán's copy of the Iona annals that served as Bede's primary source for the Vulgate chronology of his *De temporibus* and *De temporum ratione*.

Keywords

Vulgate chronology, Irish Annals, Bede's chronicles, *De temporibus*, *De temporum ratione*, Rufinus, Jerome, Adomnán, Bede, Vulgate, Septuagint, chronology, world-chronicles.

Introduction

The Venerable Bede (*c.*673–735), scholar and monk of Wearmouth and Jarrow, compiled three chronicles in his very productive life. His first was a cryptic world-chronicle embedded as chapters 17–22 in his first work on time, *De temporibus* (DT), completed in 703. His second, an enlarged world-chronicle, was similarly embedded as chapter 66 in his major work on time, *De temporum ratione* (DTR), completed in 725. His third and best-known work, his *Historia ecclesiastica gentis Anglorum* (HE), dealt principally with the ecclesiastical affairs of England and was completed in 731. While his HE has attracted a succession of critical editors, the only critical editions of his world-chronicles in DT and DTR are those published by Theodor Mommsen in 1898 in the third volume of his *Chronica minora* series among the *Auctores antiquissimi* published by the *Monumenta Germaniae Historica*. In his substantial preface to these parallel editions Mommsen discussed their distinctive pre-Christian chronology and reviewed Bede's sources.[1] The most distinctive aspect of this chronology is that it reproduces the *genuit* ages of the Hebrew patriarchs and the regnal years of the Judean kings from Saint Jerome's translation of the Hebrew Bible. Since this is known as the 'Vulgate', it is appropriate to designate this chronology as the 'Vulgate chronology'. In his discussion of this chronology, Mommsen was absolutely explicit as to who compiled it, for he commenced his discussion as follows:[2]

'In chronologia operis maioris (nam libellus minor [. . .]) Beda per duas primas mundi aetates numeris chronicorum Eusebianorum ab Hieronymo retentis substituit eos, quos idem Hieronymus posuerat biblia sacra ad Hebraicum textum vertens eosque c.273 ad ipsum Eusebium Beda refert: ita ab Adamo ad Noe computantur anni non 2242, sed 1656 [. . .].'

Thus, without enquiry of any kind, Mommsen simply asserted that it was Bede who had substituted the chronological data of Jerome's

---

[1] MGH Auct. ant. 13, 223–354: 225–46 (preface), 226–9 (sources), 230 (chronology), 247–354 (parallel editions). Charles W. Jones simply reproduced verbatim Mommsen's editions of these chronicles in his complete editions of DTR and DT in CCSL 123B, 241–544 and CCSL 123C, 579–611 (cf. ibid. p 580: 'As with DTR, this is not a new text of DT, but a transcript of the Mommsen edition of the Chronicle, chs. xvii–xxii [. . .]'). For a good account of the history of editions of DTR see Wallis (1999), xcvii–xcix. The edition of HE used here is Colgrave and Mynors (1969).

[2] MGH Auct. ant. 13, 230.

Vulgate into Jerome's Latin translation of the chronicle of Eusebius. Both Eusebius' chronicle and Jerome's translation of it had reproduced the Septuagint chronology of the patriarchal *genuit* ages and the Hebrew and Judean regnal years, which in the Book of Genesis repeatedly differs very substantially from the Hebrew Genesis. This has the result that from Adam to Noah the Septuagint totals 2242 years, whereas the Vulgate totals only 1656 years as Mommsen observed. Subsequent scholarship right down to present times has simply reiterated Mommsen's assumption of Bede's authorship of the Vulgate chronology, as the following sample demonstrates:[3]

> 1913 – E. Mac Neill: '[In DT Bede introduced] a quite new reckoning of the age of the world, differing by more than a thousand years from any older reckoning.'
>
> 1943 – C. W. Jones: 'In DT, Bede had promulgated a new *annus mundi* of his own computation.'
>
> 1957 – A.-D. von den Brinken: 'Beda läßt seine Chronologie ganz auf der 'hebraica veritas' basieren.'
>
> 1999 – F. Wallis: 'In *On Times* [DT], Bede replaced the Eusebian-Septuagint chronology of the first two World-Ages with a new chronology based on Jerome's translation of the Hebrew text of the Old Testament.'
>
> 2004 – R.-P. Pillonel-Wyrsch: 'Le décompte de Bède, largement inspire de saint Augustin s'établissai comme suit: Premier age d'Adam à Noé: 1656 ans.'

None of these authors, nor any other so far as I have observed, considered that there was any doubt regarding Bede's authorship of the Vulgate chronology. This is quite remarkable considering the fact that Jerome had completed his translation of the most relevant books of the Hebrew Bible, i.e. Genesis and Kings, by 393. Thus, if Bede really was the compiler of the Vulgate chronology, it must be accepted that it took Western Christian scholarship over three centuries to address the profound chronological conflict between the Vulgate Bible they had adopted and the Septuagint world-chronicles they had received. Moreover, while Mommsen in his preface identified sixteen sources containing entries cognate with those of DT and DTR, he did not include the

---

[3] Mac Neill (1913b), 76; Jones (1943), 133; von den Brinken (1957), 110; Wallis (1999), 358–9 (359: 'why did Bede revise the standard chronology of the *annus mundi* so radically in *On Times*, shaving nearly 1200 years off the conventional age of the world'); Pillonel-Wyrsch (2004), 16.

Irish Annals. These annals had been brought to the attention of Western scholarship as valuable historical sources by James Ussher over two centuries earlier in his *Britannicarum Ecclesiarum Antiquitates*, wherein he pointed out that they accurately dated the solar eclipse of 1 May 664, whereas in his chronicles Bede repeatedly post-dated it to 3 May. Printed editions of these Annals became available when O'Conor published partial editions of the *Annals of Tigernach* (AT), *Inisfallen* (AI), *Boyle* (AB) and *Ulster* (AU) over 1814–1826.[4] In 1887, Hennessy published an improved partial edition of AU, and over 1895–1897 Stokes published a complete edition of AT.[5] All of these annals contain world history entries cognate with those in DT and DTR, but as all these editors assumed that Bede's chronicles had served as principal sources for the world history in these Annals it seems likely that Mommsen never considered the relationship between them and DT and DTR. And indeed, this assumption remained the unanimous view of scholarship until 1972, when Morris published the results of his critical collation of passages of world history common to DTR, AT and AI and concluded:[6]

> 'It is evident that the Irish were not here using Bede, but that both used a common source [. . .] Bede and the Irish had access to the work of a scholar who had studied Jerome's commentaries closely, and used them to annotate a translation of Eusebius.'

We shall see that Morris' identification of the author as 'a scholar who had studied Jerome's commentaries closely' was prescient, but in the event his paper attracted little attention. Indeed, the only reference to it known to me dismissed it, and insisted rather that the 'common stock of the pre-Patrician chronicle is datable after 725 (the publication of Bede's *De temporum ratione*)'.[7]

However, in 1996, Dáibhí Ó Cróinín introduced me to Jones' edition of DTR and I began to collate it with AI and AT over the early Christian era, and it quickly became apparent that the entries concerning Imperial reigns were far more comprehensive in these annals than

---

[4] Ussher (1639); O'Conor (1814–26), i–iii.

[5] Hennessy (1887), reproducing AD 431–1056; Stokes (1993), which is a reprinted facsimile edition of the original publication in *Revue Celtique* 16 (1895) 374–419; 17 (1896) 6–33, 119–263, 337–420; 18 (1897) 9–59, 150–97, 267–303.

[6] Morris (1972), 85, 87; Mc Carthy (1998), 100–16 (hypotheses of annalistic dependence on Bede).

[7] Dumville (1977–9), 53 (dismissal and citation).

those of DTR. Further, when one extended the comparison over the pre-Christian era, it immediately emerged that particularly the MS Oxford, Bodleian Library, Rawlinson B 502 of AT provided a much more substantial version of the Eusebian dynastic series than DTR. In 1998, I published a summary of these and other collations and concluded:[8]

> 'they [AT and DTR] have both drawn on a common source of which they have both omitted parts. Bede's edition [DTR], which survives in more and older manuscripts, has generally preserved the better text, while AT has preserved both more of the source as well as its original chronological apparatus.'

Thus Morris and I both concluded on independent grounds that a common source lay behind the world history in both the chronicles of Bede and the Irish Annals. Now, the discovery that a common source lay behind the Irish annalistic and Bedan world-chronicles immediately raised the possibility, indeed the probability, that this was also the source of their common Vulgate chronology. To demonstrate this unequivocally, of course, what one requires is some independent manuscript survival of the common source, but alas so far there has been no sign of any such survival. Thus, the only witnesses we have at present are the Irish Annals and Bede's world-chronicles, so that if we wish to discover what this common source was like we must carefully examine and collate these witnesses, and that indeed is the principal purpose of this paper.

The Six Ages in the Irish Annals and DT/DTR

I commence by summarizing in *Figure 1* the world history regnal series found in the Irish Annals, DT and DTR, indicating for each text the appropriate paragraphs, their temporal range, and the regnal series that they reproduce.

Regarding the annalistic sources, it is clear that by far the most comprehensive world history regnal series is to be found in AT, and indeed collation of this with the parallel series in AB, AI and AU shows that these all represent acephalous and truncated abridgements of a

---

[8] Mc Carthy (1998), 116–25 (Imperial reigns), 142–8 (pre-Christian era), 150 (citation).

Chronicle[1]	Range	World history regnal series
AB §§1–128	c.Adam–c.AD 361	Hebrew, Judean, Persian, Alexandrian, and Roman.
AI §§1–562	Abraham–c.AD 575	Hebrew, Judean, Persian, Alexandrian, Roman, and Byzantine; also lacunose Assyrian, Argive, Egyptian, Athenian, Israelite, Latin, and Medean.
AT 5c–77d	c.770 BC–AD 358	Judean, Persian, Alexandrian, and Roman; also Israelite, Athenian, Macedonian, Lydian, Egyptian, Assyrian, Medean, and Chaldean.
AT 80a–187d	AD 489–720	Byzantine.
AU §§1–206	AD 81–c.387	Lacunose Roman.
AU 431–719	AD 431–720	Byzantine.
DT 17–22	Adam–AD 703	Hebrew, Judean, Persian, Alexandrian, Roman, and Byzantine.
DTR 66	Adam–AD 725	Hebrew, Judean, Persian, Alexandrian, Roman, and Byzantine.

Figure 1 Tabulation of the temporal range and regnal series of world history found in Insular chronicles.

[1] Editions and referencing: AB – Freeman (1924–7), referenced by paragraph number; AI – Mac Airt (1951), referenced by paragraph number in the pre-Patrician era; AT – Stokes (1993), which is paginated continuously through its two volumes and these page numbers followed by a quarter page letter, a = first quarter, . . . d = last quarter will be used to reference AT; AU – Mac Airt and Mac Niocaill (1983), referenced by paragraph number in the pre-Palladian section, by MS AD in the post-Palladian one; DT – CCSL 123C, 579–611, referenced by chapter number concatenated with the line number; DTR – CCSL 123B, 241–544, referenced by paragraph number.

common Vulgate source. It is also clear from the above table that between them all the Irish Annals provide a continuous regnal series extending from c.Adam (AB §1) to AD 720 (AU and AT), which, when taken together, include the Hebrew, Judean, Persian, Alexandrian, Roman, and Byzantine dynasties, which are the series essential for Biblical and Christian chronology. Turning to Bede's two chronicles it may be seen that not only does their range correspond very closely with the collective range of the Irish Annals, but also that their primary chronological apparatus likewise comprises the essential Biblical, regnal series of Hebrew, Judean, Persian, Alexandrian, Roman, and Byzantine dynasties. Furthermore, all these chronicles divide time into exactly the same six Ages as follows: Age 1 = Adam to the Flood; Age 2 = Sem to Thare; Age 3 = Abraham to Saul; Age 4 = David to Sedecias; Age 5 = Ezechiel to Augustus 41; Age 6 = Incarnation to the end of time. All these correspondences support the conclusion that the Irish Annals and Bede's chronicles share a common source, and this is further supported by a collation of their structural details against each other and other Western chronicles. Isidore also divided his chronicles into these Ages, and Jerome in his chronicle cited the first two, apparently following Eusebius. It is instructive, therefore, to compare the years of these Ages in different chronicles, and since DT was Bede's first chronicle employing Vulgate chronology and DTR reproduces virtually the same Age features, it is appropriate to tabulate DT. Thus, the following collation (*Figure 2*) compares the number of years for Ages 1–5 registered in the Irish Annals with Bede's DT and Isidore and Jerome's chronicles.

Regarding *Figure 2*, it may be seen that Jerome, Isidore, and Bede's LXX series all represent the Septuagint chronological tradition, and all these intervals reconcile to within ± 3 years. On the other hand, the Irish Annals' and Bede's HV series provide a significantly shorter chronology based essentially upon the Hebrew Bible, whose Latin translation by Jerome was earlier known as the 'Hebraica veritas', and later as the 'Vulgate'. It should be emphasised here that in both DT and DTR, while Bede cites the Septuagint totals for each Age, his actual regnal years and hence chronology conform to his Vulgate totals. Since it was in DT that Bede first employed this Vulgate chronology it is helpful to examine this work in order to establish just how Bede presented it.

Age 3 and Byzantine Chronology in DT

As was noted above, ever since Mommsen's editions it has been simply assumed that Bede was responsible for the construction of the Vulgate chronology, and discussions of it have concentrated on Ages 1–2, where the numerical discrepancies between Vulgate and Septuagint are the greatest (cf. *Figure 2*). For example, Wallis wrote: 'The dates for the period after Abraham are established by the regnal lists [. . .] The first two Ages are much more problematic'.[9] But the construction of this

Interval	Annals[1]	Bede's DT[2]	Isidore[3]	Jerome[4]
Age 1	1756	HV – 1656		
		LXX – 2242	2242	2242
Age 2	292	HV – 292		
		LXX – 942	942	942
Age 3	942	HV – 942		
		LXX – 942	940	940
Age 4	473	HV – 473		
		LXX – 485	485	486
Age 5	589	HV – 589		
		LXX – 589	586	588
Ages 1–5	HV – 3951	HV – 3951		
	LXX – 5189	LXX – 5198	5195	5198

Figure 2 A collation of the number of years for Ages 1–5 separately and collectively as found in the Irish Annals, Bede's DT, Isidore's and Jerome's chronicles. The Irish Annals and DT/DTR all cite '[H]ebreos' / 'Hebraica veritas' = HV, and '.lxx. Interpretes' = Septuagint = LXX, as the authorities for their regnal data. Where the intervals are not cited explicitly by Jerome and Isidore they have been computed from their *Annus Abrahami* and *Annus Mundi* respectively.

[1] For Annals Ages 1: AB §14; 2: AB §21; 3: AI §89.2; 4: AI §89.4; 5: AI §165.3, AT 12b, AT 36b; 1–5: HV AT 36c (Stokes misreads 3955) & AI §205 (3952–1), LXX AT 36c & AI §205 (5190–1).
[2] CCSL 123C, 600–7; for Ages 1–5 cf. DT 16.2–18 and for Ages 1: DT 17.2–3; 2: DT 18.2–3; 3: DT 19.2; 4: DT 20.2–3; 5: DT 21.2; 1–5: DT 22.3–4 (3952–1 & 5199–1).
[3] MGH Auct. ant. 11, 391–488: 428, 431, 439, 445, 454 for Ages 1–5. The intervals are computed from Isidore's AM data as: Age 2 = 3184–2242; Age 3 = 4124–3184; Age 4 = 4609–4124; Age 5 = 5195–4609.
[4] Helm (1956), 174, 174, 67a, 90a, 169 for Ages 1–5; Ages 3–5 are computed from Jerome's *Annus Abrahami* = AA as follows: Age 3 = AA 940; Age 4 = AA 1426–940 = 486; Age 5 = AA 2014–1426 = 588; Age 1–5 = AA 2014+2242+942 = 5198. Eusebius cited the same data for Ages 1–2 in his preface according to Jerome's translation, cf. Helm (1956), 15.

alternative chronology for these first two Ages required simply the substitution of the Septuagint by the Vulgate *genuit* data from Genesis, and this alternative had been considered already by Eusebius.[10] However, critical examination of all the Ages shows that the most complex analysis was required for Age 3, for which both DT and DTR affirm

[9] Wallis (1999), 355, which passes over the repeated chronological conflicts for Ages 3–5; cf. *Figures 2, 3, 5*, and *6*.
[10] Jerome's translation of Eusebius' preface states (Helm (1956), 14): *Itaque manifestum est Abraham Nini aetate generatum, iuxta eum tamen numerum, quem contractiorem editione uulgata sermo praebet Hebraeus.*

942 years by either computation. This complexity arises for the following four reasons:

a) The Vulgate version of Ex 12:40–41 asserts that the Hebrews spent 430 years in Egypt and this conflicts with its assertion in Ex 6:18–20 that 'Kohath, son of Levi, [. . .] was born in the land of Canaan, lived for 132 years, and his son Amram the father of Moses for 137 years, and Moses himself was 80 years old at the time of the departure from Egypt', and obviously the sum of these years cannot amount to 430. The Septuagint, on the other hand, provides no chronology for the Hebrews in Egypt and it must be deduced from its statement in Exodus 12:40 that they dwelt in 'Egypt and Canaan 430 years'.[11]
b) The Vulgate includes the ten-year reign of Achialon in its Hebrew regnal succession, while this reign does not appear in the Septuagint succession.
c) Neither Vulgate nor Septuagint supplies the number of regnal years for Iosue, Samuhel or Saul.
d) The Vulgate and Septuagint versions of 1 Kgs 6:1 both impose the constraint that Solomon commenced work on the Temple in his fourth year, which was in the 480th year after the Exodus. Since the reigns of Iosue, Achialon, Samuhel, and Saul *all* fall within this interval, this places a limit on their overall summation.

As a result of these conflicts, diverse numbers of years were assigned to the problematical intervals in different chronicles and these are tabulated in *Figure 3*.[12] As can be seen, Isidore followed Jerome, but the chronology of DT 19 extended Age 3 by two years by means of four modifications, and yet of all five differences Bede acknowledged only that of Achialon's ten years as: *Achialon ann ·x·; hic in ·lxx· interpretibus non habetur.*[13] But had Bede, himself as a young man aged about thirty compiling his first chronicle, actually been responsible for the computation of these data, all of which contradict the authority of the chronicles of Jerome and Isidore, then he was surely obliged to justify the remaining four differences, most especially the reduction of Samuhel and Saul's reign by eight years. Instead, at the very start of the following chapter, he immediately countermanded the assignment of thirty-two

---

[11] Wallis (1999), 166–7 (citations).
[12] Mc Carthy (1998), 142–8 (an earlier examination of these issues).
[13] CCSL 123C, 604: DT 19.18.

Interval	Bede's DT[1]	Isidore[2]	Jerome[3]	Δ
Hebrews in Egypt	145	144	144	+1
Iosue	26	27	27	−1
Achialon	10	–	–	+10
Samuhel & Saul	32	40	40	−8
Age 3	942	940	940	+2

Figure 3 Collation of the data found in Bede's DT, Isidore's and Jerome's chronicles for the conflicting Vulgate and Septuagint intervals over Age 3. The column Δ registers their differences.

[1] CCSL 123C, 603–4: DT 19.8 (Hebrews in Egypt – Jones repeats Mommsen's *CXLVII* against two MSS with *CXLV*, which value alone conforms to the Vulgate total of 942), DT 19.9 (Iosue), DT 19.18 (Achialon), DT 19.22 (Samuhel & Saul), DT 19.2 (Age 3).

[2] MGH Auct. ant. 11, 433 (§44 Hebrews in Egypt), 435 (§59 Iosue), 439 (§104 Samuhel & Saul), 432, 439 (Age 3 = AM 4124 − 3184 = 940).

[3] Helm (1956), 36a (Hebrews in Egypt), 46a (Iosue), 60a (Achialon's reign of ten years is noted but not included in the Hebrew succession), 65a (Samuhel & Saul), 67a (Age 3 = AA 940).

years to Samuhel and Saul, writing: *Salomon ann ·xl·, qui templum aedificauit anno ·ccclxxx· egressionis ex Egypto, ex quo apparet Samuhel et Saul ·xl· annis praefuisse.*[14] Since the reduction of Samuhel and Saul's reign from forty to thirty-two is the principal means by which the introduction of Achialon's additional ten years is offset, this rejection by Bede shows both that he was not responsible for the assignment of thirty-two years, nor did he understand its function. Rather it is apparent from Bede's ambivalence and his silence with respect to the other differences that he was summarising the chronology of a source that he did not properly understand. We shall see a further instance of this when we examine Age 4 below. A glance at Mommsen's *apparatus fontium* for DT shows that in it Bede repeatedly cited both Isidore's *De natura rerum* and his minor chronicle in V 38–39 of the *Etymologiae*. However, whenever the chronology of DT diverges from that of Isidore it corresponds closely with that found in the Irish Annals (cf. *Figures 2* and *5*). These details imply rather that Bede was in fact collating Isidore's minor chronicle with another chronicle that already provided Vulgate chronology, and when they diverged he followed this Vulgate source.

Notwithstanding Bede's silence regarding the construction of the chronology of his DT, so high has been his status as a chronicler that

[14] CCSL 123C, 604–5: DT 20.4–6.

he has been accorded not only full authorship of it, but also the accolade that he wrote it to 'correct' his antecedent chronographers. For example, Jones, referring to DT, asserted that 'Bede wrote primarily as a teacher, as a corrector and adaptor', while Wallis likewise considered that 'we may hypothesize, then, that in *On Times* Bede revised Eusebius' chronology in the interests of what he saw as historic accuracy'.[15] But these conjectures regarding Bede's motives simply cannot be reconciled with the Byzantine chronology of the later seventh century in DT, for this exhibits repeated conflict with the established Imperial chronology, as is demonstrated in *Figure 4*.

As can be seen, this chronology in DT, which extends over the lifetime of Bede and his older contemporaries – his abbot Ceolfrid (†716) was born *c.*642 – exhibits repeated conflict with the nomenclature and chronology of the Byzantine chronology resolved by modern scholarship. Regarding nomenclature, DT registers Constans II as *Constantinus f. Constantini*, with the result that three successive emperors are named *Constantinus*, and Leontius, whose reign ended just five years before Bede wrote DT, is registered as *Leo*. Regarding chronology, DT assigns an additional five years to Heraclius, Heraclonas' joint reign with his mother Martina for two years instead of one year with

Byzantine reigns of DT cap. 22	Byzantine chronology
22.69: Heraclius ann ·xxxvi·	610–641 Heraclius [r. 31 y].
22.70: Eraclonas cum matre sua Martina ann ·ii·	–
22.71: Constantinus filius Eracli mens ·vi·	641 Constantine III and Heraclonas.
22.73: Constantinus filius **Constantini** ann ·xxviii·	641–668 Constans II [r. 27 y].
22.75: Constantinus filius Constantini superioris ann ·xvii·	668–685 Constantine IV [r. 17 y].
22.76: Iustinianus filius **Constantini** ann ·x·	685–695 Justinian II [r. 10 y].
22.79: Leo ann ·iii·	695–698 Leontius [r. 3 y].

Figure 4 Collation of the seven penultimate Byzantine reigns in DT 22 with the Byzantine chronology of Hussey (1966), 776. Numbers in DT citations have been restored from Mommsen's capitalized versions to their MS tradition.

[15] Jones (1943), 131; Wallis (1999), 361. Mommsen, in MGH Auct. ant. 13, 226, simply stated that Bede had compiled DT by augmenting and modifying Isidore's minor chronicle: 'chronica minora Isidoriana a Beda hic illic aucta et mutata'.

Constantine son of Heraclius, places this Constantine *after* Heraclonas instead of making them joint, and assigns twenty-eight years to Constantine son of Constantine instead of twenty-seven. These errors of nomenclature and chronology extend to within five years of Bede's compilation of DT so that there can be no possible grounds on which to bestow upon him accolades for either his 'correction' or 'historic accuracy'. Rather virtually every one of these erroneous and idiosyncratic features of DT's Byzantine succession is found in the Irish Annals (cf. *Figure 7* below). I submit therefore that Bede did *not* compile either the Vulgate chronology of Ages 1–5 or the Byzantine chronology of Age 6. Rather he took these from a source that has also been transmitted in the Irish Annals. Since no earlier copy of this Vulgate source is known to exist, our next task is therefore to reconstruct its essential features by collation of the Irish Annals with Bede's world-chronicles.

Reconstruction of the Vulgate Source for Insular World History

When collating our Insular world history sources to identify the content and characteristics of the Vulgate source, we are in the fortunate position that Bede made two separate and very distinct recensions from it, both of which have survived in many early MSS. Although the two most important Irish Annalistic MSS of AT and AI are almost four centuries younger than the earliest Bedan MSS, nevertheless collation of these together throws a great deal of light on the content and organisation of their common Vulgate source. As already mentioned, the most substantial Irish Annalistic source is MS Oxford, Bodleian Library, Rawlinson B 502 of AT, and this commences in the second year of Iotham, who is the thirteenth ruler of Age 4. Before this, from the start of Age 3, the much more abridged and textually inferior MS Oxford, Bodleian Library, Rawlinson B 503 of AI provides the best Irish Annalistic representative. Since all these chronicles exhibit the same division into Ages, it is worthwhile to survey a complete Age, and so Age 4 is the first such where we may employ AT. Age 4 also has the advantage of being wholly Biblical in character and to incorporate significant Vulgate/Septuagint chronological conflicts. That is, the Bible provides an account of all the rulers of Age 4, and the Vulgate, and Septuagint give conflicting regnal years for three of the twenty reigns of this Age, viz. Athalia,

Amon and Iosias. I commence therefore by collating in *Figure 5* example entries including these three reigns taken from this Age as transmitted by AI/AT and DT.

AI/AT	DT 20
AI §89: **Finit tertia aetas mundi** . . . incipit quarta aetas (mundi quae continet) annos .ccclxxiii.	20.1: **Quarta aetas continet annos iuxta Hebreos ·ccccxliii·; ·lxx· translators ·xii· adiciunt.**
AI §90: Kl. Dauid primus rex regnauit de *tribu Iuda* annis .xl. . . .	20.2: David ann ·xl·
AI §117: K. Otholia mater Azarias annis .ui. . . .	20.12: Athalia ann ·vi·
AT 5d: K. Achaz filius Iotham rexit *Iudam* annis .xui. . . . K×13	20.17: Achaz ann ·xvi·
AT 6a: K. Osse filius Hela rexit *Israel* annis .ix., qui fuit nouissimus decim tribuum rex.	–
AT 7b: K. Achaz mortuus est.	–
AT 7b: In hoc tempore, **ut Eusebius ait**,[1] regnum defecit .*x. tribuum* . . . [a] Salmanasar rege Caldeorum et translatae sunt in montes Medorum. [cf. Jerome & Armenian chron. s.a. Achaz 11]	20.18: Israhel in Medos transfertur
AT 7b: Ezecias filius Achaz rexit *Iudam* annis .xxix. . . . K×4	20.18: Ezechias ann ·xxviiii·
AT 7c: K. Nunc incipit captiuitas .*x. tribuum*. Sexto Ezechiae anno Salminasar rex Assiriorum capta Samaria transtulit *Israel* in Assirios . . . [cf. 2 Kgs 18:10–11]	–
AT 10a: K. Ammon filius Mannasse rexit *Iudam* annis duobus **iuxta Ebreos. secundum uero .lxx. Interpretes** annis .xii. . . .	20.20: Amon ann ·ii·
AT 10a: K. Ammon a seruis suis interficitur.	–
AT 10a: K. Iosias filius Ammon rexit *Iudam* annis .xxxi. Hic mundata Iudea et Hierusalem . . . .xuiii°. anno regni sui. . . .	20.21: Iosias ann ·xxxi·
	(Continued)

AT 10c: K×17 . . . **Hoc anno ut praescripsimus** Iosias mundata Iudea, **et reliqua** . . .	–
AT 12b: **Finit quarta aetas. Incipit quinta, quae continet annos** .dlxxxix.	21.1: Quinta aetas continet ann ·dlxxxviiii·

Figure 5: A sample of entries from the Judean and Israelite (.x. *tribuum*) dynasties for Age, 4 taken from AI, AT, and DT, including the reigns of Athalia, Amon, and Iosias for which Vulgate and Septuagint chronology conflict. The dynastic identifications have been emphasized in italic and the compiler's interjections in bold, and DT number citations have been restored from Mommsen's capitalized versions to the MS tradition.

[1] Because the scribe of MS Oxford, Bodleian Library, Rawlinson B 502, 1r placed a point after the word *ait*, Stokes suffixed this interjection to the preceding obit for Achaz. However, since Eusebius does not record rulers' obits and he agrees with 2 Kgs 16:2 regarding Achaz' reign of sixteen years, but is in chronological conflict with the accounts of the Captivity in 2 Kgs 17:6 and 18:11, it is clear that this interjection in fact *prefixes* the account of the transfer of the ten tribes to the Medean mountains, and it explains the subsequent parallel account at AT 7c.

The most conspicuous feature of DT is the remarkable brevity of its entries, with regnal entries comprising simply a name and number of regnal years linked by the abbreviation *ann*, exactly as in Isidore's minor chronicle. On the other hand, the Irish Annalistic entries are relatively prosaic and their most conspicuous feature is the many *K* or *Kł* representing *Kalendae Ianuarii* distributed amongst them, and it will be demonstrated below that both the prosaic style and these calends derive from the Vulgate source. Regarding their overall structure, it is clear from their appearance in both AI/AT and DT that the Vulgate source was divided into numbered Ages, each of which commenced with a summation of years for that Age. Furthermore, from DT 20.1, it is apparent that, when the Vulgate and Septuagint chronology differed, both summations were cited, the former identified here as *Hebreos* and the latter as ·*lxx· interpretes*, or as ·*lxx· translatores*. While AI §89 does not explicitly reproduce this latter figure, space has been left for it in the MS. Both AI and AT attest that each Age began appropriately with an incipit and ended with a *finit*, which Bede has omitted. Regarding the regnal successions, both AI and AT provide multiple parallel successions (cf. *Figure 1*), of which just two are illustrated here and these are nearly always explicitly identified as *Iudam* or *Israel /.x. tribuum*. In most cases, too, the Irish Annalistic entries identify the ruler's immediate antecedent or tribe, and intermittently his obit (cf. AT 7b, 10a). In DT, on the other hand, rulers are normally identified with just a single word and neither their antecedents nor dynasty nor obit is cited. Regarding the representation of numbers, in the Irish Annals and in the MSS of Bede's chronicles these are written in lower case letters and are normally delimited with two points that are usually raised in MSS of Bede's chronicles, thus ·xl·, as indeed Jones represented in all of his own editions of Bede's works. However, Mommsen, in his editions of DT and DTR, systematically capitalized all Roman numerals and lowered both points, and he regularly suffixed the initial point to the preceding word, thus misrepresenting *David añ ·xl·* as *David ann. XL*. Unfortunately, Jones silently reproduced all these misreadings in his CCSL editions of the chronicles of DT and DTR, but in all citations from these works here the MS tradition has been restored.[16] The editors of the Irish Annals, on the other hand, have generally reproduced the MS form of the numbers accurately.

A distinctive feature of DT's treatment of the difference between Vulgate and Septuagint chronology is that, while these are always registered in the Age summations, the reigns that contribute to these differences are never themselves identified. Thus, where in Age 4 the Vulgate assigns Athalia 6, Amon 2, and Iosias 31 years, and the Septuagint assigns them respectively 7, 12, and 32 years, DT only registers the total of the Septuagint's additional twelve years. In contrast to DT, it can be seen that AT 10a has explicitly identified the second of these differences, and, since Amon's obit is given under the following calend, it is manifest that the compiler has indeed adopted the Vulgate chronology of two years. In the cases of Athalia and Iosias, while AI/AT do not identify the Vulgate/Septuagint difference, in both instances they cite the Vulgate regnal years and their calend chronology accurately reflects these. Finally, an important feature of the Vulgate source that may be identified from *Figure 5* is that not only did their compiler systematically identify the conflicts between Vulgate and Septuagint chronology, he also registered chronological conflicts found in other chronicles of antiquity. One instance of this is shown above where an interjection at AT 7b registers that Eusebius had located the termination of the Israelite dynasty at the end of Achaz' reign, asserting that the Israelites were then transferred by a Chaldean king to the Medean mountains. Since this assertion is indeed found in both Jerome's and the Armenian

---

[16] I wish to record my gratitude to Immo Warntjes, who kindly provided me with sample copies of folios from some of the principal MSS of DT, upon which this restoration is based, and for his constructive discussions regarding other MS details of DT and DTR.

translations of Eusebius' chronicle at the eleventh year of Achaz, this identification and interjection are absolutely correct.[17] However, 2 Kgs 18:10–11 asserts that it was in the sixth year of Achaz' successor, Ezechias, that the Assyrian king, Salmanasar, captured Samaria and transferred the Israelites into Assyria. Thus, as well as the conflicts in the Biblical and Eusebian narratives, there is a difference of eleven years in their chronology. The entries at AT 7b–c show that, while the compiler of the Vulgate source has registered both chronologies, he has clearly opted for the Biblical chronology, since this entry is preceded by his statement *Nunc incipit captiuitas .x. tribuum*. In DT 20.18, on the other hand, by placing *Israhel in Medos transfertur* under the reign of Achaz, Bede followed the Eusebian chronology for the start of the Israelite captivity. Hence, it can be seen from this collation that the compiler of the Vulgate source registered not only Vulgate and Septuagint chronological conflicts, but also conflicts with other authoritative sources such as Eusebius' chronicle.

Thus, in summary we conclude that in DT Bede transmitted only the skeleton of the Vulgate source, abandoning all Age incipits and finits, all calends, all obits, all dynasties except the essential series Hebrew, Judean, Persian, Alexandrian, Roman, and Byzantine, and in imitation of Isidore's minor chronicle he reduced each regnal entry to its textual minimum. This was indeed how Bede himself described the work when writing of it to Pleguin some five years later: 'These are the things which, according to the faith of Holy Scripture, I was solicitous to summarize for myself, and for those of my circle who asked me, in a concise and simple manner, as I believe and think'.[18] The Irish Annals, on the other hand, have transmitted the calends, prosaic entries, the parallel dynasties, obits, and some but not all of the interjections registering the chronological conflicts between the Vulgate and Septuagint and other authoritative sources.

Next, turning to Bede's second recension of Age 4 in DTR 66, in *Figure 6* collate it with the identical sample of AI/AT entries given in *Figure 5*. The most conspicuous development apparent from this collation is that in DTR Bede replaced DT's cryptic entries with prosaic entries that textually closely approximate the corresponding Irish Annalistic entries. In particular, his Judean regnal entries now systematically cite either the tribe or ancestor (cf. DTR 66 §§82,

---

[17] Helm (1956), 88a; Karst (1911), 182 (First Captivity).
[18] Wallis (1999), 408.

AI/AT	DTR 66
AI §89: **Finit tertia aetas mundi** . . . **incipit quarta aetas (mundi quae continet) annos** .cccclxxiii.	§81: **Quarta mundi aetas** non solum **cum inchoato gentis Iudeae imperio**, sed et cum innouata promissione quae patribus olim data est imperii Christiani sumit exordium . . .
AI §90: Kl. Dauid primus rex regnauit de *tribu Iuda* annis .xl. . . .	§82: ·īīdccccxxx· Dauid primus ex *tribu Iuda* rex an ·xl·
AI §117: K. Otholia mater Azarias annis .ui. . . .	§108: ·īīīlxxi· Athalia mater Azariae an ·ui· . . . **In** ·**lxx**· **inter-praetibus** ·uiii· an regnasse Athalia narratur.
AT 5d: K. Achaz filius Iotham rexit *Iudam* annis .xui. . . . K×13	§124: ·īīīcxxiiii· Achaz filius Ioatham an ·xui· . . .
AT 6a: K. Osse filius Hela rexit *Israel* annis .ix., qui fuit nouissimus decim tribuum rex.	–
AT 7b: K. Achaz mortuus est.	–
AT 7b: In hoc tempore, **ut Eusebius ait**, regnum defecit *.x. tribuum* . . . [a] Salmanasar rege Caldeorum et translatae sunt in montes Medorum. [cf. Jerome & Armenian chron. s.a. Achaz 11]	
AT 7b: Ezecias filius Achaz rexit *Iudam* annis .xxix. . . . K×4	§128: ·īīīccliii· Ezechias filius Achaz an ·xxuiiii·
AT 7c: K. Nunc incipit captiuitas *.x. tribuum*. Sexto Ezechiae anno Salminasar rex Assiriorum capta Samaria transtulit *Israel* in Assirios . . . [cf. 2 Kgs 18:10–11]	Huius anno sexto Salmanassar rex Assyriorum capta Samaria transtulit *Israhel* in Assyrios . . .
AT 10a: K. Ammon filius Mannasse rexit *Iudam* annis duobus iuxta **Ebreos**. secundum uero **.lxx. Interpretes** annis .xii. . . .	§133: ·īīīcccx· Amon filius Manasse an ·ii· **In Hebraica Veritate** duobus annis, **in** ·**lxx**· legitur regnasse ·xii·
AT 10a: K. Ammon a seruis suis interficitur.	–
AT 10a: K. Iosias filius Ammon rexit *Iudam* annis .xxxi. Hic mundata Iudea et Hierusalem . . . . xuiii°. anno regni sui. . . .	§136: ·īīīcccxli· Iosias filius Amon an ·xxxi· Hic mundata Iudea et Hierusalem . . . ·xuiii· anno regni sui . . .

(Continued)

	§139: **In Hebreo** ·xxxi· an regnasse *Iosias*, **in ·lxx· interpretibus** ·xxxii· legitur; sed et **Eusebius** inter regnum eius . . . Verum quid **Veritas** habeat, **Hieremias** pandit . . .
AT 10c: K×17 . . . **Hoc anno ut praescripsimus** *Iosias* mundata Iudea, **et reliqua** . . .	–
AT 12b: **Finit quarta aetas. Incipit quinta, quae continet annos** .dlxxxix.	§143: **Quinta mundi aetas** ab exterminio **coepit** regni Iudaici, quod iuxta prophetiam Hieremiae ·lxx· annis permansit.

Figure 6 The same sample of AI/AT entries for Age 4 shown in *Figure 5* is here collated with Bede's DTR 66. Again, dynastic identifications have been emphasized in italic and the compiler's interjections in bold, and DTR number citations restored to the MS tradition.

108, 124, 128, 133, 136). To these regnal entries Bede has prefixed an *Annus Mundi* (AM), which, following that of Isidore, anachronistically registers the final year of each reign.[19] That, in fact, the Vulgate source had a chronological apparatus, which registered every year, can be seen from the Irish Annalistic account of Iosias' purification of Judea and Jerusalem in his eighteenth year. This purification is related in some detail anachronistically immediately following Iosias' regnal entry, and both the Irish Annals and DTR have transmitted this account (cf. AT 10a, DTR 66 §136). However, *only* the Irish Annals have reiterated this purification appropriately seventeen calends later in Iosais' eighteenth year (cf. AT 10c), and the compiler's interjection there of *Hoc anno ut praescripsimus* [. . .] *et reliqua* shows that both entries were written by the same author. This, then, implies that this same author also inscribed the intervening seventeen calends that correctly locate this entry, all of which Bede has omitted. Likewise, as in DT, he also omitted from Age 4 all successions except the Judean succession and the obits. Furthermore, in DTR 66 §128 he silently relocated the start of the captivity of the Israelites from the reign of Achaz, where in DT it

---

[19] Mc Carthy (1998), 124 n 70 (anachronism of Isidore and Bede's AM).

agreed with Eusebius' chronology, to the sixth year of Ezechias. Thus, DTR 66 §128 now agrees with 2 Kgs 18:10–11 and Bede's relocation both corresponds textually and synchronizes with the Irish Annals. This correction is yet another verification that in 703, when Bede was compiling DT, he did not fully comprehend the chronology of his Vulgate source. Finally, Bede has moved all the contrasting summations of Vulgate and Septuagint chronology from the start of each respective Age to DTR 66 §§2–7, and instead at DTR 66 §81 he has restored an Age incipit without any corresponding finit at DTR 66 §143. However, since his incipit now incorporates an anachronistic reference to Christian rule, it is clearly Bede's own interpolation. On the other hand, in DTR 66 §§108, 133, and 139 Bede has transmitted all three interjections registering the conflict between the Vulgate and Septuagint regnal chronology for Athalia, Amon, and Iosias. Further, in DTR 66 §139 he has transmitted a protracted discussion, only partially cited here, that employs Jer 25:1 to support the Vulgate chronology against that of Eusebius. All of this common source material has been lost from the Irish Annals.

In summary we conclude from this examination of Age 4 that the Vulgate source of the Irish Annals and Bede's chronicles included:

a) A primary chronological apparatus of calends.
b) A comprehensive set of Eusebius' dynastic successions distributed over these calends.
c) Prosaic regnal entries providing an account of each ruler, regularly including their dynastic and familial affiliations, and major events involving them.
d) Substantial interjections providing critical assessments of the chronological conflicts principally between the Vulgate and Septuagint, but also other chronicles of Antiquity. These assessments favoured Vulgate chronology, and this was reflected in the calend chronology.

Of these four constituents only the Irish Annals have transmitted them all in part; in DT Bede transmitted *only* the Vulgate chronology of d), but in DTR he transmitted mostly prosaic entries and more of the interjections than have been transmitted by the Irish Annals. Thus, by collating the Irish Annals with Bede's two recensions, we can reconstruct the form and content of their Vulgate source for Age 4 in some detail. While these conclusions have been based upon a consideration of Age 4 alone, examination of the other pre-Christian Ages shows that they also apply to them. For Ages 1–2, even though our only Irish Annalistic source, AB, is evidently quite abridged and corrupted, it nevertheless clearly preserves both the calends and the Vulgate chronology (cf. *Figure 2*). For Ages 3–5, AI transmits corrupt and lacunose versions of the calends, prosaic entries and non-Biblical successions, but no interjections, while AT transmits all features a–d) for the latter part of Age 4 and all of Age 5. Of particular interest are the accumulated interjections transmitted by AT and DTR, in which the compiler explicitly cites the following authors and works: *[H]ebreos, ·lxx· Interpretes, Josephus, Josephus in Antiquitatum, Josephus in primo contra Appionem, Africanus, Julius Africanus, Eusebius, Cronicam Eusebii, Eusebius in temporum libri, Eusebius in cronicis suis,* and *Jerome in Danielis.* These interjected attributions and their associated chronological discussions are most important, for they show the compiler of the Vulgate source to have been critically collating a broad range of Christian and non-Christian sources extending at least from the Jewish Josephus' *Antiquitates* of c.AD 94 to Jerome's *In Danielem* of c.406. In these discussions, the compiler, with just one exception, favours the Vulgate chronology against that of the Septuagint and the works of Josephus (†c.100), Julius Africanus (†c.240), Eusebius, and Jerome, whenever these are in conflict. The exception is in Age 3 and has been discussed above, where the compiler points out that the Vulgate chronology of 430 years for the Hebrews in Egypt contradicts its own account of the Hebrew succession, and concludes that the Septuagint chronology must be followed. This critical collation of the Vulgate, Septuagint, and chronicles of Antiquity is very impressive, for it implies a very comprehensive reading and understanding of both Biblical and early chronicle material. The authorship of this compilation has been mistakenly attributed to Bede, who, in fact, twice edited his copy of the Vulgate source, omitting many of its most singular textual features, but reproducing and highlighting its Vulgate chronology vis-à-vis that of the Septuagint. In DT, Bede's editing resulted in just an epitome of the Vulgate source, but in DTR it must be acknowledged that Bede made his own additions. In Ages 1–5, these were occasional, such as his Vulgate AM and his Christianized incipit at DTR 66 §81 (cf. *Figure 6*), but in Age 6 Bede made numerous additions, many taken from the *Liber Pontificalis*. The textual character of the Vulgate source has rather been best preserved in the Irish Annals, most particularly in AI and AT, though, because they have lost Ages 1–2, they lack the spectacular chronological contrasts of Bede's chronicles.

Author of the Common Vulgate Source

In my 1998 paper I identified a number of singular textual and chronological correspondences between Irish Annalistic world history entries and the Latin translation of Eusebius' *Ecclesiastical History* (EH) that was completed by Rufinus of Aquileia in 402.[20] I also pointed out that in his translation Rufinus had substituted for Eusebius' Greek three passages from *De ratione paschali*, the Latin translation of the Paschal tract of Anatolius of Laodicea (†c.282).[21] The Paschal table of Anatolius' tract identified each year with *Kł* (representing *Kalendae Ianuarii*) followed by the ferial (i.e. day of the week) of 1 January, just as in the early Christian era in AT. Based upon these correspondences I proposed that 'sometime after he had completed his edition of the *Ecclesiastical history c.*402, and before his death in Sicily in *c.*410, Rufinus compiled his own chronicle employing the calends + ferial apparatus he found in *De ratione paschali*'.[22] Before proceeding it may be helpful to review the biography of Rufinus. He was born in *c.*345 near Aquileia in north-eastern Italy on the Adriatic, and studied in Rome *c.*359–368 where he became a close friend of Jerome. He returned to Aquileia and studied Christianity and asceticism under Chromatius and was baptized. About 372 he went to Egypt where he spent eight years, several of them studying in Alexandria under Didymus, a disciple of Origen. In *c.*379 he moved to Jerusalem where he spent the next eighteen years and there established a monastery on the Mount of Olives. Jerome, on the other hand, founded his monastery about twelve miles away in Bethlehem in *c.*387 and the two maintained an amicable, scholarly relationship until in *c.*393 when Jerome began to publicly and aggressively attack Origen's teaching, causing a serious breach in his relationship with Rufinus. In 397 Rufinus departed for

---

[20] Mc Carthy (1998), 129–30, 133–40 (Annalistic correspondences with Rufinus' EH).
[21] Edition by Mc Carthy and Breen (2003), 39, 92, 117–8, 138–41 (Rufinus' citations).
[22] Mc Carthy (1998), 141.
[23] Murphy (1945), 232–5 (biography).

Italy and there commenced the compilation of a substantial series of translations of Greek Christian works including numerous works of Origen. In 407 he was obliged to leave Aquileia on account of the incursions by the Goths, and by 410 he had removed to Sicily where he died shortly thereafter.[23] From this it can be seen that Rufinus was an active and indeed ardent advocate of Origen and his works, and in these circumstances I underlined the significance of eight Irish Annalistic entries, five of which are found in DTR, all of which register warm approval of Origen.[24] There is then substantial evidence that Rufinus compiled the Vulgate source, and thus to him is due the credit for the construction of the Vulgate chronology which he established using Jerome's Latin translation of the Hebrew Bible, the Septuagint, and an impressive array of the chronicles of antiquity. Regarding the matter of Rufinus' knowledge of these works it is appropriate to consider the chronology of Jerome's compilation of the Vulgate. Over the years 386–393 Jerome, having first endeavoured to establish the Biblical text from then existing editions of the Septuagint, turned in the end to the Hebrew Bible. From this, by 393, he had completed the Latin translations of the Books of Samuel, Kings, Job, Prophets, Ezra, Nehemiah, and Genesis in that order.[25] Of these the Vulgate chronology of Ages 1–4 rests principally upon the Books of Genesis and Kings. Since Rufinus and Jerome maintained amicable relations over these years, and given Rufinus' own interest in Biblical and Christian history, there are good grounds to infer that over this period Rufinus became familiar with the issues of textual and chronological conflict between the Septuagint and Hebrew Bibles. I submit therefore that the interjections making critical comparisons between the Septuagint, the Hebrew Bible, and chronicles of antiquity found in AT and DTR are ultimately the result of Jerome's endeavours to establish the Biblical text, to which undertaking Rufinus was a unique witness. I conclude that it was as a result of this that Rufinus subsequently compiled his own chronicle reflecting Vulgate chronology. On the other hand, the hypothesis that Bede had been responsible for the construction of the Vulgate chronology seems in retrospect quite implausible. For it not only required that Latin Christian scholarship took over three centuries to incorporate Jerome's Vulgate chronology into a world-chronicle, as mentioned already, but also that Bede had access to the

---

[24] Mc Carthy (1998), 138–40 (Origen entries).
[25] NPNF 6, xix–xx (chronology of the Vulgate).

chronicles of Eusebius and Josephus, which are not known to have circulated in the British Isles.

Regarding the date of Rufinus' compilation of his chronicle, the latest source he cited was Jerome's *Commentarium in Danielem*, written *c.*406. Therefore Rufinus must have been working on it within the last four years of his life, during much of which he was a fugitive from the invading Goths. Given the circumstances of collapsing Imperial authority in Italy over 407–410 and Rufinus' flight to Sicily, it seems likely that that is where his compilation ended up. As regards the question of how Rufinus' chronicle travelled thence to the British Isles we have no direct information. However, we do know that a letter from Sulpicius Severus in southern Gaul seeking chronological information was sent to Rufinus in 403, and also that Sulpicius subsequently compiled his eighty-four year *latercus* using the Latin translation of Anatolius' tract *De ratione paschali* cited by Rufinus in his translation of EH. Sulpicius' *latercus* was the Paschal tradition followed by a number of the Insular churches from the fifth to the eighth centuries, and so circumstantially it appears likely that Sulpicius and his *latercus* provided the conduit by which Rufinus' chronicle reached the British Isles.[26] Regarding the range of Rufinus' chronicle, both DT and DTR show that it commenced at Adam, while the details of the Irish Annalistic ferial series indicates that he brought it to *c.*398. But Rufinus' Byzantine succession extended only to the last complete reign in Jerome's chronicle, namely to Jovian's (†364). However, a colophon survives in AI §344 at *c.*424, which suggests that Rufinus' chronicle had received a small extension before arriving in the British Isles. It reads: 'Now ends this little work collected from the beginning of the world out of various sources, that is not from one particular exemplar'.[27] Thus it appears that the edition of Rufinus' chronicle incorporated into the Irish Annals had received an extension of about two decades.

Finally one important additional element of the chronological apparatus of the Christian era of Rufinus' chronicle should be mentioned here. At nineteen year intervals commencing at the Incarnation there were three-way synchronisms between the Septuagint AM, the Vulgate AM and the *annus ab Incarnatione* (i.e. AD). These have only survived in the Irish Annals and in AT they commence at AD 1 and

---

[26] Mc Carthy (1994), 40–1 (Sulpicius' letter & Rufinus' connections); Mc Carthy and Breen (2003), 117–8, 138–41 (Rufinus' citations of *De ratione paschali*).

[27] Anderson (1973), 5 (translation).

continue to 324 with two lacunae; for example AT 48a: *K.ui. Ab initio mundi .u.m.ccc.ix. secundum lxx, secundum Ebreos .iiii.m.c., Ab incarnatione .c.xv.* Regarding Rufinus' use of AD here, since he wrote these well over a century before Dionysius Exiguus compiled his influential Paschal table in 525, clearly Dionysius did not invent *Annus Domini* reckoning as has been repeatedly asserted in modern times. Rather I have argued elsewhere that Eusebius himself invented AD and had recorded these three-way synchronisms at nineteen-year intervals in his chronicle and that Rufinus simply transcribed these into his own chronicle.[28] This presence of AD in Rufinus' chronicle proved just as influential with Bede as did his Vulgate AM chronology, for when Bede later came to compile his HE he employed the AD intermittently throughout, and systematically in his recapitulation in HE V 24.

The Insular Route of Rufinus' Chronicle

Finally, regarding the route that Rufinus' chronicle took within the British Isles, I would first observe that, since both the Irish Annals and Bede's world-chronicles preserve unique entries and interjections, they cannot derive from the other and must derive from a common source. Furthermore, since we have concluded above that the Irish Annals retain the content and structure of Rufinus' chronicle much more substantially than does either of Bede's world-chronicles, then this common source must resemble the Annals more closely than either DT or DTR. In particular, this common source must have preserved a primary chronological apparatus of calends. Further, since the Irish Annals contain accurate phenomenological entries concerning famine, mortality, and plagues from 538 onwards that are embedded in a primary apparatus of calends extending over Ages 1–6, I conclude that the Irish annalists had Rufinus' chronicle at least by 538 and were continuing it at that time.[29] In fact I believe that Rufinus' chronicle arrived in Ireland with Sulpicius' 84-year *latercus* in the fifth century, but for the purposes of considering a source for Bede's chronicles it is sufficient here to recognise that Rufinus' chronicle was being continued in Ireland in the earlier sixth century. This Irish

---

[28] Mc Carthy (2003), 44–6 (three-way synchronisms), 49–53 (Eusebius' invention of AD); Miller (1991), 89 (three-way synchronisms).

[29] Daniel Mc Carthy, 'Chronological synchronisation of the Irish Annals,' at www.cs.tcd.ie/Dan.McCarthy/chronology/synchronisms/annals-chron *s.a.* 538 (famine), 540 & 550 (mortality), 553 (plague); cf. Baillie (1994), 215 fig. 3.

continuation did not however remain in Ireland, for Alfred Smyth's 1972 study of the early Irish Annals showed that through the later sixth, the seventh and earlier eighth centuries the Irish Annals were maintained on the island of Iona, just off the west coast of Scotland.[30] On the other hand, since Bede had compiled DT in Northumbria by 703, it is clear that a copy of Rufinus' chronicle had reached him by that year. Now, given the proximity in both space and time of Bede's DT with the Iona annals, together with Bede's substantial acknowledgement in HE III 3–5 of Iona's contribution to Christian learning, Iona is the only plausible place to seek the origin of Bede's source. But when and in what circumstances could this have happened, since Northumbria and Iona had been ecclesiastically estranged since the synod of Whitby in 664, and their conflict was not resolved until 716, that is thirteen years after Bede's compilation of DT? Well, we shall see that, notwithstanding the division at Whitby, some close and scholarly links continued between Iona and Northumbria through the later seventh century.

I first address the question of the time of arrival of Bede's copy in Northumbria. One way to pursue this is by examining the Byzantine succession, which, as noted above, is a feature of both the Irish Annals and Bede's two chronicles, in order to establish the extent of their common series. Thus, in *Figure 7* I collate the latter section of Bede's Byzantine succession from DT with the parallel series from AU, which preserve the earliest Irish Annalistic version.

Considering first of all Bede's DT reigns, I would point out that from Theodosius minor to Heraclius (M_1–I_5) Bede maintained the absolutely minimal regnal nomenclature in the style of Isidore that we saw him employ in his pre-Christian regnal successions (cf. *Figure 5*), namely a single word occasionally suffixed by *maior* or *minor* serves to identify each reign. Indeed, for M_1–7 Bede substituted Isidore's systematic regnal succession for the irregular mixture of regnal incipits and obits taken from the chronicle of Marcellinus found in AU.[31] A significant change occurs however at X_1, when Bede, in a relatively prosaic entry virtually verbatim with the Irish Annals, assigns a two-year reign to Heraclonas and his mother Martina. Similarly corresponding prosaic entries follow to X_5, the reign of Justinian son of Constantine (who reigned for 685–695), but after this the relationship between the Irish Annals and DT breaks down. For with *Leo annis tribus* Bede reverts to his minimal

---

[30] Smyth (1972), 16, 33–4, 41 (Iona c.550–740).

[31] Marcellinus' *Chronicon* is edited by Theodor Mommsen in MGH Auct. ant. 11, 37–108.

S	AU reigns	DT reigns
M1	AU 449: **Teodosius imperator uiuendi finem fecit**	DT 22.55: *Theodosius minor 26.*
M2	AU 449: **Marcianus imperator 6a 6m**	DT 22.56: *Marcianus 7.*
M3	AU 456: **Leo Senior imperator Leone Iuniore a se iam Cessare constituo . . . 17a 6m**	DT 22.58: *Leo maior 17; Leo minor 1.*
M4	AU 473: **Zenon Augustus 17a 6m**	DT 22.60: *Zenon 17.*
M5	AU 490: **Anastasius imperator 27a 2m 29d**	DT 22.61: *Anastasius 27.*
M6	AU 526: **Iustinus imperator 9a 2m**	DT 22.62: *Iustinus 8.*
M7	AU 526: **Iustinianum ex sorore sua [Iustini] nepotem**	DT 22.63: *Iustinianus 39.*
I1	AU 565: *Iustinus Minor 11*	DT 22.64: *Iustinus minor 11.*
I2	AU 576: *Tiberius Constantinus 7*	DT 22.65: *Tiberius 7.*
I3	AU 583: *Mauricius 21*	DT 22.67: *Mauricius 21.*
I4	AU 604: *Foccas 7*	DT 22.68: *Focas 8.*
I5	AU 612: *Eraclius 26*	DT 22.69: *Heraclius 36.*
X1	AU 638: Eraclas cum matre sua Martina 2	DT 22.70: Eraclonas cum matre sua Martina 2.
X2	AU 641: Constantinus filius Eraclii 6m	DT 22.71: Constantinus filius Eracli 6m.
X3	AU 642: Constantinus filius Constantini 28	DT 22.73: Constantinus filius Constantini 28.
X4	AU 672: Constantinus filius superioris Constantini regis 17	DT 22.75: Constantinus filius Constantini superioris 17.
X5	AU 689: Iustinianus Minor filius Constantini 10	DT 22.76: Iustinianus filius Constantini 10.
–	–	DT 22.79: Leo 3.
–	AU 701: Tiberius Cesar 7	DT 22.79: Tiberius dehinc 5.

Figure 7 A collation of the Byzantine succession transmitted by AU and DT showing their nomenclature and regnal years (given in Arabic numerals to conserve space). Those entries that are substantially verbatim citations from Marcellinus' or Isidore's chronicles, are identified in bold and italic respectively. The entries under siglum S in the first column identify AU's sources as M=Marcellinus, I=Isidore, or X=not identified, and serves to reference the collation.

regnal nomenclature, and this reign is not found in AU. I conclude therefore that the edition of the common source that reached Bede contained the Byzantine succession only as far as the reign of Justinian. Now, as already stated, at this time the Irish Annals were maintained on Iona, where Adomnán was abbot for the years 679 to 704, so that Justinian's reign commenced in the sixth year of Adomnán's abbacy. These details imply therefore that Bede's Vulgate source came from Iona during Justinian's reign and hence while Adomnán was abbot there.

Now as it happens Bede himself in HE has provided us with a detailed account of the transmission of documentary Biblical material from Adomnán in Iona to Northumbria sometime in the years 686–688. In HE V 15 Bede recounts how Adomnán took down a detailed account of the sacred places referenced in the Bible from a Gaulish bishop called Arculf, from which Adomnán wrote a book entitled *De locis sanctis*. Adomnán subsequently made two visits to Northumbria over 686–688, and on one of these visits, most likely the latter, he presented a copy of this book to Aldfrith, king of Northumbria for 686–705, whom he described as 'my friend' in his *Vita Columbae* II 46.[32] Bede, in his Life of St Cuthbert, twice (cap. 24) indicates that c.685 Aldfrith was educated 'among the Irish isles', or 'in the regions of the Irish', while the anonymous Life of St Cuthbert III 6 states explicitly that he was in Iona.[33] It seems likely therefore that Aldfrith had studied in Iona under Adomnán's abbacy and it was this, together with their friendship and Aldfrith's recent elevation to the Northumbrian kingship, that prompted Adomnán's present of a copy of his *De locis sanctis* to him. Bede tells us that Aldfrith circulated this book and at least by 703 this had reached Bede, for at that time he compiled his own epitome of it.[34] In HE V 15 Bede accorded great praise and respect to Adomnán's book, and his HE V 16–17 comprise simply extracts based upon it, which contribute nothing to his history of the English church and which disrupt the chronology of his account

[32] Sharpe (1995), 46–8 (Adomnán's two visits to Northumbria), 203 ('my friend').

[33] Colgrave (1940), 104, 236, 238 (Aldfrith's scholarship); Ireland (1996), 73–7 (Aldfrith's scholarship).

[34] Adomnán's *De locis sanctis* is edited by Ludwig Bieler in CCSL 175, 219–97; Bede, *De locis sanctis* is edited by Johannes Fraipont in CCSL 175, 249–80; Foley and Holder (1999), 1, 5–25 (date and translation).

[35] Ireland (1996), 73–4 (citations from Bede, Stephen of Ripon, and Alcuin praising Aldfrith's learning). I am grateful to Colin Ireland for subsequently bringing to my attention Bede's identification in cap. 15 of his *Historia abbatum* of Aldfrith's competence in Scripture, and also that Aldfrith had purchased from abbot Ceolfrith a *Codex Cosmographiorum*; cf. Blair (1970), 184–5.

of early eighth-century Northumbria. Now, Aldfrith's interest in and ability for Biblical studies are well attested, and Adomnán's present of *De locis sanctis* provided him with a detailed spatial map of the important Biblical sites.[35] Well, if Aldfrith was interested to know exactly *where* Biblical events occurred, then surely he must have been just as interested to know *when* and in what *sequence* these events occurred? And indeed, the world history found in common in the Irish Annals and Bede's chronicles provides a correspondingly detailed temporal map of Biblical events. The circumstances and synchronism of Bede's compilation in 703 of both his own epitome of *De locis sanctis* and DT suggest then that in *c.*687 Adomnán had also presented Aldfrith with at least a copy of the world history in the Iona annals, which registered the Byzantine succession up to and including Justinian's accession in 685.[36] Furthermore, it suggests that Aldfrith had likewise circulated this copy so that it reached Bede by 703 and he similarly made an epitome of its world history in his DT. The incongruity of Bede's prosaic regnal entries XI–5 as compared with all his other cryptic regnal entries in DT suggests that he knew and respected his source, and indeed Adomnán was still alive when Bede wrote DT. In fact, the regnal entry for Justinian must have been compiled when Adomnán was abbot in Iona and consequently either by him or at his direction. Furthermore, since the preceding regnal entries XI–4 are in identical style to that of Justinian, this suggests that the entire series XI–5 was the work of Adomnán. In this case, the responsibility for much of the idiosyncratic Byzantine nomenclature and chronology discussed above (cf. *Figure 4*), must lie with Adomnán. And indeed, examination of his *Vita Columbae* shows that he handled chronological matters in a very arbitrary way, most especially his chronology for the death of Columba.[37]

Bede's complete silence with regard to his Iona annals world history source is readily explained in terms of the ecclesiastical relationship then prevailing between Iona and Northumbria, for since at least the

---

[36] It seems likely that the copy also contained some Irish entries, for this would explain Bede's citation in HE III 4 of *Dearmach* as the name of Columba's monastic foundation, and his accurate figure of thirty-two years for Columba's residence in Iona over 562–593. Similarly, Bede's precise date for king Ecgfrith's assault on Ireland in HE IV 26 repeats that of the Irish Annals *s.a.* 685.

[37] Daniel Mc Carthy, 'The chronology of St. Colum Cille', at http://www.celt.dias.ie/publications/tionol/dmcc01.pdf (Adomnán's distortion of the chronology of Columba's death); Sharpe (1995), 8, referring to the *Vita*: 'it is unhappily the case that Adomnán is very sparing of facts in the form of dates [. . .] and what he does say is not always in agreement with the evidence of the Irish annals or of Bede'.

Synod of Whitby in 664 Iona and its Paschal tradition had been classified by the Northumbrian church as schismatic. In citing Biblical chronology from a chronicle from Iona, Bede was playing with ecclesiastical fire, and it comes as no surprise that members of Wilfrid's party should subsequently criticize him. While Bede's *Epistola ad Pleguinam* in 708 represents that this criticism arose simply from his substitution of Vulgate for Septuagint chronology, any knowledge of his actual source for this Vulgate chronology must surely have helped fan the criticism directed against him.[38] On the other hand, Bede's conspicuous and emphatic acknowledgement accorded to Adomnán for his *De locis sanctis* in HE V 15–17 may be read as expressing his gratitude for Adomnán's role in providing him with a chronicle authoritatively reflecting the chronology of Jerome's Vulgate. Since no schismatic controversy was attached to the spatial distribution of Biblical sacred sites, Bede could more safely express his appreciation for Adomnán's literary contribution to Northumbrian Christian scholarship by praising his *De locis sanctis*. But for Bede to have publicly acknowledged the actual source of his Vulgate chronology in DT, and later DTR, would have been to invite a Roman ecclesiastical anathema.

While the foregoing analysis has shown that Bede received his copy of Rufinus' world-chronicle via Iona, it must also be acknowledged that much later, in the eleventh century, an Irish annalist with the ancestor to AT as preserved in MS Oxford, Bodleian Library, Rawlinson B 502 collated this with Bede's DTR. This collation disclosed some of Bede's own additions and this annalist transcribed these none too carefully, occasionally identifying Bede as his source.[39] It was these attributions when combined with their inadequate collation of the textual and chronological features of the world-chronicle entries that prompted editors and scholars of the Irish Annals to mistakenly assume that Bede's world-chronicles had served as source for all of the Annals' world history.[40] This misguided reflex persisted unchallenged until Morris' paper of 1972.

---

[38] Bede, *Epistola ad Pleguinam* (ed. by Charles W. Jones in CCSL 123C, 613–26); Wallis (1999), 405–15, 405 (English translation, criticism).

[39] Stokes (1993), 7, 11, 13, 36, 37, 41 (attributions to Bede); Grabowski and Dumville (1984), 112, 118–9, 122 (AT citations of DTR).

[40] Mc Carthy (1998), 100–16 (review of ascriptions of Bede's priority).

Summary and Conclusions

This paper commenced with Mommsen's critical editions of Bede's world-chronicles in DT and DTR and pointed out that Mommsen had simply assumed Bede's authorship of their Vulgate chronology, which assumption has been reiterated by scholars ever since. However, critical examination of Bede's handling of the chronological complexities of Age 3 in DT strongly suggests that rather than constructing a new chronology he was merely epitomizing the Vulgate chronology of an existing chronicle. Certainly identification of DT's idiosyncratic Byzantine regnal succession in the later seventh century precludes all possibility of maintaining that Bede had compiled DT to 'correct' the chronology of his predecessors. Rather, critical collation of Age 4 in DT, DTR and the Irish Annals shows that as well as its Vulgate chronology their common source included a number of very distinctive features, namely, Ages, calends, parallel regnal successions, prosaic regnal entries, regnal obits, and interjections in which the compiler provided critical assessments of chronological conflicts of a wide range of Biblical and early chronicle sources. These features have been best transmitted by the Irish Annals, most particularly by AT as preserved in MS Oxford, Bodleian Library, Rawlinson B 502. However in DTR Bede preserved a number of interjections and occasional entries lost from the Irish Annals, so that the best view of the common source is obtained by combining AT with DTR.

Regarding the author of this common source, based upon citations from his Latin translation of the *Ecclesiastical History* and entries referring warmly to Origen and his works, he has been identified as Rufinus of Aquileia, who was a unique witness to Jerome's work in establishing the text of the Latin Bible. The date of Rufinus' completion of his compilation is bounded by his citation of Jerome's *Commentarium in Danielem* of *c.*406 and his death in Sicily in 410. As regards the transmission of his chronicle to the British Isles, the indications are that it came via Gaul and travelled from thence with the Paschal *latercus* of Sulpicius Severus. By at least the earlier sixth century Rufinus' chronicle was being continued in Ireland, but was removed to Iona later in that century. In *c.*687 Adomnán, abbot of Iona, presented to Aldfrith, king of Northumbria, a copy of the world chronicle of the Iona annals extending to the reign of Justinian, as well as a copy of his own *De locis sanctis*. By 703 these copies had reached Bede, who compiled epitomes of them both. In 725 Bede incorporated much more substantial citations from his Iona source in chapter 66 of his DTR. Thus it was Adomnán's copy of the Iona annals that served as Bede's primary source for the Vulgate chronology of his DT and DTR.

The purpose of this enquiry has been to clarify the evolution of Vulgate chronology and to correctly identify who was responsible for its construction, and to establish the outline of its transmission path across Western Europe. It has not been the intention to diminish Bede's status as a chronicler, which in any case is secured by his HE, though modern accolades celebrating his supposed 'historic accuracy' cannot stand. Rather the results of this enquiry allow us to appreciate much more clearly the significance of Rufinus' contribution to the scholarship of early Western Christianity, and the extent to which his influence permeated the Christian learning of the British Isles. They also allow us to discern the key role played by the Irish monasteries and monastics of the early medieval period in the transmission of these works to Anglo-Saxon Britain. Finally, these results enable us to establish much more precisely the extent and nature of Bede's own contribution to the world-chronicle of DTR.

MASAKO OHASHI

THE *ANNUS DOMINI* AND THE *SEXTA AETAS*: PROBLEMS IN THE TRANSMISSION OF BEDE'S *DE TEMPORIBUS*[1]

Abstract

In chapter 14 of *De temporibus*, Bede implies that he composed this text in AD 703. The latter part of the work (chapters 17–22) contains the *Chronica minora*, where chapter 22 covers the *Sexta aetas* (the age after the coming of Christ) up to Bede's own time. In some early manuscripts, this chapter begins: 'the Sixth Age contains 709 completed years', while the correct number of years according to chapter 14 should be 703. Was this due to a simple scribal error? It is argued in this paper that the most likely scenario is that this error occurred rather due to confusion between the chronological systems featuring in the Victorian and the Dionysiac Easter tables at a time of transition from the Victorian to the Dionysiac reckoning.

Keywords

Bede, *De temporibus*, *Chronica minora*, chronology, *Annus Domini*, *Annus Passionis*, *Annus Mundi*, Easter table, Dionysius Exiguus, Victorius of Aquitaine, Incarnation era.

Introduction

Counting years for chronological purposes is nowadays a simple activity in the West, since *Anni Domini*, unlike consulate or regnal years,[2]

---

[1] This paper is a revised version of my previous article: Ohashi (2003).
[2] Blackburn and Holford-Strevens (1999), 675–6, 763–5.

are counted continuously from the supposed year of Chirst's birth to the present year. When Dionysius Exiguus employed this system in his Easter table, he simply used it as a parameter within this table, not as a chronological system in its own right.[3] Two centuries later, Bede in his *Historia ecclesiastica*[4] introduced *Anni Domini* as a relative chronology for historical events, and this later became the standard for counting years. It was not in the *Historia* but in *De temporibus* where he first dealt with *Anni Domini*. Based on the *Argumenta* of Dionysius,[5] Bede explains how to reckon the *Annus Domini* for the 'present year', that is AD 703.[6] In *De temporum ratione*, he explains certain elements of the Dionysiac calculation with reference to his *annus praesens*, designated as *Annus Domini* 725.[7]

The main chronology used by him in the two chronicles attached to *De temporibu*s (*Chonica minora*) and De temporum ratione (*Chronica maiora*) are *Anni Mundi* (total years from Adam). After he had published *De temporibus*, however, Bede faced a serious problem concerning this reckoning. He was accused of heresy because he calculated the years from Adam to Christ's birth as 3952 years, while other authors traditionally reckoned 5000–5500 years for the period.[8] The differences in calculation lay within the First to the Fifth Age of the world: in the Augustinian division of Universal History, the period from the creation of the world to Christ's birth is divided into five ages, that is the *Prima aetas* from Adam to Noah, the *Secunda aetas* from Noah to Abraham, the *Tertia aetas* from Abraham to David, the *Quarta aetas* from David to the Captivity, and the *Quinta aetas* from the Captivity to the coming of Christ.[9] It has been suggested that Bede created (or adopted) a shorter reckoning because of the fear of millenarianism which occurred and continued in the West at that time, or rather as criticism of that

---

[3] Declercq (2000), 150; Declercq (2002), 237.

[4] Bede's *Historia ecclesiastica gentis Anglorum* is edited and translated by Colgrave and Mynors (1969). This edition is used in the following.

[5] Dionysius Exiguus, *Argumentum I* (Krusch (1938), 75).

[6] Bede, *De temporibus* 14 (CCSL 123C, 598–9). The same work was first published without the *Chronica minora* by Jones (1943), 295–303.

[7] Bede, *De temporum ratione* 47, 49, 52, 54, 58 (CCSL 123B, 427–33; 434–5; 441; 443; 447). Besides these instances, Bede uses the *Annus Domini* only twice in the *Chronica maiora* (*De temporum ratione* 66: CCSL 123B, 521, 533; Wallis (1999), 225, 235).

[8] Bede's reaction to this accusation is recorded in his letter to Plegwin (AD 708). Bede, *Epistola ad Pleguinam* (CCSL 123C, 617–26), translated by Wallis (1999), 405–15. Concerning the traditional calculation of *Anni Mundi* cf. Declercq (2000), 39–41.

[9] Bede, *De temporibus* 17–22 (CCSL 123C, 601–11).

mindset.[10] As Bede's letter to Plegwin written in AD 708 implies, some people assumed that the sixth millennium would soon come to an end.[11] Because of St Augustine of Hippo's criticism of millenarianism and the persisting fear that the world would come to an end at the close of the sixth millennium, the reckoning of years before the coming of Christ was a sensitive issue for Bede.[12] On the other hand, a problem of the chronology of the *Sexta aetas* prevalent in Bede's days seems not to have received full attention.

This problem, though relatively minor in its appearance, is found in some manuscripts of *De temporibus*. According to Theodor Mommsen's critical edition of Bede's *Chronica minora* in the *Monumenta Germaniae Historica* published in 1898, which was later reproduced in Charles W. Jones' *Corpus Christianorum* edition of *De temporibus*, chapter 22 of *De temporibus* begins: *Sexta aetas continet annos praeteritos DCCVIIII* ('the Sixth Age contains 709 completed years').[13] This means that 709 years have passed from Christ's birth to the time of Bede composing this passage. The figure clearly differs from the chronological data given elsewhere in this text, since Bede implies in chapter 14 that he wrote *De temporibus* in AD 703.[14] How did the discrepancy occur, and why did such inconsistencies survive in manuscripts during the Middle Ages? This paper examines the chronological background to the problem leading to a new interpretation of so-called chronological 'errors' in medieval sources.

Problems in Early Manuscripts

First, it is necessary to offer a survey of the manuscripts of *De temporibus* before turning to the problems they contain. Given the great

---

[10] See Mc Carthy in this volume. Concerning the possible origin of the reduction of *Anni Mundi* in an Irish milieu see Stevenson (1995), 178. The *Laterculus Malalianus* (a Latin text composed at Canterbury according to Stevenson) is partly based on the sixth-century *Chronographia* of John Malalas. Although the extant Greek text of the *Chronographia* records 5500 years from Adam to Christ's birth, the *Laterculus* gives 5967 years, insisting that 6000 years had already passed at the time of Christ's passion (Stevenson (1995), 122–5). It is suggested that this unique chronology was created to counter some shorter chronology of Irish origin (Stevenson (1995), 178).

[11] Bede, *Epistola ad Pleguinam* 15 (CCSL 123C, 624; Wallis (1999), 413).

[12] Hunter Blair (1970), 265–7; Wallis (1999), 353–66.

[13] Bede, *De temporibus* 22 (MGH Auct. ant. 13, 280; CCSL 123C, 607); the translation is mine.

[14] Bede, *De temporibus* 14 (CCSL 123C, 598–9).

number of medieval manuscripts of *De temporibus*, an investigation of all of them is, at this point, not possible.[15] Concerning the number given at the beginning of chapter 22 these manuscripts can be classified in four groups.

The principal manuscripts of Mommsen's edition of Bede's *Chronica minora* are MSS E, F, H, M and P. Mommsen prints the number in question as *DCCVIIII* on the basis of two of these manuscripts (MSS F and P), unfortunately without exact folio reference for P (MS Vatican, Biblioteca Apostolica, Pal. Lat. 1448).[16] MS F (MS Berlin, Staatsbibliothek, Phillipps 1831, 105r) clearly gives the reading, as *Plate 1* illustrates.

In his introduction Mommsen states that the number 709 was inserted by some scribe who was a contemporary of Bede.[17] MS F is thought to have been copied around 800, and this would thus be the oldest surviving manuscript of *De temporibus*. Moreover, there are at least four other ninth-century manuscripts containing the number *DCCVIIII* at this point: MS Berlin, Staatsbibliothek, Phillipps 1832, 14r;[18] MS München, Bayerische Staatsbibliothek, Clm 14746, 95v;[19]

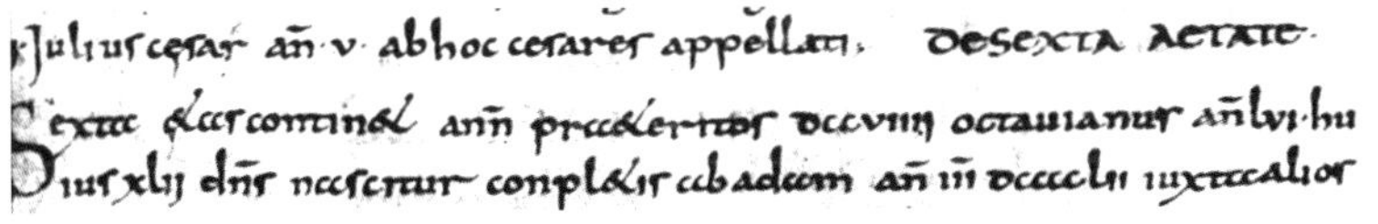

Plate 1 Detail of MS Berlin, Staatsbibliothek, Phillipps 1831, 105r, showing the number in question.

---

[15] Jones listed 93 manuscripts for *De temporibus* 1–16 and 25 for *De temporibus* 17–22 (CCSL 123C, 580–3). Wesley Stevens adds another four for *De temporibus* 1–16 (Stevens (1985), 34).

[16] In this MS, Bede's *De temporibus* including the chronicle can be found on fols 82v–91r, with the number in question occurring on f 91r. I am grateful to Dr Immo Warntjes for providing these details.

[17] MGH Auct. ant. 13, 226 n 3: *Numerus P. Chr. 709, qui ponitur c.173, minus fidus est venitque opinor a librario auctoris aequali*.

[18] Interestingly, chapter 14 of *De temporibus* in this manuscript calculates the *Annus Domini* for the *annus praesens* as *DCCCXI* (811) on folio 12r, while the beginning of chapter 22 still reads *DCCVIIII* (709). This manuscript was compiled in *c*.873, containing a detailed commentary on *De temporum ratione* in the margin, which is published in parallel by Jones in CCSL 123B.

[19] This manuscript contains only the *Chronica minora*, and gives the number as *DCC nouem*, not *DCCVIIII*. This may imply that the scribe noted some problem of the number itself.

MS Paris, Bibliothèque Nationale, Nouvelle acquisition latine 1615, 139v;[20] MS St. Gallen, Stiftsbibliothek, 878, 274.[21]

Three manuscripts omit the number itself: E (MS Einsiedeln, Stiftsbibliothek, 167, 388), H (MS St. Gallen, Stiftsbibliothek, 251, 29) and M (MS Paris, Bibliothèque Nationale, Lat. 4860, 89r).[22] The reason for the omission may simply be that the date given was obsolete. *De temporibus*, as well as *De temporum ratione*, in its original form was divided into two parts: the explanation of time and the calculation of Easter forming the first, the chronicle the second part. In the Middle Ages, it sometimes happened that the latter part was copied separately as the *Chronica minora* (as it happened with the *Chronica maiora* of *De temporum ratione*), with some scribes even continuing that chronicle to their own time of writing.[23] Because of the unfinished, open character of these continuations, no *annus praesens* (i.e. the year in the Sixth Age) could be fixed.

Before Mommsen published his edition, the number in question was accepted by scholars as *DCCVIII* (708). In his note on Bede's historical writings, Charles Plummer suggests that *V* was mistakenly added into *DCCIII*.[24] Plummer does not specify the edition he used, but it is probable that he followed Giles' edition of Bede's complete works published in 1843.[25] Migne's *Patrologia Latina*[26] gives the same reading as Giles, and neither editor comments on the number. Both Giles and Migne followed the *Opera Bedae Venerabilis presbyteri anglosaxonis* (Basel 1563) published by Johannes Heerwagen (Hervagius)[27] who used a fourteenth-century manuscript, now lost, for his edition.[28] Still, some manuscripts survive containing this number: MS Oxford, Bodleian Library, Auc. F. 3.14, 32r;[29] MS Oxford, Bodleian

---

[20] Chapter 14 gives the *Annus Domini* as *DCCIII* (f 137v).

[21] This and the other St Gall manuscripts mentioned in this article can now be consulted in the world wide web at http://www.e-codices.unifr.ch/de/list/csg.

[22] Again, I am grateful to Dr Warntjes for providing the details of this MS.

[23] Mommsen edited the continuation of Bede's *Chronica minora* in MGH Auct. ant. 13, 344–5 from MSS E, H and M; all three of them transmit the chronicle independent of *De temporibus*.

[24] Plummer (1896), i cxlvi.

[25] Giles (1843), vi 136: *Sexta aetas continet annos praeteritos DCCVIII*.

[26] PL 90, 290.

[27] Not examined. Cf. Stevens (1985), 34.

[28] Jones (1939), 5–19, especially 14–8; Jones (1943), vii.

[29] Chapter 14 gives *DCCUI* (f 30r). It is possible that this is a simple scribal error for *DCCIII*.

Library, Auc. F. Infra I.2, 191rb,[30] and MS Berlin, Staatsbibliothek, lat. Fol. 307, 8ra.[31] Thus the number *DCCVIII* was one of the traditional readings in this passage.

The number corresponding to that given in chapter 14 (*DCCIII*) also survives in some manuscripts: MS München, Bayerische Staatsbibliothek, Clm 21557, 98r; MS St. Gallen, Stiftsbibliothek, 248,[32] 98a, and MS St. Gallen, Stiftsbibliothek, 250, 160.

An analysis of the numeral at the beginning of chapter 22 of *De temporibus*, therefore, leads to four groups of manuscripts: manuscripts omitting the number; manuscripts recording *DCCIII*; manuscripts recording *DCCVIII*; and finally manuscripts recording *DCCVIIII*. There is no problem concerning the first two groups, but it is difficult to understand the latter two as being simple mistakes.

Background to the Differences: Some Hypotheses

It is highly probable that, when he commented on the number *DCCVIII*, Plummer did know about the other 'erroneous' number *DCCVIIII* found in some of the earliest manuscripts. If he knew about it, would he have understood both numbers as simple mistakes? Certain aspects suggest that these two numbers were no scribal errors, but deliberately chosen. Since the text of *De temporibus* is short, it is rather difficult to imagine that a scribe, having read the first part of the work with AD 703 mentioned in chapter 14, accidentally added *V* to *DCCIII*, while later scribes would then have copied this mistake without any criticism or consideration. It appears even less probable that another scribe accidentally added *VI* to *DCCIII*. As Mommsen in his edition points out concerning *DCCVIIII*, this number seems to have already been employed by some of the earliest scribes, presumably contemporaries of Bede or his successors.

Investigating these two problematic numbers, we should deal with them in context. Bede gives slightly different information about the year of publication in chapters 14 and 22, and this may provide a clue to the problem. In chapter 14, he refers to the 'present year' as 'the fifth

---

[30] Chapter 14 gives *DCCIII* (f 190rb).

[31] This manuscript contains the *Chronica minora* only.

[32] Jones suggests that *De temporibus* ends with chapter 15 in this manuscript, but it actually also contains the part of chronicle relevant to the present study.

year of the Emperor Tiberius', 'the first indiction' and 'AD 703'.³³ At the end of Chapter 22, on the other hand, he states: *Tiberius dehinc quintum agit annum ind. prima. Reliquum sextae aetatis Deo soli patet.* ('Tiberius reigns in the fifth year, the first indiction. The remainder of the Sixth Age is known only to God.').³⁴ Since the *Chronica minora* was sometimes copied separately, users of this text, if they did not have full information about the Dionysiac table, could have understood the figure *DCCIII* as a mistake.

Before introducing the Dionysiac table, many people in the West followed for the 532-year Victorian table which had been published by Victorius of Aquitaine at the request of Archdeacon Hilarus (later Pope Hilarus) in AD 457.³⁵ Probably because of its convenience and papal authority, the Victorian table was particularly prominent in Gaul, as the decrees of the Council of Orléans of 541³⁶ and Gregory of Tours' *Historia* of the late sixth century³⁷ illustrate. It was followed there right into the eighth century.³⁸ As the letter of Cummian

---

³³ Bede, *De temporibus* 14 (CCSL 123C, 598–9).

³⁴ Bede, *De temporibus* 22 (CCSL 123C, 611); the translation is mine.

³⁵ Victorius of Aquitaine, *Cyclos* (Krusch (1938), 27–52). Following the traditional Roman calculation, Victorius fixed Easter Sunday between the sixteenth and the twenty-second day of the moon (*lunae* XVI–XXII), while the Alexandrian-Dionysiac tables fixed it between the fifteenth and the twenty-first day of the moon (*lunae* XV–XXI). There were some other differences between the two systems causing discrepancy of Easter dates. Moreover, Victorius sometimes noted double-dates for Easter Sunday in his table. Cf. especially Jones (1934), 411–3.

³⁶ *Concilium Aurelianense* (541. Mai. 14) §1 (CCSL 148A, 32): *Placuit itaque Deo propitio, ut sanctum pascha secundum laterculum Victori ab omnibus sacerdotibus uno tempore celebretur.*

³⁷ Gregory of Tours, *Historiarum libri decem* X 23 (MGH SS rer. Merov. 1,1 514–5): *Dubietas pascae fuit ob hoc, quod in cyclum Victuri luna XV. pascham scripsit fieri. Sed ne christiani ut Iudei sub hac luna haec solemnia celebrarent, addidit: Latini autem luna XXII. Ob hoc multi in Gallis XV. luna celebraverunt, nos autem XXII.* The year in question is AD 590.

³⁸ The continuations of Fredegar's chronicle written in the early eighth century refer to the Victorian Cycle and its chronological data, including the *Annus Passionis* reckoning: *Continuationes Chronicarum Fredegarii* 16 (MGH SS rer. Merov. 2, 176): *certe ab initio mundi usque ad passionem domini nostri Iesu Christi sunt anni 5228 et a passione Domini usque isto anno praesente, qui est in cyclo Victorii ann. 177. Kl. Ian. die dominica, ann. 735; et ut istum miliarium impleatur, restant ann. 63.* The phrase '5228 years from the beginning of the World to Christ's Passion' reflects the Victorian *Annus Passionis* chronology. The data given for the 'present year' (*ann. 177* and *Kl. Ian. die dominica*) refer to year *CLXXVII* of the Victorian table (Victorius, *Cyclos* (Krusch (1938), 35)), which does not correspond to AD 735 (as Fredegar suggests), but to AD 736. Some manuscripts written in the Frankish kingdom in the eighth and early ninth century contain Victorian material. For Frankish texts of the eighth century based on the Victorian reckoning see especially *Dial. Burg.* and *Quaest. Austr.* in Borst (2006), 353–74, 466–508.

addressed to the abbot of Iona reveals, the table was also accepted by the southern Irish in the early 630s.[39]

Concerning Britain, however, the direct evidence for the use of the Victorian table is rather limited. Although Bede severly criticizes the Victorian reckoning in *De temporum ratione*,[40] he never refers to its use in England or other places. Only a few sources may imply the existence and the use of the Victorian table in Britain: Aldhelm's letter to the British king Geraint of Cornwall[41] and Ceolfrith's letter to King Nechtan of the Picts.[42] Both authors criticizes the Victorian table, a criticism that made sense only if the table was in actual use. Moreover, a quotation from the letter of Victorius to Hilarus (which usually accompanied Victorius' table) can be found in the preface to the *Vita S. Cuthberti* written by an anonymous author at Lindisfarne c.700.[43] These sources are proof that the table was actually known and probably used before the exclusive introduction of the Dionysiac table. And the application of the Victorian table seems to lie behind the 'erroneous' numbers.

There are six columns in the Victorian table: the numerals *I–DXXXII*, a consul list, the week day for 1 January, the day of the moon for 1 January, the Julian date for Easter Sunday, and the age of the moon for Easter Sunday. It is the first of these elements that concerns us here. Victorius started his own table with the year of Christ's Passion; the numeral *I* refers to the very year of the Passion. The numerals up to *DXXXII* refer to the number of years from the Passion, counted inclusively. Since the 532-year table was designed to be a true cycle,[44] users would return to the beginning of the table to use it continuously. Concerning the *Annus Passionis* (AP), i.e. the number of years from the Passion, 532 years had to be added to the numeral given

---

[39] Cummian, *De controversia paschali* (Walsh and Ó Cróinín (1988), 56, 68).

[40] Bede, *De temporum ratione* 51 (CCSL 123B, 437–41; Wallis (1999), 132–5), where Bede directly names and attacks Victorius. On the different attitudes towards the Victorian calculation between *De temporum ratione* and the *Historia ecclesiastica* see Ohashi (2005).

[41] Aldhelm, *Epistola ad Gerontium* (MGH Auct. ant. 15, 480–86; translated by Lapidge and Herren (1979), 152–70).

[42] A copy of this letter survives only in Bede, *Historia ecclesiastica gentis Anglorum* V 21 (Colgrave and Mynors (1969), 534–51).

[43] *Vita Cuthberti* I 1 (Colgrave (1940), 60–2); the citation is from Victorius of Aquitaine, *Epistola ad Hilarum* (or *Prologus*) 1 (Krusch (1938), 17–8).

[44] Bede, *De temporum ratione* 47 (CCSL 123B, 427–8; Wallis (1999), 126); *Historia ecclesiastica gentis Anglorum* V 21 (Colgrave and Mynors (1969), 546–7).

in the first column in such cases. We may assume at this point that some Victorian tables now lost would be accompanied by notes in the margins, as found in many medieval manuscripts of the Dionysiac table;[45] regnal years and death notices of kings, e.g., were there recorded.[46] What implication, then, do these facts have concerning the numbers given in chapter 22 of *De temporibus*?

At the end of chapter 22, Bede refers to the year of writing as 'the fifth year of the Emperor Tiberius' and 'the first indiction'. Since the Victorian table in its original form lacks the indiction, users of this table, when reading this passage, would in all likelihood focus on the information about the regnal year of Tiberius III and the number of years of the Sixth Age. And it can be suggested that those who were using only the Victorian table thought the number *DCCIII* was mistaken. We now take the fifth year of Tiberius III as corresponding to AD 702–703. Users of the Victorian table, however, would rather have thought in *Anni Passionis* terms, connecting the fifth year of Tiberius with AP 675–676 by adding 532 years to the numbers found in the Victorian table (*CXLIII–CXLIIII*); the *Annus Passionis* could be used to reckon the years from Christ's birth, that is the number of years of the *Sexta aetas*.

As Bede in *De temporum ratione* 47 (on 'the years of the Lord's Incarnation') suggests, the Synoptic Gospels imply that Jesus was crucified in the thirty-fourth year of his life.[47] This means that adding 33 years to the years from the Passion (*Anni Passionis*) results in the number of years from Christ's birth. If users of the Victorian table followed this method to reckon those years, or rather the number of years of the *Sexta aetas* (i.e. if their method of converting *Anni Passionis* into *Anni Domini* was AP+33=AD) they will inevitably reckon an additional six years compared to the Dionysiac *Anni Domini*: The first year of the Victorian table, that is the 'year of the Passion' (AP 1), corresponds to AD 28, and this means that Christ should have become 34 years old in that year according to the Synoptic Gospels.

---

[45] Medieval annals were usually based on records in the margins of Dionysiac Easter tables. Cf. McCormick (1975).

[46] For instance, a copy of the Victorian table found in the Sirmond MS (MS Oxford, Bodleian Library, Bodley 309, 113r–120r) contains references to emperors from Nero to Justinian I in the margins of fols 113v, 116v, and 118r–120r. This manuscript is thought to be based on the sources used by Bede. Cf. Jones (1937); Ó Cróinín (1983c), repr. in Ó Cróinín (2003), 173–90; Ó Cróinín (2003a).

[47] Bede, *De temporum ratione* 47 (CCSL 123B, 430–1; Wallis (1999), 127–8).

The fifth year of Tiberius was AP 675–676; as a result, those who tried to convert this into years from Christ's birth (total years of the *Sexta aetas*) would arrive at 708–709 by adding 33 years. Since the references in the margins of Easter tables were sometimes noted in ambivalent positions (i.e. between two years, or referring to different years in different tables), both numbers *DCCVIII* and *DCCVIIII* could be deduced from different manuscripts of the Victorian table;[48] in these cases, Bede's statement that the fifth year of Tiberius coincided with the first indiction appears to have been totally neglected.

As Mommsen's edition and some early manuscripts reveal, the number *DCCVIIII* seems to have been copied especially in the Carolingian period, when many copies of manuscripts of Bede's writings were made for monastic schools. This means that even though some scribes might have carelessly copied Bede's computistical writings, others may have carefully made their copies by comparing some earlier manuscripts which have since been lost. Thus the problematic numbers in *De temporibus* can be understood as the result of careful reading and editing by Carolingian scribes. How, then, did they choose *DCCVIIII* instead of the correct *DCCIII*?

When publishing *De temporum ratione* in 725, Bede basically followed the same style as in his first computistical text, dividing the work into two parts. In the *Chronica maiora* he does not give any number for the total number of years of the *Sexta aetas*, although he followed his previous writing for the other five ages.[49] Does this fact reflect the re-calculation of the year by some copyists of manuscripts of the *Chronica minora*? Seen from this perspective, there are indeed some interesting passages in the main text of *De temporum ratione*. In chapter 47, Bede explains how to calculate the *Annus Domini* by quoting (as in *De temporibus*) Dionysius' *Argumenta*. Then he turns to discussing the year of Christ's Passion. Bede states:

Habet enim, ni fallor, ecclesiae fides dominum in carne paulo plus quam ·xxxiii· annos usque ad suae tempora passionis uixisse, [. . .].

'If I am not mistaken, the faith of the Church holds that the Lord lived in the flesh, up until the time of the Passion, a little more than 33 years [. . .].'[50]

---

[48] The variation in the reckoning of the beginning of the year could also cause such confusion.

[49] Bede, *De temporum ratione* 66 (CCSL 123B, 463–4; Wallis (1999), 157–8).

[50] Bede, *De temporum ratione* 47 (CCSL 123B, 430–1; Wallis (1999), 127–8).

Then, a little later:

Sancta siquidem romana et apostolica ecclesia hanc se fidem tenere et ipsis testatur indiculis quae suis in cereis annuatim scribere solet ubi, tempus dominicae passionis in memoriam populis reuocans, numerum annorum triginta semper et tribus annis minorem quam ab eius incarnatione Dionysius ponit adnotat. Denique anno ab eius incarnatione iuxta Dionysium septingentesimo primo, indictione quarta decima, fratres nostri qui tunc fuere Romae hoc modo se in natale domini in cereis Sanctae Mariae scriptum uidisse et inde descripsisse referebant: A passione domini nostri Iesu Christi anni sunt ·dclxuiii·.

'That the holy Roman apostolic Church holds this belief is testified by the legends that it customarily inscribes, year by year, upon its [Paschal] candles. Recalling the time of the Lord's Passion to the memory of the people, it records a number of years which is always 33 less than what Dionysius laid down [as having passed] since the Incarnation. Hence, in the 701st year of His Incarnation (according to Dionysius), in the 14th indiction, our brothers who were then in Rome related that they saw, and copied down, the following inscription from the candles in St. Mary's: "From the Passion of Our Lord there are 668 years."'[51]

Here Bede clearly suggests that the years from the Passion can be calculated by reducing 33 years from the *Annus Domini*. Vice versa, this means that adding 33 years to the *Annus Passionis* leads to the year from Christ's Incarnation. The referrence to his brothers' experience in Rome suggests that Bede, facing some controversy about the issue, tried to convince those who were following a different calculation to accept the one outlined by him.

We now turn our attention to the Victorian calculation of the *Annus Passionis*. Even though Bede refers in the same chapter to Victorius,[52] he does not mention the Victorian chronology. This probably implies that Bede intended to ignore the Victorian chronology because

---

[51] Bede, *De temporum ratione* 47 (CCSL 123B, 431; Wallis (1999), 128). The visit to Rome is recorded in the anonymous *Vita Ceolfridi* (written in Bede's monastery of Wearmouth-Jarrow) and also in Bede's *Historia abbatum*: *Vita Ceolfridi Abbatis* 20 (Plummer (1896), i 395); Bede, *Historia abbatum* 15 (Plummer (1896), i 380).

[52] Bede, *De temporum ratione* 47 (CCSL 123B, 428; Wallis (1999), 126).

of problems posed by this chronology in his earlier *De temporibus*.[53] But it is highly probable that those reading the term *annus passionis* would naturally connect this not to Dionysiac but to Victorian chronology. Some extant manuscripts illustrate this: MS St. Gallen, Stiftsbibliothek, 110 (saec. viii/xi), which contains some computistical formulae, reads (p 512): *Si uis inuenire quotus annus sit a passione domini, sume annos incarnatione ipsius. A quibus subtrahe xxuii, quod remanet totus annus est a passione Christi.* ('If you want to find out how many years have passed from the Passion of the Lord, take the years from His Incarnation. From these subtract 27, and what remains is the total number of years from the Passion of Christ'). A similar explanation can be found in another ninth-century manuscript: MS Berlin, Staatsbibliothek, Phillipps 1831[54] (f 90r): *Si uis scire an[nos] a passione domini, summe an[nos] ab incarnatione Christi dcclvii. De ipsis subtrahe xxvii. Remanent dccxxx. Isti sunt anni a passione domini.* ('If you want to know the years from the Passion of the Lord, take the 757 years from the Incarnation of Christ. From these subtract 27. 730 remain. These are the years from the Passion of the Lord'). The formulae here actually calculate the *Annus Passionis* of Victorius from the *Annus Domini* of Dionysius. Thus those scribes (or authors) were in fact dealing with two different Easter tables.

There is, moreover, a curious passage related to Dionysiac chronology in chapter 47 of *De temporum ratione*. According to Bede, compu-

---

[53] Most revealing for Bede's silence about Victorian chronology is *De temporum ratione* 51 (CCSL 123B, 437–41, especially 441; Wallis (1999), 132–5, especially 135): In this chapter Bede criticizes the Victorian criteria concerning the first day of the first lunar month. Victorius set the limit between 5 March and 2 April, while the Alexandrian-Dionysiac calculation fixed it between 8 March and 5 April. Here Bede suggests that Victorius may calculate the Paschal full moon (*luna* XIIII) before the vernal equinox (21 March). This fact had already been criticized in the letter of Abbot Ceolfrith addressed to Nechtan, king of the Picts between 706 and 716 (Bede, *Historia ecclesiastica gentis Anglorum* V 21; Colgrave and Mynors (1969), 540–3). At the end of chapter 51, Bede quotes a work of Victor, bishop of Capua between 541 and 554, discussing the date of Easter of AD 550. According to Bede's quote, Victorius had calculated the date as 17 April, *luna* XV (irregular lunation sometimes found in the table), while the Alexandrian-Dionysiac calculation (followed by Victor) fixed it on 24 April. Just before the quotation, Bede states that the year in question was 'the 13th indiction' and 'the ninth year of the proconsul Basilius'. In Bede's quote proper the year is characterized as 'the 13th indiction'. It is noteworthy that Bede (and Victor in the quotation) refer only to data not found in the Victorian table, which could prove useless for Victorian adherents who followed the discussion.

[54] MS F in Mommsen's (and Jones') edition(s).

tistical elements for AD 566 in the Dionysiac table would correspond to those of AD 34 (the year of the Passion) since, he insists, the luni-solar calendar recurs after 532 years.[55] Additionally, Christian tradition relates that Christ was crucified on the eighth Kalends of April (25 March), and rose on the sixth Kalends of April (27 March). Bede, with a great sense of irony, draws the following conclusion:

> *Et ideo circulis beati Dionysii apertis, si quingentesimum sexagesimum sextum ab incarnatione domini contingens annum, quartam decimam lunam in eo ·viiii· kal. Apr. quinta feria repereris, et diem paschae dominicum ·vi· kal. Apr. luna decima septima, age Deo gratias quia quod quaerebas, sicut ipse promisit, te inuenire donauit.*

'And so, with the cycles of Dionysius open before you, should you find in the 566th year from the Incarnation of the Lord that the 14th moon falls on Friday the 8th Kalends of April [25 March], and Easter on Sunday the 6th Kalends of April [27 March], then give thanks to God, for He has granted that you find what you were looking for, just as He promised!'[56]

But in reality users of the Dionysiac table would not have found the dates mentioned above, but rather the fourteenth day of the moon on 21 March and Easter Sunday on 28 March.[57] A late ninth-century glossator, who probably was a school master, comments on this sentence: *hironice dixit hoc, dum nullo modo inuenitur* ('he says this ironically, because it cannot be found in this way').[58] Faith Wallis, in her commentary on this chapter, suggests: 'Bede himself doubted the historical foundation of Dionysius's AD chronology'.[59] The gloss might explain why some scribes chose the problematic numbers in chapter 22 of *De temporibus*. Following Bede's doubt on the Dionysiac AD chronology, those who found the numbers *DCCVIIII* or *DCCVIII* in their copies of *De temporibus* may not have dared to change them.[60]

---

[55] Bede, *De temporum ratione* 47 (CCSL 123B, 428; Wallis (1999), 126).
[56] Bede, *De temporum ratione* 47 (CCSL 123B, 431–2; Wallis (1999), 128).
[57] Dionysius Exiguus, *Cyclus paschalis* (Krusch (1938), 71).
[58] MS Berlin, Staatsbibliothek, Phillipps 1832, 46r (margin); ed. in CCSL 123B, 432.
[59] Wallis (1999), 338.
[60] The doubt about the Dionysiac chronology later led to some attempts at re-calculations of this era by authors such as Abbo of Fleury; cf. von den Brincken (1960); idem (1961); Verbist (2003). The same doubt had already been recorded in the ninth century when many MSS of Bede's computistical writings were compiled. Cf. Jones (1939), 81.

Conclusion

Since there is a six-year interval between Victorian chronology (Christ's Passion occurring in the year corresponding to AD 28) and the information provided by the Synoptic Gospels when connected to Dionysiac chronology (Christ's Passion occurring in AD 34), those computists using the Victorian table could have converted wrongly the Victorian *Anni Passionis* into the Dionysiac *Anni Domini*. This is particularly true for the conversion period from the Victorian to the Dionysiac reckoning in the late seventh and the eighth century. Some curious chronological dates, which cannot immediately be explained as simple scribal errors, are sometimes found in texts and manuscripts written at that time.[61] This study, therefore, on the problematic numerals found in manuscripts of *De temporibus*, might give some clue as to how to approach and examine other such cases which remain unsolved because historians regard them as simple scribal errors.

---

[61] For instance, the *Annales Lundenses* (MGH SS 29, 192) report the death of St Cuthbert of Lindisfarne *s.a.* AD 693, while Bede and other English sources give AD 687. Two Frankish annals (the *Annales Prumienses* and the *Annales Bertiniani*) give different dates for the marriage of Pippin III and his wife: The former records AD 744, while the latter has AD 749 (MGH SS 15,2, 1290; MGH SS 1, 136).

KERSTIN SPRINGSFELD

EINE BESCHREIBUNG DER HANDSCHRIFT ST. GALLEN, STIFTSBIBLIOTHEK, 225

Abstract

MS St. Gallen, Stiftsbibliothek, 225 was probably compiled in 773. It contains 21 hitherto unexplored pages dealing with natural science and computistics: three tables (Easter table, Easter Sundays, lunar age), two schematic chapters (horologium and Julian calendar with multiplication-tables), eight computistical *argumenta* (dealing with the bissextile day, the *annus mundi*, the *annus domini*, the indiction, the *concurrentes*, common and embolismic lunar years, the *annus passionis*, the seasons), and two inserted chapters on ecclesiastical feastdays and Greek names.

This material shows parallels with the Cologne textbook on time-reckoning of 805, as well as the encyclopaedia on time-reckoning from Aachen of 809; yet, this manuscript does not contain a cohesive work, but only disconnected computistical bits.

It is evidently influenced by Irish computistical thought, but it also transmits Roman traditions through its lunar calculations, which were later promoted by Alcuin in the Carolingian kingdom. A pseudo-calculation of Irish-Frankish origin is mentioned in the context of the bissextile day. Likewise, a table listing all possible Easter Sundays depending on the weekday of the Easter full moon also derives from Irish-Frankish Easter calculations. This calculation, using certain regulars for March and April, is explained in the *Calculatio Albini* of 776, which is an adaptation of Irish material, presumably by Alcuin.

Keywords

Alcuin, Bede, Dionysius Exiguus, Isidore of Seville, Charlemagne, Palladius, horologium, computistics, March and April regulars, Alexandrian and Roman lunar calculations, Easter calculations, Irish-Frankish Easter calculations and computistics, Easter table, time-reckoning.

Diese von der komputistischen Forschung bisher wenig beachtete Handschrift[1] wurde der Öffentlichkeit erstmals 2004 auf der Ausstellung 'Karl der Große und seine Gelehrten. Zum 1200. Todesjahr Alkuins (gest. 804)' in der Stiftsbibliothek St. Gallen zugänglich gemacht.[2]

Es handelt sich um eine in frühalemannischer Minuskel stilvoll geschriebene St. Galler Sammelhandschrift aus dem letzten Drittel des 8. Jahrhunderts, wahrscheinlich aus dem Jahr 773.[3] Sie ist im Katalog *libri scottice scripti*, also der irischen Handschriften, von 835 aufgeführt. St. Gallen hatte als irische Gründung eine reiche Sammlung irischer Schriften.[4]

Die Handschrift ist inhaltlich breit gefächert. Für die Hagiographie ist sie besonders bedeutend, da sie die älteste erhaltene Fassung der Lebensgeschichte der Zürcher Stadtheiligen Felix und Regula (*Passio sanctorum Felicis et Regulae*) enthält.[5] Die Palette der behandelten Wissensgebiete reicht außerdem[6] von Synonymik (Isidor von Sevilla, *Differentiae*) über Exegetik (Eucherius von Lyon, *Formulae spiritalis*

---

[1] Auf die Handschrift aufmerksam machte mich Prof. em. Dr.-Ing. Winfried Görke, ehemals Leiter des Instituts für Rechnerentwurf und Fehlertoleranz, Fakultät für Informatik der Universität Karlsruhe. Mit ihm fand ein interessanter wissenschaftlicher Austausch statt, für den ich mich an dieser Stelle sehr herzlich bedanken möchte. Eine Vielzahl von Anregungen und Hinweisen verdanke ich Immo Warntjes, dem ich ebenfalls sehr herzlich danken möchte.

[2] Während des Alkuin-Symposiums 'Alkuin von York (um 730–804) und die geistige Grundlegung Europas' (Internationales Kolloquium in der Stiftsbibliothek St. Gallen, 30. September bis 2. Oktober 2004) in St. Gallen nahm ich Einsicht in die Handschrift und erhielt Kopien der Seiten 114–137. Mein Dank gilt den Mitarbeitern der Stiftsbibliothek. Vgl. die elektronische Version der Handschrift in der Virtuellen Handschriftenbibliothek der Schweiz (http://www.e-codices.unifr.ch/de/csg/0255).

[3] Tremp, Schmuki und Flury (2004), 38. Handschriftenseite 126 abgebildet 41, Literatur 47. Zum Entstehungsjahr 773 vergleiche unten die Bemerkung zur Ostertafel auf pp. 114–115 der Handschrift.

[4] Cordoliani (1955a), 164; Cordoliani (1955b), 289.

[5] Die Editionen bei Etter (1988).

[6] Vgl. das ausführlichere Inhaltsverzeichnis im Anhang zu diesem Artikel.

intelligentiae), Liturgie (ein Hieronymus zugeschriebener Traktat über Psalmen und Hymnen sowie ein *Ordo Romanus* mit rituellen Anweisungen für liturgische Amtshandlungen im Laufe eines Kirchenjahrs) und Theologie (Bericht über die Auffindung des Kreuzes Christi[7] und die *Revelationes* des Pseudo-Methodius) über Grammatik (Deklination lateinischer Wörter, ein Text über griechische Namen) bis hin zur Komputistik und den Naturwissenschaften, ja sogar der Medizin (Verhaltensmaßregeln für jeden Monat des Jahres zur Erhaltung der körperlichen Gesundheit).

Das Inhaltsverzeichnis pp. 2–31 und der Hauptteil pp. 114–461 sind von einer Hand geschrieben, eine gleichzeitige andere Hand schrieb pp. 33–113, von späterer Hand sind pp. 461–478.[8] Winithar, der erste namentlich bekannte Schreiber und vermutlich auch Leiter des St. Galler Skriptoriums, versah die Handschrift mit Kapitelnummern.[9]

Der komputistisch-naturwissenschaftliche Teil der Handschrift pp. 114–134 beginnt mit Kapitel VI, das aus drei Tabellen besteht, und zwar aus zwei Zyklen der Ostertafel sowie einer Tafel zur Ermittlung der Ostersonntage. Ebenfalls recht schematisch sind die komputistischen Kapitel VII und VIII zu Körperuhr und Kalender. Fließtext enthalten die kurzen komputistischen Kapitel VIIII bis XV. Kapitel XVI ist eine Jahreszeitenfigur mit anschließenden allegorischen Deutungen. Etwas länger sind Kapitel XVII zu den kirchlichen Festen und Kapitel XVIII zu den griechischen Namen. Kapitel XVIIII listet in Zweiwortsätzen Tierstimmen auf. Das letzte Kapitel dieses Teils der Handschrift ist eine doppelseitige Tafel der Mondalter der Monatsersten im 19jährigen Zyklus.

Bei der Ostertafel handelt es sich genauer um zwei 19jährige Zyklen für die Jahre 760–778 und 779–797 (Faksimiles erscheinen im folgenden als *Plate 1* und *Plate 2*). Die besondere Kennzeichnung des Jahres 773 im ersten Zyklus durch ein Kreuz (auf p 115 in der entsprechenden Zeile ganz links; siehe *Plate 1*) legt die Vermutung nahe, daß die Handschrift in diesem Jahr geschrieben ist. In den späteren komputistischen Kapiteln tauchen aber auch die Jahresangaben 743 und 751 auf.

Die beiden Zyklen sind Teil der 532jährigen großen Ostertafel der dionysisch-bedanischen Zeitrechnung. Der aus dem heutigen Rumänien (Skythien) stammende Dionysius Exiguus war seit 496 römischer Abt. Er erhielt im Jahr 525 von Papst Johannes I. den Auftrag, die römische Osterberechnung an die alexandrinische anzupassen. Diesem Auftrag

---

[7] Holder (1889).

[8] Bruckner (1936), 71.

[9] *CLA* 7, 51 (No. 928). Zu Winithar siehe Ochsenbein (2000).

leistete Dionysius Folge, indem er eine alexandrinische Ostertafel, die des alexandrinischen Bischofs Kyrill für die Jahre 437–531, übernahm und fortsetzte. Er stellte seiner Ostertafel den letzten Zyklus des Kyrill voran, von dem 525 noch sechs Jahre übrig waren, und berechnete die Osterfeste für die nächsten fünf 19jährigen Zyklen, also von 532–626. Durch seine Tafel, in der er die Jahre nach Christi Geburt rechnet, führte Dionysius Exiguus die Inkarnationszählung im Abendland ein.[10] Der Angelsachse Beda von Jarrow mit dem Beinamen Venerabilis übernahm in seinen komputistischen Schriften (703 und 725) die Osterberechnungsweise des Dionysius und setzte seine Ostertafel bis ins Jahr 1063 fort.[11] Im fränkischen Reich begann sich die dionysisch-bedanischen Zeitrechnung seit der Regierungszeit Karlmanns, des Onkels Karls des Großen, durchzusetzen.[12] Durch die Bestrebungen Karls des Großen und seines angelsächsischen Beraters Alkuin wurden Bedas komputistische Lehrbücher zu Standardwerken für das Karolingerreich.[13]

Die St. Galler Ostertafel folgt dem Vorbild Bedas, umfaßt allerdings zwei Spalten mehr, nämlich insgesamt zehn: Inkarnationsjahr, Indiktion (alter römischer Steuerzyklus von 15 Jahren), Epakte (Mondalter am 22. März), Konkurrente (Zahl des Wochentages am 24. März), Jahr des jüdischen Mondzyklus (Beginn drei Jahre nach dem 19jährigen Zyklus), Datum des Fastensonntags *Quadragesima*, dessen Mondalter, Datum des Ostervollmonds, Datum des Ostersonntags und dessen Mondalter.[14] Bei Beda fehlen die beiden Spalten zur *Quadragesima*.[15] So erklären sich vielleicht auch die Fehler in den St. Galler Tafeln: Es wurde nämlich der Beginn der Fastenzeit immer stur sechs Wochen vor dem Ostersonntag

---

[10] Die komputistischen Schriften des Dionysius Exiguus sind ediert von Krusch (1938), 59–86; dort 63–8 der einführende Text *Libellus de cyclo magno paschae DCCCII annorum* als Brief an den Bischof Petronius; 69–74 der Zyklus, also die eigentliche Ostertafel, vorangestellt die Jahre 513–531 nach der Tafel des Kyrill, noch datiert nach der Ära des Christenverfolgers Diokletian, dann die fünf 19jährigen Zyklen des Dionysius für die Jahre 532–626.

[11] Beda, *De temporum ratione*, ediert in CCSL 123B, 239–544, Ostertafel rekonstruiert 551–62.

[12] Krusch (1884), 137–41.

[13] Springsfeld (2002), 300–1. Jüngst zusammenfassend Borst (2006), 84–5, 90–1.

[14] Für 779–797 ist allerdings die Reihenfolge etwas anders als für 760–778, und zwar ab der fünften Spalte, nämlich: erst Datum des Ostervollmondes, Datum und Mondalter des Ostersonntages, dann Datum und Mondalter des Fastensonntags. Siehe *Plate 1* und *Plate 2*.

[15] Nach Cordoliani (1955b), 301 hat sonst von den St. Galler Handschriften nur MS St. Gallen, Stiftsbibliothek, 251 (um 830), 22–25 eine Ostertafel mit Datum und Mondalter des Beginns der Fastenzeit; diese enthalte außerdem auch die Daten für *Rogationes* und Pfingsten. Vgl. http://www.e-codices.unifr.ch/de/csg/0251.

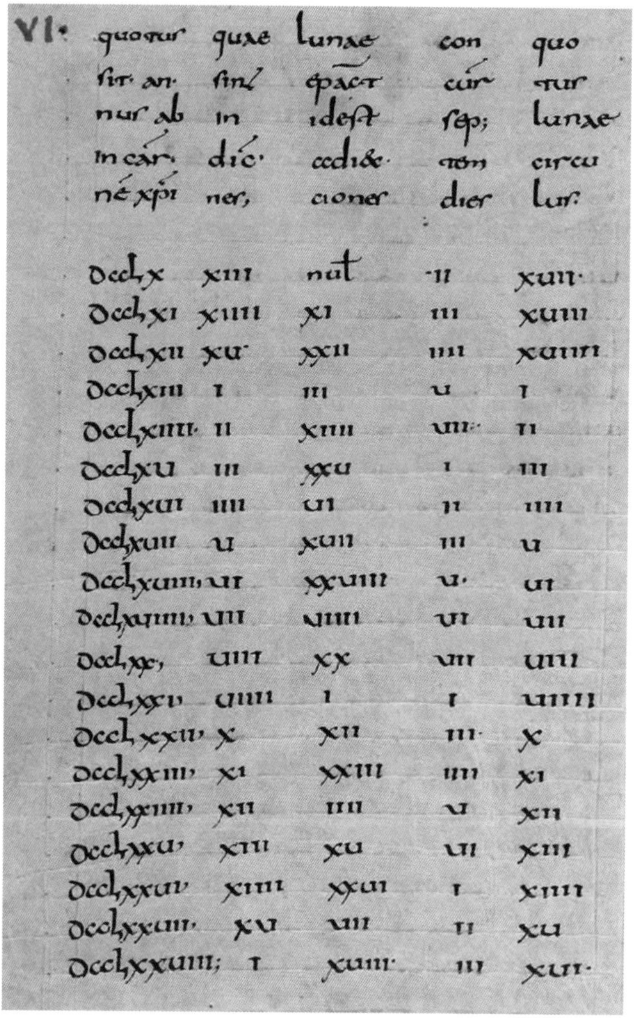

VI· quotus sit annus ab incarnatione xpi	quae sint indictiones	lunae epactae id est adiectiones	concurrentes septem dies	quotus lunae circulus
dcclx	xiii	nul	ii	xuii
dcclxi	xiiii	xi	iii	xuiii
dcclxii	xu	xxii	iiii	xuiiii
dcclxiii	i	iii	u	i
dcclxiiii	ii	xiiii	uii	ii
dcclxu	iii	xxu	i	iii
dcclxui	iiii	ui	ii	iiii
dcclxuii	u	xuii	iii	u
dcclxuiii	ui	xxuiii	u	ui
dcclxuiiii	uii	uiiii	ui	uii
dcclxx	uiii	xx	uii	uiii
dcclxxi	uiiii	i	i	uiiii
dcclxxii	x	xii	iii	x
dcclxxiii	xi	xxiii	iiii	xi
dcclxxiiii	xii	iiii	u	xii
dcclxxu	xiii	xu	ui	xiii
dcclxxui	xiiii	xxui	i	xiiii
dcclxxuii	xu	uii	ii	xu
dcclxxuiii	i	xuiii	iii	xui

Plate 1 Faksimile der Ostertafel pp. 114–115. *(Continued)*

vermerkt, ohne die Schalttage zu beachten, so daß einige Daten Samstag statt Sonntag angeben.[16] Diesen systematischen Fehler diskutierte übrigens als erster der Alkuin-Schüler Hrabanus Maurus im Jahr 820.[17]

[16] Dieser Fehler tritt auf, wenn der Beginn der Fastenzeit vor dem 24. Februar, dem Sitz des Sonnenschalttages, liegt: nämlich in den Jahren 760, 764 und 772 sowie 780, 788 und 796; die Daten geben Samstag statt Sonntag an.

[17] Hrabanus Maurus, *De computo* 83 (CCCM 44, 302): *Notandum tamen quod quadragesimalis terminus, quando ante sexto kalendas martias fuerit, et bissextus eodem anno*

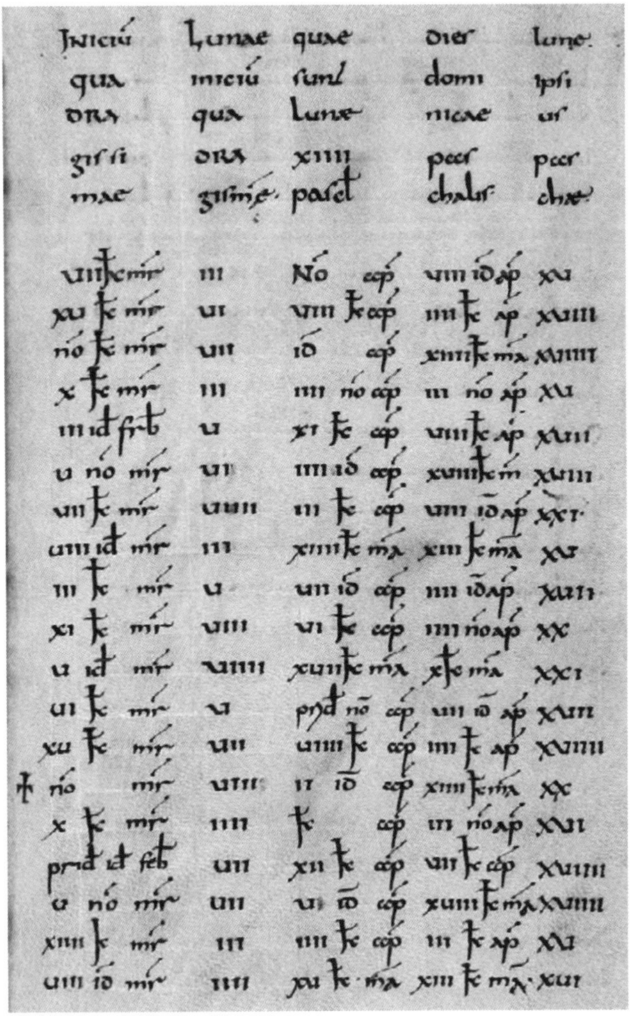

Plate 1 *(Continued)*

Im Wesentlichen korrekt wurde das Mondalter bei Beginn der Fastenzeit angegeben: jeweils 12 weniger als das Mondalter des Ostersonntages (6 Wochen = 42 Tage = ein Mondmonat und 12

euenerit, non obseruat praescripta loca terminorum propter bissextum in luna quem februario addi debere diximus, et ob hoc in sequentem diem transferendus est, ut concordet cum termino paschali in feria. Verbi gratia, in presenti anno terminus quadragesimalae, qui conscriptus est in undecimo kalendas martias, translatus est in decimo kalendas martias propter bissextum.

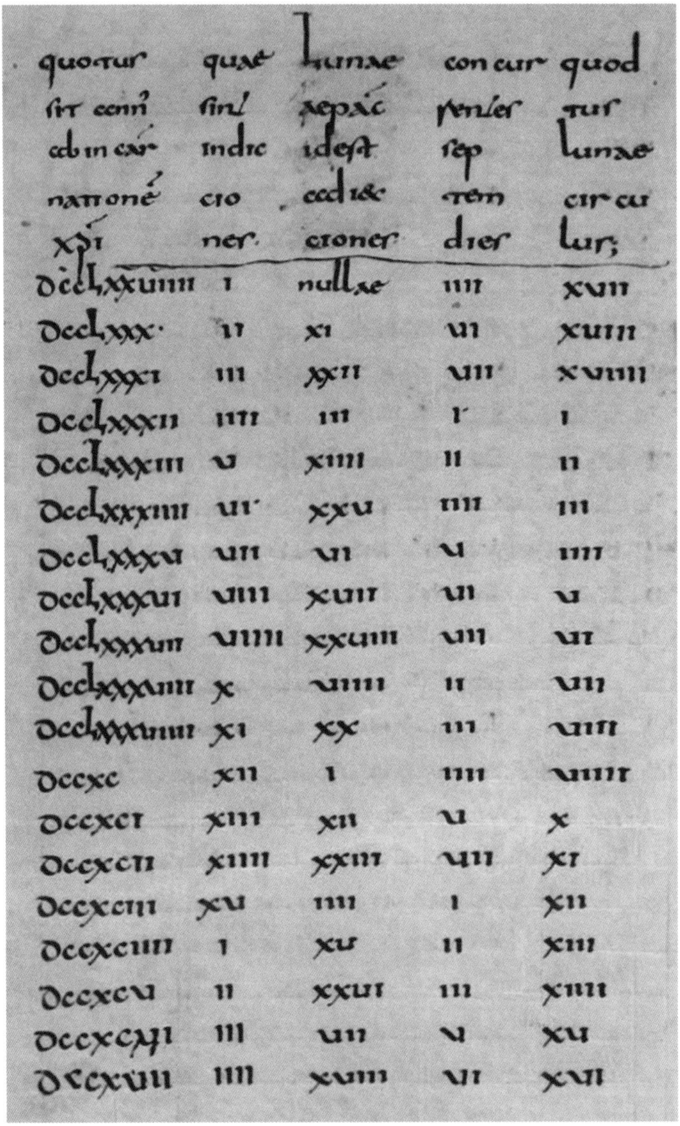

Plate 2 Faksimile der Ostertafel pp. 116–117. *(Continued)*

Tage). Es ist unabhängig vom Schalttag, da in Sonnenschaltjahren der Februarmondmonat statt mit 29 mit 30 Tagen gezählt wurde, um eine Verschiebung zu vermeiden. Die *initia quadragesimae* sind

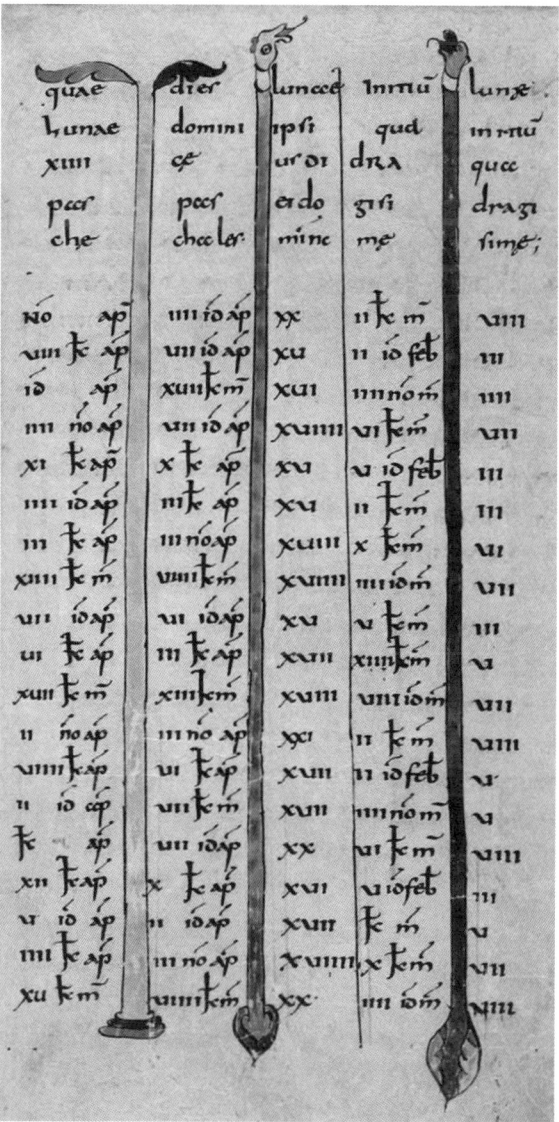

Plate 2 *(Continued)*

schon in der 84jährigen Ostertafel aus dem fünften Jahrhundert, welche die Iren noch bis ins achte Jahrhundert hinein benutzten, mit julianischem Kalenderdatum und Mondalter angegeben. Vielleicht war seitdem die Verzeichnung der *initia quadragesimae* in Ostertafeln eine irische Tradition?[18]

Wie häufig in Ostertafeln haben sich einige (Abschreib-?)Fehler eingeschlichen, für den Zyklus 779–797 deutlich mehr als für 760–778.[19] Außerdem gibt der Schreiber die Indiktion für 779 mit I statt II an, obwohl bereits bei 778 korrekt I stand, und setzt den Fehler bis in die letzte Zeile fort. Noch bemerkenswerter ist jedoch, daß der Schreiber im rechten Teil des Zyklus 779–797 auf p 117 unterhalb der Spaltenüberschriften eine Leerzeile eingefügt hat, so daß sich die Zeilen im rechten Teil um eine verschieben, was die Benutzung der Tafel erschwert haben dürfte (siehe *Plate 2*). Vielleicht wurde der erste Zyklus von einer älteren Tafel abgeschrieben und der zweite berechnet?

Nicht auf einen bestimmten Zeitraum bezogen sondern allgemeingültig ist die folgende Tafel der Ostersonntage der irofränkischen Tradition (ein Faksimile ist hier gedruckt als *Plate 3*; in moderner Notation ist diese Tafel wiedergegeben in *Figure 1*). Es handelt sich praktisch um eine ewige Ostertafel, der man ausgehend vom Wochentag des Ostervollmondes das Datum und das Mondalter des Ostersonntages entnehmen kann.

Genauer erklärt enthält die Tafel in der vierten Spalte die 19 Daten der Ostervollmonde, in den darauf folgenden sieben Spalten die entsprechenden Daten der Ostersonntage, die vom Wochentag des Ostervollmonds abhängen. Über diesen sind die Wochentage des Ostervollmonds Sonntag bis Samstag (*feria I* bis *VII*) sowie noch darüber

---

[18] Diese Ostertafel (MS Padua, Biblioteca Antoniana, I 27, 76r–77v) wurde entdeckt von Dáibhí Ó Cróinín (siehe Mc Carthy und Ó Cróinín (1987–8), 227–42) und überzeugend rekonstruiert von Mc Carthy (1996), 285–320; ein Faksimile findet sich in Warntjes (2007), 80–2.

Auch in MS Vatican, Biblioteca Apostolica, Pal. Lat. 1448, 71v gibt es eine bedanliche Ostertafel für 779–797. Diese enthält allerdings keine Angaben für die Fastenzeit. Sie ist von hoher Qualität und hat nur einen Fehler bei einem Mondalter für den Ostersonntag (im Jahr 18 steht 18 statt 19). Auf der folgenden Seite enthält diese Handschrift übrigens das Widmungsgedicht für Alkuins heute verlorenen *Libellus annalis*, gedruckt als *Carmen LXXII* in MGH Poetae 1, 294–5; vgl. Springsfeld (2002), 26; dort noch falsche Zählung der Handschriftenseiten.

[19] pp. 114–115: Fehler in Zeile 17 zum Jahr 776, Ostervollmond, VI Id. Ap. statt V. pp. 116–117: in der letzten Zeile steht für 797 statt DCCXCVII nur DCCXVII; weiter fünf Fehler bei den Ostersonntagen: Zeile 1 zum Jahr 779: IIII Id. Ap. statt III Id. Ap., Zeile 2 zum Jahr 780: VII Id. Ap. statt VII Kal. Ap., Zeile 6 zum Jahr 784: III Kal. Ap. statt III Id. Ap., Zeile 12 zum Jahr 790: III Non. Ap. statt III Id. Ap., Zeile 14 zum Jahr 792: VII Kal. Mai. statt XVII Kal. Mai., sowie ein falsches Fastendatum in Zeile 13 zum Jahr 791: II Id. Feb. statt Id. Feb.; schließlich drei Fehler beim Mondalter zu Beginn der Fastenzeit: Zeile 11 zum Jahr 789: VII statt VI, Zeile 12 zum Jahr 790: VIII statt VIIII, Zeile 16 zum Jahr 794: III statt IIII.

das entsprechende Mondalter des Ostersonntags 21 bis 15 (*luna XXI* bis *XV*) notiert. In der letzten Zeile stehen die Mondalter für den Beginn der Fastenzeit, die immer 12 weniger als die Mondalter der Ostersonntage betragen, also 9 bis 3 (*luna VIIII* bis *III*). Findet der Ostervollmond an einem Sonntag (*feria I*) statt, hat der Ostersonntag eine Woche später das Mondalter 14+7=21 (*luna XXI*), der Sonntag zu Beginn der Fastenzeit das Mondalter 21−12=9 (*luna VIIII*); für einen Montag (*feria II*) ergeben sich die Mondalter 20 und 8 (*luna XX* und *VIII*), für einen Dienstag (*feria III*) 19 und 7 (*luna XVIIII* und *VII*) usw. bis sich für einen Samstag (*feria VII*) Mondalter 15 und 3 (*luna XV* und *III*) ergeben. Man benutzt die Tafel wie folgt: Im 1. Jahr des 19jährigen Zyklus findet z.B. der Ostervollmond am 5. April statt. Ist dieser Tag ein Sonntag, gilt die nächste Spalte mit *feria I* und der Ostersonntag ist am 12. April. Ist er ein Montag, gilt die Spalte darauf mit *feria II* und der Ostersonntag ist am 11. April; ist er ein Dienstag, fällt der Ostersonntag auf den 10. April usw. bis zum Samstag für den Ostervollmond und dem korrespondierenden Ostersonntag am 6. April.

Im Jahr 773 (dem möglichen Abfassungsjahr der Handschrift) mußte man, wollte man nicht das Datum in der Ostertafel ablesen, es in der 6. Spalte für *feria II* = Montag suchen, da der Ostervollmond am Montag, dem 12. April, stattfand und Ostern deswegen am 18. April mit dem Mondalter 20 gefeiert werden mußte.

Die Tafel enthält außerdem noch in der ersten Spalte die Konkurrenten, in der zweiten die Abfolge der Mondschalt- und -gemeinjahre übereinstimmend mit Beda und in der letzten die bedanischen Epakten. Die dritte Spalte enthält spezielle Regularen für den Monat, in dem der Ostervollmond stattfindet, d.h. entweder 4 für März oder 7 für April. Diese März-/April-Regularen, genannt *regulares minores*, dienen der Bestimmung des Wochentages des Ostervollmondes. Man addiert sie zusammen mit der aktuellen Konkurrente zum Monatstag des Ostervollmondes, zieht anschließend noch Vielfache von 7 ab und erhält so den Wochentag des Ostervollmondes und schließlich durch Nachsehen in der entsprechenden Spalte das Datum des Ostersonntages.

In der ersten Zeile der Tabelle sind noch weitere Regularen für die Ostermonate März und April notiert: sie betragen 36 und 35 und heißen *regulares maiores*. Mit ihrer Hilfe kann man das Datum des Ostervollmondes bestimmen. Dazu muss man zunächst wissen, in welchem Monat der Ostervollmond stattfindet. Das erkennt man an den Epakten. Betragen diese mehr als 15 oder weniger als 5, findet er im April statt, sonst im

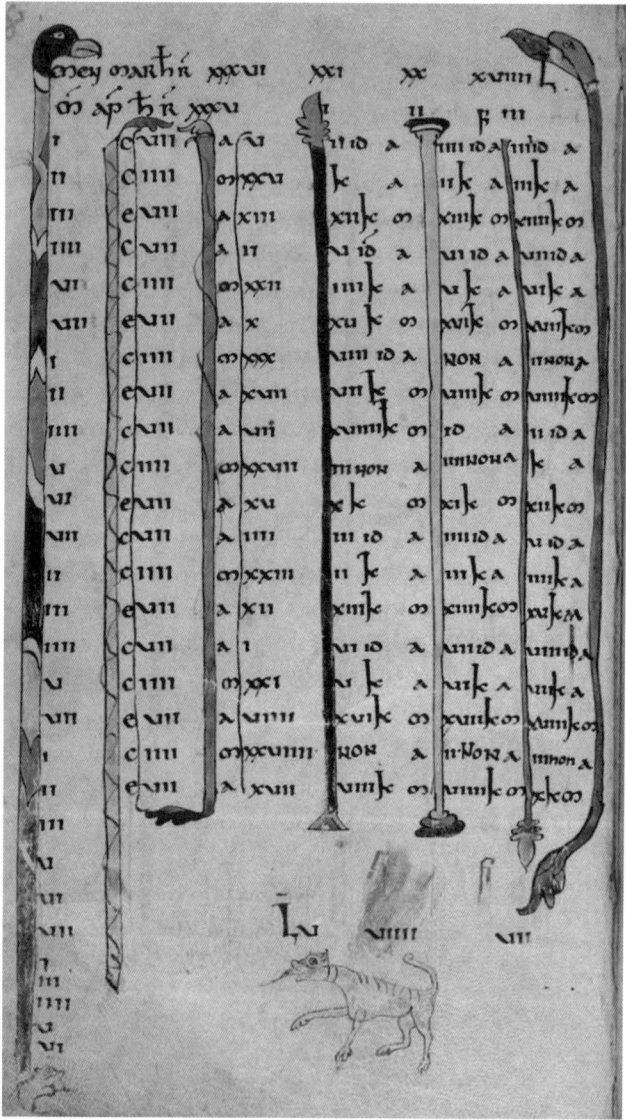

Plate 3 Faksimile der Tafel der Ostersonntage pp. 118–119. *(Continued)*

März. Man subtrahiert dann von den Regularen die Epakten, um den Monatstag des Ostervollmondes zu erhalten. Z.B. beträgt im ersten Jahr des 19jährigen Zyklus die Epakte 0 bzw. 30, d.h. der Ostervollmond findet im April statt; man rechnet 35−30=5 und erhält den 5. April. Im zweiten

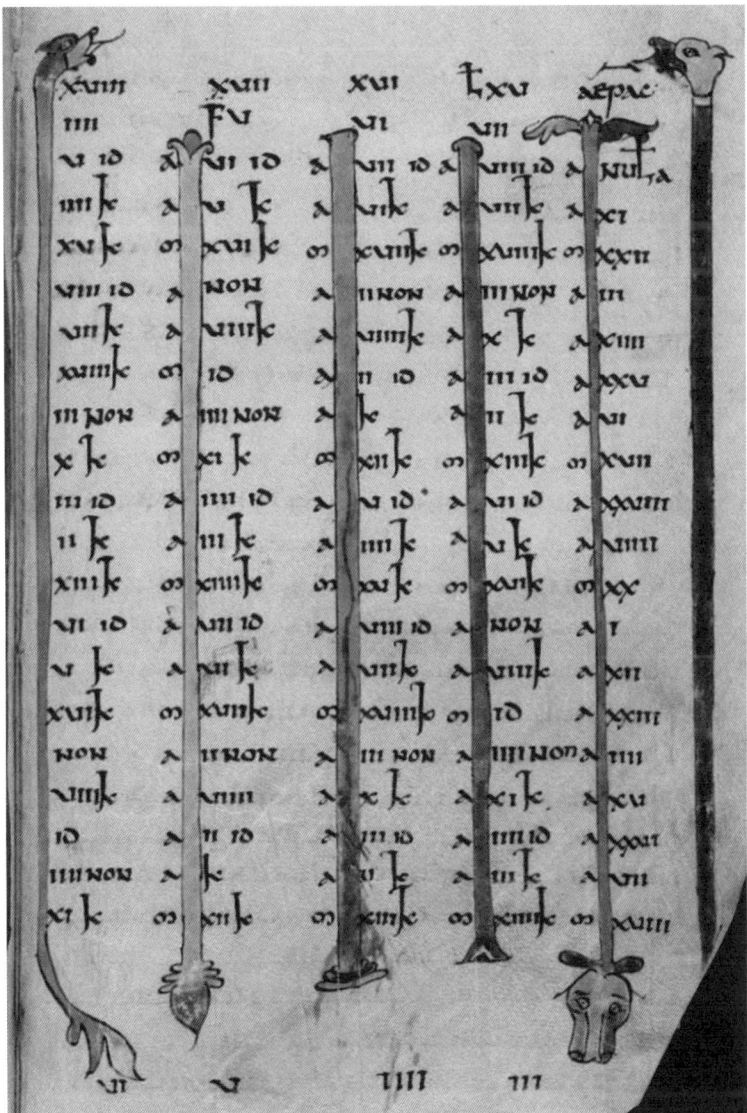

Plate 3 *(Continued)*

Jahr beträgt die Epakte 11, also liegt der Ostervollmond im März, man rechnet 36−11=25 und erhält so den 25. März und so weiter.[20]

[20] Zur Berechnung Springsfeld (2002), 176–80, 183; dort allerdings ein Druckfehler, es muß heißen:
  Epakte $\leq$ 5 oder Epakte $>$ 15 $<=>$ Ostervollmond im April,
    5 $<$ Epakte $\leq$ 15 $<=>$ Ostervollmond im März.

COMPUTUS AND ITS CULTURAL CONTEXT IN THE LATIN WEST, AD 300–1200

März - Regularen	36	luna XXI	XX	XVIIII	XVIII	XVII	XVI	XV	Epakten
April - Regularen	35	I (So.)	II (Mo.)	III (Di.)	IIII (Mi.)	V (Do.)	VI (Fr.)	VII (Sa.)	
I C 7	April 5	12. April	11. April	10. April	9. April	8. April	7. April	6. April	nulla
II C 4	März 25	1. April	31. März	30. März	29. März	28. März	27. März	26. März	11
III E 7	April 13	20. April	19. April	18. April	17. April	16. April	15. April	14. April	22
IIII C 7	April 2	9. April	8. April	7. April	6. April	5. April	4. April	3. April	3
VI C 4	März 22	29. März	28. März	27. März	26. März	25. März	24. März	23. März	14
VII E 7	April 10	17. April	16. April	15. April	14. April	13. April	12. April	11. April	25
I C 4	März 30	6. April	5. April	4. April	3. April	2. April	1. April	31. März	6
II E 7	April 18	25. April	24. April	23. April	22. April	21. April	20. April	19. April	17
III C 7	April 7	14. April	13. April	12. April	11. April	10. April	9. April	8. April	28
V C 4	März 27	3. April	2. April	1. April	31. März	30. März	29. März	28. März	9
VI E 7	April 15	22. April	21. April	20. April	19. April	18. April	17. April	16. April	20
VII C 7	April 4	11. April	10. April	9. April	8. April	7. April	6. April	5. April	1
II C 4	März 24	31. März	30. März	29. März	28. März	27. März	26. März	25. März	12
III E 7	April 12	19. April	18. April	17. April	16. April	15. April	14. April	13. April	23
IIII C 7	April 1	8. April	7. April	6. April	5. April	4. April	3. April	2. April	4
V C 4	April 9	28. März	27. März	26. März	25. März	24. März	23. März	22. März	15
VII E 7	April 9	16. April	15. April	14. April	13. April	12. April	11. April	10. April	26
I C 4	März 29	5. April	4. April	3. April	2. April	1. April	31. März	30. März	7
II E 7	April 17	24. April	23. April	22. April	21. April	20. April	19. April	18. April	18
III									
V									
VI									
VII	luna	VIIII	VII	VI	V	IIII	III		
I									
III									
IIII									
V									
VI									

Figure 1 Tafel der Ostersonntage pp. 118–119.

Diese Berechnung, bei der man den Monatstag des Ostervollmondes im März oder April erhält, erklärt vielleicht die Tatsache, daß die Daten der Ostervollmonde 'modern' notiert sind und nicht wie sonst im Mittelalter üblich in der römischen Schreibweise, d.h. statt '2. Iden des April' steht *A V*, also 5. April. Dies gilt aber nicht für die Osterdaten, die wieder römisch notiert sind. Diese konsekutive Zählweise der Monatstage (von 1 bis 30/31) erscheint zwar selten in komputistischen Tabellen und Texten, ist aber nicht gänzlich ungewöhnlich; bspw. wird sie mit Vorliebe von Dicuil benutzt, in seinem zwischen 814 und 816 geschriebenen astronomischen Werk.[21] Ein früheres Zeugnis aus Gallien ist die Tafel von Périgueux mit fünf 19jährigen Zyklen, die 631 beginnen.[22]

Auch diese Tafel enthält Schreibfehler, allerdings nur wenige.[23]

Die Tafel der Ostersonntage abhängig vom Wochentag des Ostervollmondes taucht in Form eines Kreisdiagramms auch im Zusammenhang mit dem Kölner Lehrbuch von 805[24] auf, und ähnliche Tafeln enthalten auch die Enzyklopädien zur Zeitrechnung von 809 und 818.[25] Bei ihnen fehlt allerdings der Bezug zu den speziellen März-/April-Regularen. Diese Regularen entstammen der irofränkisch Osterberechnung. Ihr Gebrauch ist in der sogenannten *Calculatio Albini magistri*[26] erklärt. Sie stammt wahrscheinlich von Alkuin, obwohl nur eine der zahlreichen Handschriften, die diesen Text enthält, ihn als Verfasser angibt. Es handelt sich um eine im Frankenreich vorgenommene Bearbeitung aus dem Jahr 776 von irischem Material, das Beda übergangen

[21] Esposito (1907).

[22] Krusch (1884), 129–30.

[23] Bei den Ostervollmonden für das Jahr 8 fehlt ein I, es müßte A XVIII statt XVII heißen, ebenso für das Jahr 13 M XXIII statt XXIIII. Dafür ist in der Sonntags-Spalte für Jahr 9 ein I zuviel bei den XVIIII Kal. Ap., die gibt es nicht, gemeint sind die XVIII. Kalenden. Die erste Zeile hat zwei falsche Einträge, von denen der zweite verbessert ist: in der Montags-Spalte steht IIII Id. Ap. statt III Id. Ap. und in der Dienstags-Spalte ist über der römischen 3 von III Id. Ap. ein weiteres I zum korrekten Datum IIII Id. Ap. ergänzt; danach geht es richtig weiter mit V Id. Ap.

[24] Das Kölner Lehrbuch von 805 nur in MS Köln, Diözesan- und Dombibliothek, 83² (Köln 798 und 805), 45r–58v, ediert in Borst (2006), 891–950. Berühmt ist diese Handschrift wegen dreizehn astronomisch-komputistischer Diagramme, von denen hier das erste auf f 79v gemeint ist, abgebildet in van Euw (1998b), 137. Vgl. die elektronische Version der Handschrift in der Virtuellen Handschriftenbibliothek der Erzbischöflichen Diözesan- und Dombibliothek Köln (http://www.ceec.uni-koeln.de).

[25] 7-Bücher-Computus oder *Libri computi* I 10 und 11, bzw. 3-Bücher-Computus oder *Liber calculationis* II: eine Tafel für die Daten der Ostersonntage abhängig vom Wochentag des 1. Januar, die andere vom Wochentag des 1. März; ediert in Borst (2006), 1132–45, 1394. Siehe auch Springsfeld (2002), 189.

[26] Ediert in PL 101, 999–1002. Text und Übersetzung in Springsfeld (2002), 322–8; vgl. auch op. cit. 80–8 und 171–94.

hatte: das Dionysius Exiguus zugeschriebene *Argumentum XIV*,[27] verfaßt für das Jahr 532, dem Beginn der dionysischen Ostertafel. Die *Calculatio* war aufgrund der einfachen Berechnungsmethode des Osterfestes äußerst beliebt und ist u.a. auch in der Aachener Enzyklopädie der Zeitrechnung (*Libri computi*) enthalten.[28]

Kapitel VII beinhaltet eine sogenannte Körperuhr oder Wanderuhr.[29] Das Kapitel verzeichnet stereotyp für jeden Monat und jede Tagesstunde die Schattenlängen, die der menschliche Körper abhängig von Jahres- und Tageszeit wirft. Der eigene Körper dient als Schattenstab (Gnomon) bei dieser einfachen und sehr groben Form der Zeitbestimmung nach dem Vorbild des Palladius.

Dieser lateinische Agrarschriftsteller (4. Jh.) gibt in seinem berühmten landwirtschaftlichen Lehrbuch *De agricultura*,[30] gegliedert in zwölf Bücher für die 12 Monate des Jahres, jeweils zu Beginn jedes Buches für jede Tagesstunde die Schattenlängen in Fuß (*pedes*) an, so daß man durch Messen des eigenen Körperschattens mit dem eigenen Fuß die ungefähre Uhrzeit ermitteln konnte. Seine Werte für die Schattenlängen sind für den Winter auf Alexandria, für den Sommer auf Rom bezogen, das Verhältnis von Körperlänge zu Fußlänge beträgt 6:1 (siehe *Figure 2*).

		Jan/ Dez	Feb/ Nov	Mär/ Okt	Apr/ Sep	Mai/ Aug	Jun/ Jul
hora I und XI	*pedes*	29	27	25	24	23	22
hora II und X	*pedes*	19	17	15	14	13	12
hora III und IX	*pedes*	15	13	11	10	9	8
hora IV und VIII	*pedes*	12	10	8	7	6	5
hora V und VII	*pedes*	10	8	6	5	4	3
hora VI	*pedes*	9	7	5	4	3	2

Figure 2 'Körperuhr' des Palladius – Längen des Körperschattens.

[27] Ediert in Krusch (1938), 78–9. Nach Borst und Jones sind die *Argumenta XI–XVI* im Jahr 675 von einem irischen Komputisten eingeschoben (Jones (1943), 106–7; Borst (1998), 177–8). Zur Meinung, der Einschub sei früher und im westlichen Mittelmeerraum verfaßt, siehe den Beitrag von Immo Warntjes in diesem Tagungsband.

[28] *Annalis libellus* oder Veroneser Jahrbüchlein 19 und 7-Bücher-Computus oder *Libri computi* IIII 19; ediert in Borst (2006), 704–11. Kopien des 9. Jahrhunderts enthalten u.a. MS St. Gallen, Stiftsbibliothek, 184 und 397. Vgl. http://www.e-codices.unifr.ch/de/csg/0184 bzw. 0397.

[29] Springsfeld (2002), 255–6.

[30] Ediert in Rodgers (1975).

		Jan/ Dez	Feb/ Nov	Mär/ Okt	Apr/ Sep	Mai/ Aug	Jun/ Jul
hora I und XI	*pedes*	32	30	28	26	24	22
hora II und X	*pedes*	21	19	17	15	13	11
hora III und IX	*pedes*	19	17	15	13	11	9
hora IV und VIII	*pedes*	17	15	14	11	9	7
hora V und VII	*pedes*	15	13	11	9	7	5
hora VI	*pedes*	13	11	9	7	5	3

Figure 3 'Körperuhr' des 7-Bücher-Computus (*Libri computi*) – Längen des Körperschattens.

		Jan/ Dez	Feb/ Nov	Mär/ Okt	Apr/ Sep	Mai/ Aug	Jun/ Jul
hora I und XI	*pedes*	29	27	25	23	21	19
hora II und X	*pedes*	19	17	15	13	11	9
hora III und IX	*pedes*	17	15	13	11	9	7
hora IV und VIII	*pedes*	15	13	11	9	7	5
hora V und VII	*pedes*	13	11	9	7	5	3
hora VI	*pedes*	11	9	7	5	3	1

Figure 4 'Körperuhr' in St. Gallen, Stiftsbibliothek, 225, p 120 – Längen des Körperschattens.

Körperuhren sind in Handschriften ab 800 recht verbreitet.[31] Allerdings sind die Schattenlängen nicht immer gleich, sondern wurden auf andere Standorte umgerechnet, wahrscheinlich auf den geographischen Breitengrad des jeweiligen Bearbeiters. Wahrscheinlich auf Aachen umgerechnet ist die Körperuhr der Enzyklopädie zur Zeitrechnung von 809 (*Libri computi*; siehe *Figure 3*).[32]

---

[31] Dazu Jones (1939), 87, der angibt, Zinner (1925), 399 nenne mit den Nummern 1167–1195 neunundzwanzig Handschriften, davon fünf mit Diagrammen. Jones ergänzt noch die teilweise autographische Handschrift Abbos von Fleury, MS Berlin, Staatsbibliothek, Phillipps 1833 (Fleury, um 1000), 38v, deren Kreisfigur die gleichen Schattenlängen wie der Text aus St. Gallen (siehe unten) enthält. Auch die *Annalis Libellus*-Handschrift London, British Library, Royal 13 A XI (England um 1100) enthält fols 13v–14r zwei Texte sowie eine Kreisfigur, von denen der erste Text die mathematische Konstruktion der Körperuhr erklärt, der zweite Text identisch mit dem aus St. Gallen ist und auch die Kreisfigur die gleichen Schattenlängen enthält. Mein Dank gilt Dr. Eric Graff, der mir diese Handschriftenseiten zugänglich machte.

[32] 7-Bücher-Computus, Einschub c in MS Madrid, Biblioteca Nacional, 3307 (Murbach etwa 820), 50v; ediert als *Libri computi* IIII 28a in Borst (2006), 1232–4.

Den wieder anderen Schattenlängen unserer Handschrift entsprechen die eines Diagramms im Zusammenhang mit dem Kölner Lehrbuch zur Zeitrechnung von 805.[33]

Vergleicht man die drei Tabellen so fällt auf, daß die umgerechneten Werte systematischer aussehen als die des Palladius. Sie sind vermutlich bei der Umrechnung vereinfacht worden. Dies ging sicher auf Kosten der Genauigkeit, machte aber die Körperuhr leichter memorierbar und gestaltete sie damit anwendungsfreundlicher.

Kapitel VIII über den Aufbau des römischen Kalenders mit Kalenden, Nonen und Iden gehört zum komputistischen Standardmaterial.[34] Zunächst wird für die Monate die Anzahl der Tage genannt und angegeben, wieviele Monatstage jeweils vor den Nonen, Iden und Kalenden liegen. Dann werden die Tage für das ganze Jahr durchgezählt und angegeben, wieviele Jahrestage jeweils vor den Stichtagen liegen: z.B. an den Kalenden des Februar 32, an denen des März 60 usw. bis zu den Kalenden des Januar 366.

Die Multiplikationstafeln, die sich anschließen, sollten das Rechnen mit Wochentagen und Mondmonaten im Kalenderjahr erleichtern, deswegen geht die Tafel für die Sieben bis $50 \times 7 = 350$ und die für Neunundfünfzig bis $6 \times 59 = 354$, was einem Mondjahr mit $12 \times 29,5$ Tagen entspricht.

Ebenfalls zum komputistischen Standardmaterial gehört der erste Teil von Kapitel IX über die Länge des Jahres: 12 Monate, 52 Wochen und ein Tag bzw. gegebenenfalls wegen des Sonnenschalttages zwei Tage, 365 Tage bzw. im vierten (Sonnenschalt-)Jahr 366 Tage, 8760 Stunden (365×24) und 35040 Punkte bzw. Viertelstunden (8760×4). Der Schalttermin 24. Februar ist der seit Julius Cäsar übliche. Er steht bei Dionysius und Beda und auch schon bei dem spanischen Gelehrten Isidor von Sevilla.

Kein Allgemeingut sondern eine irofränkische Eigenheit[35] ist die daran anschließende Begründung für den Sonnenschalttag: Es wird die Zahl der Stunden im Jahr 8760 durch 7 geteilt, ergibt 1251, Rest 3. Diese drei

---

[33] MS Köln, Diözesan- und Dombibliothek, 83² (Köln 798 und 805), 84r, das elfte der dreizehn astronomisch-komputistischen Diagramme, abgebildet in van Euw (1998b), 145 und http://www.ceec.uni-koeln.de. Die gleichen Werte enthält auch ein halbkreisförmiges Diagramm gedruckt in PL 90, 953–5.

[34] Z.B. enthalten im 7-Bücher-Computus (*Libri computi*) als Kapitel II 14 und 18, ediert in Borst (2006), 1162–3, 1170 mit Verweis auf die Rheinische Anleitung von 760/792 (*Lectiones computi*) I 11, ediert in Borst (2006), 554; Vorlage für 7-Bücher-Computus (*Libri computi*) II 18 war Beda, *De temporum ratione* 22 (CCSL 123B, 351–2); Jones vermerkt als Bedas Quelle die anatolische Tafel aus *De ratione paschali* 10 (erste Edition von Krusch (1880), 323–4; neue Edition von Mc Carthy und Breen (2003), 50, 98).

[35] Springsfeld (2002), 204–5.

Stunden würden den Schalttag bewirken. Denn je drei Stunden im 1., 2., 3. und 4. Jahr ergäben zusammen 12 Stunden, die den Schalttag bildeten.

Obwohl bei den 8760 Stunden im Jahr der Tag noch mit 24 Stunden gerechnet wird, hat der Schalttag offensichtlich nur 12 Stunden, die Nacht wird also nicht mitgezählt. Diese Rechnung hatte schon Beda scharf kritisiert. Er schreibt in Kapitel 39 seines Hauptwerks zur Zeitrechnung *De temporum ratione*, sogar dem gemeinen Volk sei bekannt, daß 'der Tag mit seiner Nacht' 24 Stunden habe.[36]

Schon das pseudo-dionysische *Argumentum XVI* rechnet den Tag mal mit 12, mal mit 24 Stunden, genau wie der fränkische Traktat *De bissexto*, in dem um 790 irisches Material bearbeitet wurde. Der merowingische *Computus* von 727 und das fränkische Lehrbuch von 737 rechnen ebenfalls mit drei Stunden pro Jahr für den Schalttag.[37]

Die für den mittelalterlichen Komputisten schwierige Division durch 7 wird in Kapitel IX unserer Handschrift nicht genauer ausgeführt. Dafür wäre allerdings auch eine weitergehendere Multiplikationstafel als die aus dem vorangegangenen Kapitel nötig, wie sie sicher der Verfasser des Traktats *De bissexto* bei seiner Rechnung benutzt hat: Er subtrahiert von 8760 nacheinander 7000, 700, wieder 700, 280, 63 und 14 und erhält ebenfalls den Rest 3.[38]

Unklar ist, warum überhaupt durch 7 geteilt wird, da die Zahl der Wochentage nichts mit dem Sonnenschalttag zu tun hat. Begründet wird dies im *Argumentum XVI* mit der Genesis: Gott schuf die Welt in sechs Tagen, am siebten aber ruhte er. Deshalb sei vollständig einsehbar (?), daß man die Stunden des Tages durch 7 teilen müsse, der Rest bewirke den Schalttag. Diese eigentümliche Erklärung des Schalttages findet sich auch in irischen komputistischen Handbüchern, nicht in *De ratione conputandi*, aber im Münchener Computus, verfaßt 718, und in dem von Immo Warntjes neu entdeckten *Computus Einsidlensis*.[39]

---

[36] Beda, *De temporum ratione* 39 (CCSL 123B, 252).

[37] *Argumentum XVI* in Krusch (1938), 80. Es wird mühsam vorgerechnet, daß 365 Tage 4380 Stunden haben, ebensoviele Stunden außerdem 365 Nächte, also insgesamt 8760 Stunden. Bei Division durch 7 bliebe ein Rest von 3 Stunden, welche in vier Jahren den Schalttag ergäben. *De bissexto* in PL 101, 994C–D; vgl. dazu Springsfeld (2002), 204. Kapitel 10 des merowingischen *Computus* von 727 in Krusch (1938), 55, jetzt neu ediert als Kapitel 14 dieses Textes in Borst (2006), 366–7. Dieselbe 'altfränkische Theorie' habe Krusch in dem alten Berner Victorius-Codex (MS Bern, Burgerbibliothek, 645) f 50r gefunden.

[38] Springsfeld (2002), 204; Springsfeld (2003), 226.

[39] Münchener Computus in MS München, Bayerische Staatsbibliothek, Clm 14456 (Regensburg 810–823), 8r–46r: Division durch 7 fols 22v–23r. *Computus Einsidlensis* in MS Einsiedeln, Stiftsbibliothek, 321 (647), 82–125: Division durch 7 p 108; zu diesem Text siehe Warntjes (2005), 61–4; Bisagni und Warntjes (2008).

In Kapitel X geht es um das Weltjahr. Die Juden hätten beim Auszug aus Ägypten 3689[40] Jahre nach Erschaffung der Welt durch das Töten eines Lammes am Freitag 25. März bei Vollmond (*luna XIV*) das erste Passahfest gefeiert. Bis zu Christi Passion am Freitag 25. März bei Vollmond seien 5228 Jahre vergangen. Der Osterzyklus des Victorius sei von diesem Termin ausgehend für 532 Jahre berechnet worden.

Hier wird leicht vereinfacht Victorius von Aquitanien wiedergegeben. Ihm hatte der Archidiakon Hilarus im Jahr 457 im Namen der Kurie den Auftrag gegeben, den Osterzyklus neu zu berechnen. In der Vorrede zu seiner Ostertafel berechnet Victorius die Jahre seit Erschaffung der Welt auf 5658. Dem Kirchenvater Hieronymus folgend setzt er von Adam bis zur Sintflut 2242 Jahre an und von der Sintflut bis Abraham 942 Jahre. Während Hieronymus dann aber von Abraham bis Christi Geburt noch 2015 Jahre zählt, also auf das Weltjahr 5199 für die Geburt des Erlösers kommt, rechnet Victorius bis in seine Gegenwart im Jahr 457 noch 2474 Jahre dazu, kommt also auf das Weltjahr 5658 und somit für Christi Geburt auf das Weltjahr 5201. Seine Weltjahresepoche weicht also von der des Hieronymus um zwei ab.[41]

Victorius schreibt etwas genauer, die Juden seien im Jahr 3689 nach Erschaffung der Welt am Abend des 24. März, ein Donnerstag mit *luna XIII*, durch den Tod des Lammes gerettet worden. Am nächsten Tag, Freitag mit *luna XIV*, zu Beginn des Jahres 3690 hätten sie das erste Passahfest begangen, das man niemals am Ende eines Jahres feiere. Christi Passion habe 5228 Jahre nach Erschaffung der Welt stattgefunden, am 25. März, ein Freitag mit *luna XIV*, zu Beginn des Jahres 5229.

Kapitel XI–XIII zu Indiktion und Konkurrente gehen auf die dionysischen *Argumenta I, II* und *IV* zurück, die 703 auch Beda aufgriff.[42]

Kapitel XI berechnet nun das Inkarnationsjahr mit Hilfe der Indiktionsangabe des 15jährigen römischen Steuerzyklus:

$49 \times 15 = 735$; $735 + 12$ Regularen + aktuelle Indiktionszahl (hier 4) $= annus\ praesens$ (hier 751).

Es folgt noch der Hinweis, daß man nach Ablauf des aktuellen Indiktionszyklus nicht mehr 49 sondern 50×15 rechnen und dazu immer 12 Regularen und die aktuelle Indiktionszahl addieren müsse.

Umgekehrt berechnet Kapitel XII die Indiktion 4 aus dem Inkarnationsjahr 751 durch die Division modulo15: Es sei $15 \times 40 = 600$,

---

[40] *III milia DCXXXVIIII* = 3639 ist ein Schreibfehler, das *L* wurde vergessen.

[41] Victorius von Aquitanien, *Prologus* 7 und 9, ediert in Krusch (1938), 22–3, 24–5.

[42] *Argumentum I–IV* ediert in Krusch (1938), 75–6. Beda, *De temporibus* 14 (CCSL 123C, 59).

[43] Modern notiert: k=(j+[j/4]+4Regularen)mod7.

751−600=151;   15×9=135,   151−135=16;   16+3 Regularen=19, 19−15=4.

Diese beiden Kapitel gehören offensichtlich zusammen, da sie mit dem gleichen *annus praesens* 751 rechnen.

Kapitel XIII nennt die Regel zur Berechnung der Konkurrente, also des Wochentages des 24. März im 28jährigen Sonnenzyklus: Inkarnationsjahr + dem ganzzahligen Anteil bei Division durch 4 + 4 Regularen, schließlich modulo 7.[43] Als Beispieljahr wird als *annus praesens* 743 angegeben, die Konkurrente aber nicht errechnet.[44]

Kapitel XIIII ist ein Isidorzitat über Mondgemein- und -schaltjahr.[45] Das Gemeinjahr habe in 12 Mondmonaten 354 Tage, oft würden zwei Gemeinjahre aufeinander folgen. Das Schaltjahr dagegen stehe immer allein und habe in 13 Monaten 384 Tage. Es sei das Jahr Mose, in dem befohlen worden sei, im zweiten Monat Pascha zu feiern. Der Ausdruck *Embolismus* für Schaltjahr komme aus dem Griechischen, die Lateiner würden es *superaugmentum*, also Zuwachs, nennen. Weil den Gemeinjahren 11 Mondtage (Schreibfehler *XII*) zu fehlen scheinen, habe man die Schalt- und Gemeinjahre erfunden. Wenn vom vorangegangenen Ostervollmond[46] bis zum folgenden 384 Tage vergingen, sei ein Schaltjahr, wenn 354 ein Gemeinjahr. Das Kapitel endet mit einem Halbsatz, der nicht bei Isidor steht: 'wie leicht einleuchtet und die folgende Regel zeigt'.[47] Da aber auf das Kapitel keine 'Regel' zu Schalt- und Gemeinjahren folgt, ist hier offensichtlich aus einer Handschrift abgeschrieben worden, die danach noch etwas zu Schalt- und Gemeinjahren enthielt, etwa deren Abfolge im 19jährigen Zyklus oder eine Ostertafel mit einer entsprechenden Spalte.[48]

---

[44] Man muß rechnen: $(743+[743/4]+4) \mod 7 = (743+185+4) \mod 7 = 932 \mod 7 = 1$; der 24. März des Jahres 743 war in der Tat ein Sonntag.

[45] Isidor von Sevilla, *Etymologiae* VI 17, 21–24 (ediert in Lindsay (1911)).

[46] Die Handschrift schreibt *luna XV* wie Dionysius, bei Isidor steht aber *luna XIIII*. MS St. Gallen, Stiftsbibliothek, 225, 125 (vgl. http://www.e-codices.unifr.ch/de/csg/0255): *Si enim a XV luna paschae precedentis usque ad XIIII sequentes CCCLXXXIIII dies fuerint, embolismus annus est; si CCCLIIII, communis annus est.* Dionysius Exiguus, *Epistola ad Bonifacium et Bonum* (ediert in Krusch (1938), 84): *A quinta decima luna paschalis festi anni, verbi gratia, praecedentis usque ad XIIII sequentis paschae, si communis annus est, CCCLIIII dies habebit: si embolismus, CCCLXXXIIII [dies]. [...] A XV itaque luna praeteriti festi, usque ad decimam quartam praesentis, quot sunt dies, diligentius inquiramus.* Isidor von Sevilla, *Etymologiae* VI 17, 24 (Lindsay (1911)): *Si enim a quarta decima luna paschae praecedentis usque ad quartam decimam sequentis CCCLXXXIV dies fuerint, embolismus annus est; si CCCLIV, communis [est].*

[47] MS St. Gallen, Stiftsbibliothek, 225, 125 (vgl. http://www.e-codices.unifr.ch/de/csg/0255): *Ut autem facilius eluceat subiecta regula demonstrat.*

[48] Wie z.B. Tafel auf p 118 unserer Handschrift, Spalte 2. Auch in der Ostertafel in Isidor von Sevilla, *Etymologiae* VI 17, 5–9 (Lindsay (1911)) sind sie angegeben. Beda, *De*

Ein anderes Thema behandelt Kapitel XV. Um das Passionsjahr[49] zu erhalten, müsse man vom Inkarnationsjahr 27 subtrahieren. Auch Victorius hatte als Jahr der Passion Christi das Weltjahr 5228 (also Ära 5200+28=5228) bzw. das Inkarnationsjahr 28 genannt, gleichzeitig das Jahr 1 seiner Ostertafel.[50]

Im nächsten Satz des Kapitels wird allerdings das Weltjahr aus dem Inkarnationsjahr 743 auf das Jahr 5942 berechnet: $15 \times 395^{51} = 5925$; 5925+6 Regularen+Indiktionszahl 11 (die übrigens zum Jahr 743 paßt)=5942. Hier liegt also die im Mittelalter verbreitete und bei Iren beliebte Ära des Hieronymus zugrunde (5199+743=5942), obwohl sonst alles zu Victorius paßt.[52]

Da in Kapitel XV und XIII der gleiche *annus praesens* 743 benutzt wird, ist das Isidorexzerpt aus Kapitel XIIII wohl eingeschoben worden.

An die Weltjahrberechnung schließt sich ohne Überschrift ein Auszug aus Isidors Naturgeschichte über die Jahreszeiten an:[53] Der Frühling bestehe aus Feuchtigkeit und Feuer, der Sommer aus Feuer und Trockenheit, der Herbst aus Trockenheit und Kälte, der Winter aus Kälte und Feuchtigkeit. Deswegen hätten die Jahreszeiten ihre Namen als Mischung aus ihren gemeinsamen Eigenschaften erhalten, wie die folgende Figur zeige. Es folgt tatsächlich eine Jahreszeitenfigur,[54] welche die Kapitelnummer XVI trägt (siehe *Plate 4*). Sie besteht aus einem Kreis mit vier größeren und vier kleineren, einander überschneidenden, im Kreisinneren liegenden Halbkreisen. In den größeren Halbkreisen stehen im oberen die Wörter 'Frühling' und 'Osten', im rechten 'Sommer' und 'Süden', im unteren 'Herbst' und 'Westen' und im linken 'Winter' und 'Norden'. In den kleineren, die jeweils zwei der größeren schneiden, stehen die Wettereigenschaften: im Frühlings- und im

---

temporum ratione 45 (CCSL 123B, 420) verzeichnet ihre Abfolge im Text. Weiteres Beispiel: *Annalis libellus* oder Veroneser Jahrbüchlein 14 zu Schalt- und Gemeinjahren und den Epakten (Springsfeld (2002), 341–2; Borst (2006), 696–8).

[49] Das Passionsjahr enthalten nach Cordoliani (1955b), 315 in der St. Galler Stiftsbibliothek nur die Handschriften 251 und 902, 225 erwähnt er nicht. Vgl. http://www.e-codices.unifr.ch/de/csg/0251 bzw. 0902.

[50] Victorius von Aquitanien, *Prologus* 9 und *Cyclos s.a. I* (Krusch (1938), 24, 27).

[51] MS St. Gallen, Stiftsbibliothek, 225, 125, Schreibfehler *CCCLXXXV*. Vgl. http://www.e-codices.unifr.ch/de/csg/0225.

[52] Der 7-Bücher-Computus (*Libri compoti*) II 1 (Borst (2006), 1145–6) verwendet für die Weltjahrberechnung die bedanische Ära 3952: $(317 \times 15) + 4 +$ Indiktion 2=4761, entspricht Inkarnationsjahr 809; siehe dazu auch Springsfeld (2002), 188.

[53] Isidor von Sevilla, *De natura rerum* VII 4–5 (Fontaine (1960), 201–3).

[54] Vgl. eine ähnliche Figur abgedruckt zu Isidor, *De natura rerum* VII 4 in Fontaine (1960), 202 (*bis*).

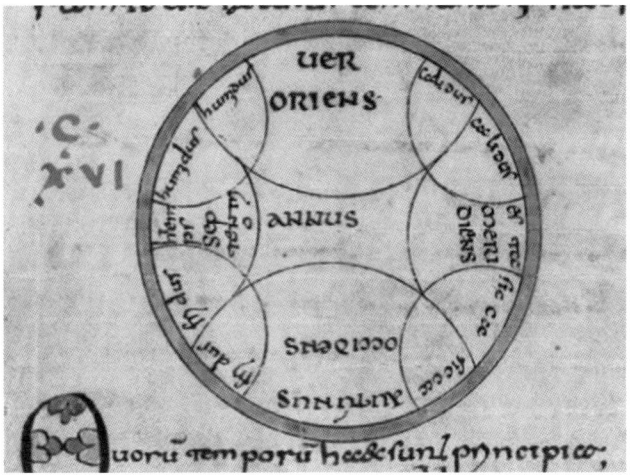

Plate 4 Faksimile der Jahreszeitenfigur p 126.

Sommerhalbkreis jeweils 'warm', im Sommer- und im Herbsthalbkreis 'trocken', im Herbst- und im Winterhalbkreis 'kalt' und im Winter- und Frühlingshalbkreis 'feucht'.

Es folgen die Daten für den Beginn und die Dauer der Jahreszeiten: Der Frühling beginne am 22. Februar und dauere 91 Tage, Sommer 24. Mai 91 Tage, Herbst 23. August 93 Tage,[55] Winter 25. November 90 Tage. Von da an wiederholten sich die 365 Tage des Jahres. Dies sei die Verschiedenheit der Jahreszeiten gemäß der Natur.

In den nachfolgenden Sätzen wird dann noch eine allegorische Deutung der Jahreszeiten Winter, Sommer und Frühling gegeben. Der Winter stehe für die Mühsale der Welt, der Sommer für die Verfolgung des Glaubens, der Frühling aber mit dem Osterfest für die Erneuerung des Glaubens und den Frieden der Kirche.

Kapitel XVII behandelt die kirchlichen Feste in Anlehnung an Isidor.[56] Zunächst geht es um das Jubeljahr oder Erlaßjahr, das alle 50 Jahre begangen wird (Lev 25:10). Dann wird unterschieden zwischen dem *Pascha* der Juden, gefeiert in Erinnerung an die Befreiung aus der ägyptischen Gefangenschaft, und dem christlichen Osterfest. *Pentecoste*, wie das Fest bei den Juden heiße, stamme vom griechischen Wort *pente* = fünf ab. Das Jubeljahr zeige den Weg in die ewige Ruhe der Erholung an. Das jüdische Fest der *Scenopegia* sei von den Griechen als Weihe des

---

[55] Nach *LX* unleserlich, bei Isidor steht *LXIII* = 93.
[56] Isidor von Sevilla, *Etymologiae* VI 18, 1–17 (Lindsay (1911)).

Tabernakels interpretiert worden. Es werde im Monat September gefeiert. *Enceniae* sei die Weihe des neuen Tempels und werde im Oktober gefeiert. Griechisch *Epiphania* bedeute Erscheinen und sei der Tag der Taufe des Herrn und des Verwandelns von Wasser in Wein (bei der Hochzeit in Kanaa).[57] *Noeminia* feiere man an den Kalenden eines Monats, bei den Hebräern aber und den Griechen, die nach dem Mondkalender rechneten, bei Neumond. *Parascae* diene der Vorbereitung des Sabbats, den der Herr befohlen habe von Arbeiten frei zu halten.

Kapitel XVIII über die griechischen Namen und Kapitel XVIIII über die verschiedenen Stimmen der Lebewesen haben gar nichts mehr mit Komputistik zu tun, dafür aber die letzte Tabelle mit der Kapitelnummer XX, die den komputistischen Teil der Handschrift beschließt. Sie enthält die Mondalter der Monatsersten im 19jährigen Zyklus und damit die Grundlage der gesamten Osterberechnung. Solche Mondaltertafeln finden sich praktisch in jeder komputistischen Handschrift.[58] In unserer Handschrift steht diese Tafel auf Vor- und Rückseite, was sehr unpraktisch ist, weil man blättern muß (Faksimiles sind hier gedruckt als *Plate 5, Figure 5* gibt sie in moderner Notation wieder).

Die Tafel erklärt sich wie folgt. Ausgehend von einem Neumond am 24. Dezember erhält man durch Weiterzählen für den 1. Januar das Mondalter 9. Danach zählt man diesen Mondmonat, einen vollen mit 30 Tagen, bis zu seinem Ende am 22. Januar. Man erhält den nächsten Neumond am 23. Januar und für den 1. Februar das Mondalter 10. Dieser Mondmonat ist ein hohler mit 29 Tagen, er endet also am 20. Februar mit dem Mondalter 29. Der nächste Neumond findet am 21. Februar statt, der 1. März hat dann das Mondalter 9. Dieser volle Mondmonat endet am 22. März, Neumond ist am 23. März. Der 1. April hat dann das Mondalter 10, usw. Das erste Jahr endet am 31. Dezember mit dem Mondalter 19 (Neumond am 13. Dezember), so daß der 1. Januar des 2. Jahres des 19jährigen Zyklus das Mondalter 20 hat. So kommt man schließlich zeilenweise bis zum 31. Dezember des 19. Jahres mit dem Mondalter 8.

Noch einfacher erscheint die Konstruktion der Tafel, wenn man innerhalb einer Spalte bleibt, also beispielsweise nur den Januar betrachtet. Man beginnt im 1. Jahr mit dem Mondalter 9, im 2. Jahr erhält man das Mondalter 20, weil sich die Mondphasen jedes Jahr um 11 Tage verschieben, das Mondalter also pro Jahr um 11 steigt. Im 3. Jahr müßte

---

[57] Taufe des Herrn heute als eigenes Fest am Sonntag nach Epiphanie.
[58] Als Beispiel die Mondaltertafel des 7-Bücher-Computus (*Libri computi*) IIII 2 in MS Madrid, Biblioteca Nacional, 3307, 40r (ediert in Borst (2006), 1198–1200).

der 1. Januar eigentlich das Mondalter 31 haben, da aber ein Mondmonat höchstens 30 Tage hat, muß man 30 abziehen: also 31−30=1, es war also Neumond. Im 4. Jahr hat der 1. Januar das Alter 12, usw. bis ins 19. Jahr des 19jährigen Zyklus, dann hat der 1. Januar das Mondalter 27. Vom 19. zum 1. Zyklusjahr kommt man schließlich durch Addition von 12 wegen des *saltus lunae* (Mondsprung), dem Ausgleichstag zwischen 19 Mondjahren mit 235 Mondmonaten bzw. 6936 Tagen und 19 Sonnenjahren mit 6935 Tagen. Man rechnet also in jedem Jahr spaltenweise das Mondalter + 11 und zieht gegebenenfalls einen vollen Mondmonat ab.

Die St. Galler Tafel enthält allerdings einige Abweichungen von dieser Regel, und zwar die in der Tafel mit (!) gekennzeichneten Zahlen: im November des 7. Jahres steht nicht (13+11=)24 sondern 23, im Juli des 8. Jahres nicht (19+11=)30 sondern 29 und im Dezember des 19. Jahres nicht (25+11=36, 36−30=)6 sondern 7. Außerdem wurde im Januar des 8. Jahres die 27 zur 26 verbessert, es folgt im 9. Jahr ein offensichtlicher Schreibfehler (*IIII* statt *VII*); weiter wurde im August des 19. Jahres die 3 zur 2 verbessert.

Wodurch entstehen die drei Abweichungen, die (wahrscheinlich) keine Schreibfehler sind? Beda erklärt 725 seine Mondalterrechnung so, dass sich fünf Ausnahmen von der Regel '+11' ergeben. Sie sind in *Figure 6* fettgedruckt. Vier dieser Ausnahmen entstehen aufgrund der Lage der Schaltmonate in den Jahren 8, 11 und 19:

1. Der 1. Mai im 8. Zyklusjahr hat nicht das Mondalter (17+11=)28 sondern 27.
2. Der 1. Juli im 8. Zyklusjahr nicht (19+11=)30 sondern 29.
3. Der 1. März im 11. Zyklusjahr nicht (18+11=)29 sondern 28.
4. Der 1. Mai im 19. Zyklusjahr nicht (18+11=)29 sondern 28.

Die fünfte Ausnahme im Jahr 19 ergibt sich wegen des Mondsprunges:

5. Der 1. Dezember im 19. Zyklusjahr nicht (25+11=36, 36−30=)6, sondern 7.[59]

---

[59] Beda, *De temporum ratione* 20 (CCSL 123B, 347–8). Weil in den vier anderen Mondschaltjahren der Schaltmonat jeweils am 2. Monatstag beginnt, tritt der Ausgleich schon im folgenden Monat ein. Eine Kontrolle ergibt sich durch Bedas Angabe der Neumonddaten der sieben 30tägigen Schaltmonate in Kapitel 45 (CCSL 123B, 422): 1. im 2. Zyklusjahr am 2. Dezember; 2. im 5. Zyklusjahr am 2. September; 3. im 8. Zyklusjahr am 6. März; 4. im 10. Zyklusjahr am 4. Dezember; 5. im 13. Zyklusjahr am 2. November; 6. im 16. Zyklusjahr am 2. August; 7. im 19. Zyklusjahr am 5. März. Vgl. Springsfeld (2002), 131–2.

Plate 5 Faksimile der Mondaltertafel pp. 133–134. (*Continued*)

Die St. Galler Tafel stimmt mit Beda in den Ausnahmen Nr. 2 und 5 überein. Die fehlende Ausnahme Nr. 3 für das Jahr 11 ist wahrscheinlich wieder ein Schreib- oder Verständnisfehler, da beim Mondalter 30 für den 1. Februar nur das Mondalter 28 für den 1. März in Frage kommt. Anders für das 8. Zyklusjahr: Da die St. Galler Tafel im Jahr 8 am 1. Mai noch das Mondalter 28 hat (Beda

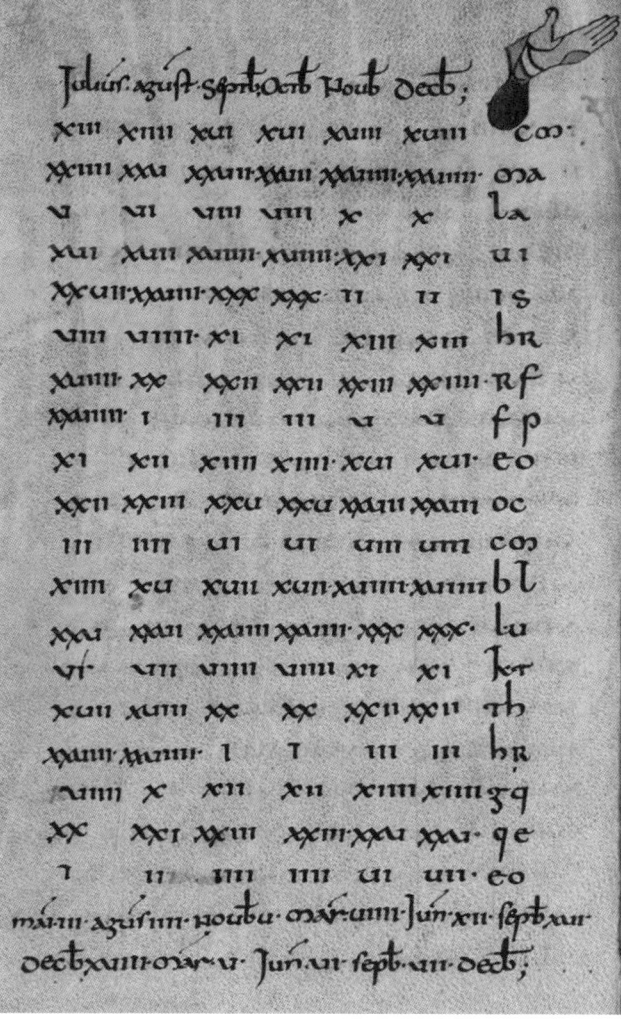

Plate 5 (*Continued*)

erst 27, Ausnahme Nr. 1), am 1. Juli aber 29 wie Beda (Ausnahme Nr. 2), beginnt der Schaltmonat hier wohl einen Monat später (4. April) als bei Beda (6. März).

Das Mondalter 23 für den 1. November im Jahr 7 muß wieder ein Fehler sein (richtig 24), da der Mondschaltmonat, wie eben gezeigt, erst im folgenden Jahr, nicht schon in diesem eingeschaltet wurde.

	Jan	Feb	Mär	Apr	Mai	Jun	Jul	Aug	Sep	Okt	Nov	Dez
1	9	10	9	10	11	12	13	14	16	16	18	18
2	20	21	20	21	22	23	24	25	27	27	29	29
3	1	2	1	2	3	4	5	6	8	8	10	10
4	12	13	12	13	14	15	16	17	19	19	21	21
5	23	24	23	24	25	26	27	28	30	30	2	2
6	4	5	4	5	6	7	8	9	11	11	13	13
7	15	16	15	16	17	18	19	20	22	22	23 (!)	24
8	26	27	26	27	28	29	29 (!)	1	3	3	5	5
9	4 (!)	8	7	8	9	10	11	12	14	14	16	16
10	18	19	18	19	20	21	22	23	25	25	27	27
11	29	30	29	30	1	2	3	4	6	6	8	8
12	10	11	10	11	12	13	14	15	17	17	19	19
13	21	22	21	22	23	24	25	26	28	28	30	30
14	2	3	2	3	4	5	6	7	9	9	11	11
15	13	14	13	14	15	16	17	18	20	20	22	22
16	24	25	24	25	26	27	28	29	1	1	3	3
17	5	6	5	6	7	8	9	10	12	12	14	14
18	16	17	16	17	18	19	20	21	23	23	25	25
19	27	28	27	28	29	30	1	2	4	4	6	7 (!)

Figure 5 Mondalter der Monatsersten im 19jährigen Zyklus nach St. Gallen, Stiftsbibliothek, 250, pp. 133–134.

	Jan	Feb	Mär	Apr	Mai	Jun	Jul	Aug	Sep	Okt	Nov	Dez
1	9	10	9	10	11	12	13	14	16	16	18	18
2	20	21	20	21	22	23	24	25	27	27	29	29
3	1	2	1	2	3	4	5	6	8	8	10	10
4	12	13	12	13	14	15	16	17	19	19	21	21
5	23	24	23	24	25	26	27	28	30	30	2	2
6	4	5	4	5	6	7	8	9	11	11	13	13
7	15	16	15	16	17	18	19	20	22	22	24	24
8	26	27	26	27	**27**	**29**	**29**	1	3	3	5	5
9	7	8	7	8	9	10	11	12	14	14	16	16
10	18	19	18	19	20	21	22	23	25	25	27	27
11	29	30	**28**	30	1	2	3	4	6	6	8	8
12	10	11	10	11	12	13	14	15	17	17	19	19
13	21	22	21	22	23	24	25	26	28	28	30	30
14	2	3	2	3	4	5	6	7	9	9	11	11
15	13	14	13	14	15	16	17	18	20	20	22	22
16	24	25	24	25	26	27	28	29	1	1	3	3
17	5	6	5	6	7	8	9	10	12	12	14	14
18	16	17	16	17	18	19	20	21	23	23	25	25
19	27	28	27	28	**28**	30	1	2	4	4	6	7

Figure 6 Mondalter der Monatsersten im 19jährigen Zyklus nach Beda.

Das Fehlen von Bedas Ausnahme Nr. 4 läßt sich wieder durch die andere Lage des Mondschaltmonats erklären: Im 19. Zyklusjahr legt Beda den Mondschaltmonat vom 5. März bis zum 3. April, unsere Handschrift aber erst vom 3. Mai bis zum 1. Juni, wie es später auch der Ire Dicuil in seiner Mondaltertafel macht.[60]

Sinnvoll erklären lassen sich also zwei Abweichungen der St. Galler Handschrift zu Beda, nämlich die im 8. Zyklusjahr (unsere Handschrift schiebt den zusätzlichen Monat erst am 4. April statt am 6. März ein), und die im 19. Zyklusjahr (erst am 3. Mai statt am 5. März). Aber auch in diesen beiden Fällen halte ich einen Fehler für wahrscheinlicher, weil es üblich war, die Schaltmonate vor dem Osterfest einzufügen.[61]

Daß St. Gallen auch mit Beda in der Ausnahme Nr. 5 übereinstimmt (Mondsprung am 24. November, also 1. Dezember Mondalter 7 statt 6), ist umso bemerkenswerter, als es für den Mondsprung drei Termine gab: 1. der 22. März, der in der irischen Komputistik des späten siebten und frühen achten Jahrhunderts befolgt wird; 2. einen alexandrinischen am 30. Juli; 3. einen römischen am 24. November. Darüber wurde einige Jahre nach Entstehung dieser Handschrift zur Zeit Karls des Großen sogar heftig gestritten, und zwar zwischen Alkuin, dem wichtigsten Berater Karls in Dingen der Wissenschaft, Schule und Kirche, und anderen, namentlich nicht bekannten Beratern Karls. Alkuin nannte sie 'ägyptische Knaben', weil sie der alexandrinischen Mondalterberechnung folgten, er dagegen die römische Tradition vertrat.[62] Durch Verzeichnen des Mondsprunges am 24. November und den römischen Mondaltern für August bis November 2, 4, 4, 6, statt der alexandrinischen 3, 5, 5, 7 (für den August wurde sogar – wie oben erwähnt – das alexandrinische Mondalter 3 in das römische Mondalter 2 radiert) folgt die St. Galler Tafel also der römischen Tradition Alkuins.

Die letzten beiden Spalten der Tafel enthalten außerdem noch Mondbuchstaben, die ebenfalls zur Berechnung des Mondalters verwendet wurden. Ihren Gebrauch erklärt Beda, der sie nicht erfunden

---

[60] In Dicuils Mondaltertafel (*Liber de astronomia* I 7; Esposito (1907), 391) ist für das Jahr 19 in der einzigen bekannten Handschrift Valenciennes, Bibliothèque Municipale, 404 (386), 73 (online einsehbar unter: http://bookline-03.valenciennes.fr/bib/fondsvirtuels/ microfilms/accueil.asp) das Mondalter 28 am 1. Mai zu 29 korrigiert und über dem Mondalter 30 am 1. Juni ausdrücklich *emb.* für den Mondschaltmonat verzeichnet.

[61] Gegenbeispiel dazu Dicuil, *Liber de astronomia* I 7; siehe vorige Anm.

[62] Springsfeld (2002), 38–61, 142–7.

hat, aber als erster erklärt.⁶³ Zwei synodische Monate werden zu 59 Tagen zusammengefaßt und mit der dreifachen Buchstabenreihe A–U, A.–U., .A–.T kodiert. Mit dieser Folge werden die Jahrestage im Kalender belegt. Eine Tafel ordnete jedem Mondalter einen Buchstaben für jedes Jahr des 19jährigen Zyklus zu.

In unserer Handschrift fehlt allerdings diese zugehörige Tafel. Hinzukommt, daß zwar die Buchstaben richtig wiedergegeben sind, aber die Punktierungen fehlen. So waren diese beiden Spalten mit den Mondbuchstaben also eigentlich nicht nutzbar. Dies kam aber in komputistischen Handschriften häufiger vor.⁶⁴

Unter der Tafel stehen auf beiden Seiten jeweils noch zwei Zeilen. In diesen werden, allerdings merkwürdig gruppiert, erst noch einmal die Mondregularen aus Zeile 1 der Tafel angegeben, außerdem aber noch die Sonnenregularen der Monatsersten.

Die Sonnenregularen betragen von Januar bis Dezember 2, 5, 5, 1, 3, 6, 1, 4, 7, 2, 5, 7. Sie geben den Wochentagsunterschied der Monatsersten zum 24. März, dem Sitz der Konkurrenten, an. Addiert man sie zur aktuellen Konkurrente des 28jährigen Sonnenzyklus, erhält man den Wochentag der Monatsersten.⁶⁵

Die Reihenfolge in unserer Handschrift ist allerdings merkwürdig: Auf p 133 stehen unten erst die Mondregularen für Januar, April, Juli und Oktober 9, 10, 13 und 16, danach ohne Kennzeichnung für die gleichen Monate die Sonnenregularen 2, 1, 1, 2. Dann werden für Februar, Mai, August und November erst wieder die Mondregularen 10, 11, 14 und 18 angegeben, danach noch der Februar genannt, allerdings ohne die Sonnenregulare 5. Dann geht es auf p 134 unten weiter mit den Sonnenregularen für Mai, August und November 3, 4 und 5. Schließlich folgt das gleiche für März, Juni, September und Dezember, erst die Mondregularen 9, 12, 16 und 17, dann die Sonnenregularen 5, 6, 7 und die für Dezember 7 fehlt wieder.

---

⁶³ Beda, *De temporum ratione* 23 mit Tafel (CCSL 123B, 353–5). Ein Mondalteralphabet schon bei Philocalus, eine Verbesserung vielleicht gegen Ende des 7. Jahrhunderts durch den gelehrten Mönch Aldhelm von Sherborne, vielleicht aber auch in jenem Jahrbüchlein angelsächsischer Mönche um 680, von dem Bedas Kirchengeschichte erzählt, dann wahrscheinlich in der Vorlage für Willibrords Kalender um 685, jedenfalls um 705 in dessen Echternacher Kopie. Borst (1998), 405–6.

⁶⁴ Z.B. MS Köln, Diözesan- und Dombibliothek, 103, 2r, ebenfalls ohne Punktierung. Vgl. http://www.ceec.uni-koeln.de.

⁶⁵ Vgl. Beda, *De temporum ratione* 21 (CCSL 123B, 349–50).

Es scheint, daß die Regularen hier vierteljahresweise zusammengefaßt worden sind, wobei der Sinn dieser Gruppierung unklar ist.

Diese letzte Tafel beschließt den komputistisch-naturwissenschaftlichen Teil. Wie gezeigt wurde, enthält diese St. Galler Handschrift von 773 viel komputistisches Standardmaterial, wie z.B. diese Mondaltertafel, aber auch Osterzyklen, Regeln zur Indiktion und Konkurrente, Bemerkungen zum Aufbau des Julianischen Kalenders, der Jahreslänge, dem Sitz des Sonnenschalttags und den Mondschalt- und -gemeinjahren, Multiplikationstafeln, die Jahreszeitenfigur und die zugehörigen Bemerkungen nach Isidor, außerdem eine Körperuhr nach Palladius. Seltener in komputistischen Handschriften vertreten sind die Bemerkungen zum Weltjahr nach Victorius. Die Tafel der Ostersonntage mit den Angaben der speziellen März-/April-Regularen verweist auf die irische Tradition. Die *Calculatio*, in der ein Komputist (Alkuin?) im Frankenreich 776 diese irische Osterberechnungsweise aufgreift und erklärt, war zwar weit verbreitet, dies nicht zuletzt, da sie Teil der Aachener Enzyklopädie der Zeitrechnung von 809 war, mir sind jedoch nur Tafeln der Ostersonntage ohne Verweis auf die März-/April-Regularen bekannt, nämlich die der Aachener Enzyklopädie. Ebenso in irofränkischer Tradition steht die Begründung des Schalttages durch die Division der Stunden des Jahres durch 7 und die Rechnung des Schalttages mit 12 Stunden. Noch einmal betont werden soll die Tatsache, daß die Mondalterberechnung fast identisch mit der Bedas ist und daß sie durch den Mondsprung am 24. November der römischen Tradition folgt, der Tradition, die Alkuin im Streit mit den 'ägyptischen Knaben' 797/798 vertritt und die er im Frankenreich zu verbreiten suchte.

Die Handschrift 225 bezeugt also für das Kloster St. Gallen im letzten Drittel des 8. Jahrhunderts gleichzeitig komputistisches Standardmaterial ähnlich den Regeln Bedas sowie Elemente der irischen Osterberechnungsweise.

ANHANG: INHALTSVERZEICHNIS DER HANDSCHRIFT ST. GALLEN, STIFTSBIBLIOTHEK, 225

Paginae	Inhalt
2–31	Hauptregister über den ganzen Band und 21 Spezialindices über die einzelnen bis p 439 reichenden Teile
32	leer
33–56	Auszug aus Isidors *Differentiae*
56–61	*Quibus modis retenetur homo bono trib. opere.* Text ähnlich zu Isidors *Differentiae*
62–113	Isidors *Allegoriae*
114–134	Komputistik und anderes
114–117	Kap. VI Osterzyklen 760–778 und 779–797
118–119	Tafel der Ostersonntage nach der *Calculatio Albini magistri*
120	als Kap. VII *Horologium*, d.h. Schattenlängen der Tagesstunden bzw. Körperuhr des Palladius
120–121	Kap. VIII *Incipit compotus*. Über die Längen der Kalendermonate und die jeweilgen Stichtage der Kalenden, Nonen und Iden, danach Multiplikationstafeln für 7 und 59
122	Kap. VIIII *De bissexto*. Zur Jahreslänge mit einer irofränkischen Begründung des Sonnenschalttages
122	Kap. X zum Weltjahr
122–123	Kap. XI berechnet das Inkarnationsjahr 751 aus der Indiktion
123–124	Kap. XII berechnet die Indiktion aus dem Inkarnationsjahr 751
124	Kap. XIII *Concurrentes septimane dies qualiter invenire*. Zur Berechnung der Konkurrenten, Bsp. 743
124–125	Kap. XIIII *De communibus et embolismis*. Isidorzitat zu Mondgemein- und -schaltjahr
125–126	Kap. XV zum Passionsjahr, Bsp. 743
126–127	daran anschließend Isidorzitat aus der Naturgeschichte inkl. Jahreszeitenfigur mit der Kapitelnummer XVI und anschließen-

	den allegorische Deutungen zu den Jahreszeiten
127–129	Kap. XVII *De sollemnitatibus*. Zu den kirchlichen Festen in Anlehnung an Isidor, *Etymologiae* VI 18, 1–17.
129–132	Kap. XVIII *De grecis nominibus*. Über griechische Namen. In eine Textlücke auf der rechten Blattseite von 129 ist eine Hand gemalt.
132	Kap. XVIIII *De voce varium animantium*. Über Stimmen von Tieren, darunter sind zwei Hände gemalt
133–134	Kap. XX Tabelle der Mondalter der Monatsersten im 19jährigen Zyklus, oben rechts ist eine Hand gemalt.
135	vier Zeilen zur Aufgabe des Lehrers[66]
135–137	Kap. XXI *Incipit tempora per sanitate corporum propter quale observatione debent homini*. Medizinische Anleitungen zur Erhaltung der Gesundheit des Körpers
137–140	Kap. XXII *In nomine domini incipit ordo librorum catholicorum qui ponuntur in anno circulo in ecclesiae romana*.[67]
140	Kap. XXIII *De declinatione nominum*. Über die Deklination von Namen
141–146	Kap. XXIIII *Incipit eiusdem de salmis et hymnis*, überschrieben mit *Hieronimus dixit*.
146–166	Kap. XXV *Inventio crucis*. Über die Kreuzesauffindung[68]
166–249	als Kap. XXVI und XXVII Auszüge aus den *Formulae spititalis intelligentiae* des Eucherius von Lyon
249–384	Eucherius, *Instructiones* (Buch 1 und Anfang von Buch 2)
384–439	Pseudo-Methodius, *Revelationes*, apokalyptischer Text

[66] Ein Pädagoge sei derjenige, dem die Jugendlichen zugewiesen werden. Der Name stamme aus dem Griechischen und bedeute, 'der die Jungen führt', d.h. lehre und den Mutwillen dieses Alters zügele.

[67] Liturgischer Text zum Kirchenjahr ähnlich zum *Ordo Romanus* (Kollektion B) in MS Köln, Diözesan- und Dombibliothek, 138, 1v–40v. Vgl. http://www.ceec.uni-koeln.de.

439–461	Exzerpte aus Isidors *Differentiae* (Buch II, Kap. 4–16)
461–473	*Interrogatio: Homo propter quid dicitur. Responsio: Homo dictus ab humo.* Ende: *in die factus est mundus.* Katechismus nach Isidors *Differentiae* von späterer Hand.
473–478	Ebenso von späterer Hand die Lebensgeschichte der Zürcher Stadtheiligen Felix und Regula

[68] Holder (1889).

BRIGITTE ENGLISCH

KAROLINGISCHE REFORMKALENDER UND DIE FIXIERUNG DER CHRISTLICHEN ZEITRECHNUNG

Abstract

The Carolingian renaissance is generally regarded as formative in many areas, one of them being computistical studies, while Alcuin appears to be the major force behind the reforms of that time. From this perspective it is surprising that his treatises dealing with the *bissextus* and the *saltus lunae* do not show much innovation. One possible reason for this is the discrepancy between the Julian calendar and certain visible celestial phenomena, of which especially the spring equinox and the full moon were of immediate relevance for the calculation of Easter. In this context solar or lunar eclipses could potentially lead to a severe crisis, since it would in some cases contradict this calculation. Yet, the science of that epoch was not advanced enough to solve this problem; following Charlemagne's demand for a *norma rectitudinis*, activity in this field was restricted to the recording of these celestial phenomena, as well as reiteration of and references to the conventional computistical texts, which did not exceed the knowledge transmitted through Bede's works. Therefore, the Carolingian renaissance should be regarded as formative in computistical studies for the reason that Bede's works were made authoritative to a degree that differences between theory and reality were not solved before the Gregorian calendar reform.

Keywords

Alcuin, Charlemagne, Bede, Victorius of Aquitaine, calendar reform, deviation of the calendar, Lorsch, Easter Sunday, full moon, lunar and solar eclipse, computistic, time-reckoning, *Ratio de luna, De saltu lunae ac bissexto, Annalis libellus, Admonitio generalis*, bissextum, saltus lunae, *Propositiones ad acuendos iuvenes*.

DIE FIXIERUNG DER CHRISTLICHEN ZEITRECHNUNG

Der Angelsachse Alkuin, der vielleicht bedeutendste Träger der Reformen Karls, wird im Mittelpunkt meines Aufsatzes stehen. Er kam im Jahr 782, vermutlich etwa im Alter von 50 Jahren, an dessen Hof[1] und machte dabei nicht nur Karls Bestreben nach der Herstellung eines nach außen wie nach innen funktionierenden Reiches zu seinem eigenen, sondern auf sein Know-how und vielleicht auch auf seinen Mut, pragmatische Entscheidungen zu fällen, ist es zurückzuführen, daß Karl eine der größten Schwierigkeiten seiner Epoche bewältigte, an der seine Vorgänger insgesamt gescheitert waren: Europa eine einheitliche Zeitordnung zu verschaffen.

Umso erstaunlicher ist es, daß Alkuins Bild in der Forschung ausgesprochen ambivalent ist. Zum einen ist man geneigt, seiner Charakteristik durch Karls Biographen Einhard zuzustimmen, der ihn mit Ehrerbietung als *virum doctissimum*[2] bezeichnet. In diesem Zusammenhang wird zumeist auf seine Vorgeschichte hingewiesen, so das Faktum, daß er seine Ausbildung in York, dem bedeutendsten Bildungszentrum Angelsachsens, erhielt und er selbst seit 778 der Kathedralschule und der Bibliothek Yorks vorstand. All dies dürfte Alkuin als Intellektuellen, als sachkundigen Mann ausgewiesen haben, insbesondere in Fragen der Zeitrechnung, da seine Lehrer, Egbert und Aelbert, Schüler Bedas waren,[3] d. h. des Autors, dessen umfassende Lehrschriften zum Osterfest bereits ihm frühen Mittelalter höchste Reputation genossen. Dies macht insbesondere der zuletzt von Dietrich Lohrmann untersuchte Briefwechsel mit Karl dem Großen deutlich,[4] der belegt, daß man am Karlshof bei allen 'kniffligen' Fragen zu Mathematik, Astronomie und Osterfestberechnung bei Alkuin Rat suchte.

Zum anderen ist aber gerade die Komputistik, die christliche Festberechnung, das Gebiet, auf dem Alkuin – zumindest nach modernen Maßstäben – faktisch am wenigsten leistete. Dies beruht weitaus weniger auf der Anzahl als auf dem Inhalt seiner Werke. So lässt sich die Autorschaft Alkuins nach den Forschungen Kerstin Springsfelds zwar nur für die komputistische Schrift *Ratio de luna*[5] tatsächlich

---

[1] Springsfeld (2002), 20.
[2] Einhard, *Vita Karoli magni* 25 (MGH SS rer. Germ. 25, 30): *in ceteris disciplinis Albinum cognomento Alcoinum, item diaconem, de Brittania Saxonici generis hominem, virum undecumque doctissimum, praeceptorem habuit, apud quem et rethoricae et dialecticae, praecipue tamen astronomiae ediscendae plurimum et temporis et laboris inpertivit. Discebat artem conputandi et intentione sagaci siderum cursum curiosissime rimabatur.*
[3] Bieler (1961), 164; Springsfeld (2002), 17.
[4] Lohrmann (1993).
[5] Alkuin, *Ratio de luna* ist ediert in PL 101, 981C–984B.

nachweisen, weitaus folgenreicher ist aber, daß er dort keineswegs originäre Gedanken präsentiert, sondern nur das Wissen Bedas zum Teil derart minutiös aufbereitet, daß weite Passagen auf den ersten Blick lediglich als zweckentfremdete Gedankenspielerei erscheinen. Und selbst wenn man die Alkuin von Arno Borst noch zugeschriebenen Werke *De saltu lunae ac bissexto*[6] und das sogenannte *Annalis libellus*[7] mit in die Überlegungen einbezieht, zeigt sich ein ähnliches Bild, denn Borst selbst beurteilt das eine Werk als mit schneller Feder verfaßte Kompilation wie alle anderen[8] und das andere lediglich als Erläuterung zu Bedas Werk, welches zudem voller Fehler stecke.[9] Möglicherweise könnte dies auch der Grund für Borst gewesen sein, die von ihm angenommene Entwicklung eines abstrakten Kalenderkonstrukts – unter dem nicht ganz zutreffenden Terminus einer karolingischen Kalenderreform publiziert – in das Kloster Lorsch zu verlegen, also als Leistung zu klassifizieren, die sich gänzlich unabhängig von Alkuin vollzogen habe.[10] Allgemein hat sich – angesichts dieser für die Berechtigung von Einhards Werturteil problematischen Quellenlage – die stillschweigende Übereinkunft herausgebildet, Alkuins Leistung im Karlsreich als die eines '*spiritus rector*' der Bildungsreform zu interpretieren.[11] Dessen vornehmlichste Leistung habe darin bestanden, den Lehrkanon zum Quadrivium und zur Komputistik gemäß den karolingischen Bildungsbestrebungen aufzubereiten und für ihre Verbreitung zu sorgen.[12] Einen kleinen Lichtblick in dieser eher rezeptiv geäußerten Gelehrsamkeit bieten allein die Alkuin zugeschriebenen *Propositiones ad acuendos juvenes* (Aufgaben zur Schärfung des Geistes der Jugend),[13] die vielleicht wichtigste

---

[6] Alkuin, *De saltu lunae ac de bissexto* ist ediert in PL 101, 985C–999; zu diesem Werk siehe Springsfeld (2002), 64–79.

[7] Das Werk ist teilweise gedruckt in Monachi ordinis S. Benedicti abbatiae Montis Casini (1873), 80–9, 96; Gesamttext und Übersetzung findet sich in Springsfeld (2002), 329–75; kritisch ediert nun in Borst (2006), 676–772; zur handschriftlichen Überlieferung siehe Springsfeld (2002), 96–104; Borst (2006), 664–75; vgl. dazu Borst (1998), 55–7, der dort Alkuin als Autor postuliert, während er 317–8 richtiger betont, daß der Autor namenlos bleibt.

[8] Borst (1998), 55; vgl. dazu Springsfeld (2002), 64–5.

[9] Borst (1998), 55; vgl. dazu Springsfeld (2002), 104.

[10] Borst (1998), 231–311.

[11] Springsfeld (2002), 23.

[12] Springsfeld (2002), 26.

[13] Die maßgebliche Edition des Textes stammt von Folkerts (1978), der damit die alte Migne-Ausgabe ersetzte. Die kritische Textausgabe wurde nochmals abgedruckt und mit einer Übersetzung versehen in der Ausgabe von Folkerts und Gericke (1993).

Sammlung von Rechen- und Scherzaufgaben in lateinischer Sprache,[14] auf deren Bedeutung auch in komputistischer Hinsicht allerdings hier nur hingewiesen werden kann.[15]

Es stellt sich mithin die Frage, ob sich nicht Gründe dafür aufzeigen lassen, warum ein gemäß dem Urteil der Zeitzeugen wissenschaftlich befähigter Mann wie Alkuin sich darauf beschränkte, die Lehren eines Vordenkers seiner Spezialdisziplin zu verbreiten. Es wird in diesem Zusammenhang zu betrachten sein, welcher Natur die von Karl angestrebten Prämissen waren, denen Alkuin auf dem Gebiet der Bildung und der Zeitrechnung zu genügen hatte und ob sich, läßt man die einzelnen von Alkuin stammenden und von ihm verfaßten Werke Revue passieren, gewissermaßen ein gemeinsamer Nenner als zentrales Thema herauskristallisieren läßt. Darüber hinaus werden wir uns der Frage widmen, in welcher Situation sich die Komputistik im Frankrenreich bei dem Regierungsantritt Karls des Großen befand und ob, neben den rein technischen Schwierigkeiten, auch konkrete Probleme bei der Berechnung des Osterfestes bestanden. Vor diesem Hintergrund wird zu zeigen sein, daß Alkuin im Namen Karls etwas gelang, was für viele Jahrhunderte die Einheit des christlichen Abendlandes sicherte: Europa eine Zeitordnung zu geben.

Das Osterfest war einer der zentralen Krisenpunkte des frühmittelalterlichen Christentums. Daß sich hier so tiefgreifende Auseinandersetzungen entwickelten, hängt damit zusammen, daß das Osterfest, anders als Weihnachten, nicht an einem festen Termin begangen wird, sondern es in den Zeitrahmen zwischen dem 22. März und dem 25. April fallen kann. Die Ursache hierfür war das Bemühen, beim zentralen Fest der Christenheit nicht ein Kalenderdatum zu wiederholen, sondern die astronomisch-kalendarische Situation zu rekonstruieren, zu der die Auferstehung Christi stattfand. Man wollte so die von Gott gegebene Situation der Auferstehung Christi in die Gegenwart transportieren, um alljährlich die Erlösungsgewißheit für jeden einzelnen nachvollziehbar zu gestalten. Dies war insofern problematisch, als das Osterfest gemäß der Evangelien an den jüdischen Mondkalender angebunden war. Diese Ausrichtung spiegelt sich in der auch heute noch geltenden Osterregel, den Gedenktag der Auferstehung am 'ersten Sonntag nach Frühlingsvollmond' zu begehen, die

---

[14] Folkerts (1978), 35; Folkerts und Gericke (1993), 284.
[15] So beschäftigt sich auch die Studie von Singmaster (1998) vornehmlich mit der möglichen Deszendenz und der Entschlüsselung mancher Aufgaben mit modernen mathematischen Methoden.

gemeinhin als Beschluß des Konzils von Nicäa aus dem Jahre 325 angesehen wird.[16]

Damit war aber die Festlegung des Osterfests von der kontinuierlichen Bestimmung des Laufs von Sonne und Mond abhängig. Die hieraus erwachsende Kalenderrechnung, die Komputistik, dominierte als komplexer Aufgabenbereich im Mittelalter einen bedeutsamen Bereich der intellektuellen Aktivität. Ihre Bewältigung war vornehmlich deshalb schwierig, weil man die oben aufgeführte Regel durchaus verschieden interpretieren konnte. So herrschte z.B. Uneinigkeit darüber, wie man den Vollmond definierte, ob er nun am Mondalter 14 oder 15, d. h. am 14. oder 15. Tage nach dem ersten Sichtbarwerden einer feinen Sichel nach Neumond eintrat.[17] Auch der Umfang eines vollen Osterzyklus, dessen Parameter für jedes Jahr in sogenannten Ostertafeln tabellarisch aufgelistet wurden, bereitete Probleme. Die zentrale Eigenschaft dieser Berechnungsweise war nämlich ihre zyklische Struktur, d.h. die Verkettung des 28jährigen Sonnenzyklus und des 19jährigen Mondzyklus. Dies bedeutete, daß das Funktionieren des Systems letztendlich auf der Vollständigkeit eines beide Parameter umfassenden 532jährigen Zyklus beruhte.[18] Diese Zeiteinheit ist aber nicht nur für den einzelnen Menschen außergewöhnlich lang und schwer zu überblicken, sondern läßt auch langfristig keine äußeren Korrekturen zu. Demzufolge hatten sich in der Spätantike und im frühen Mittelalter verschiedene Osterzyklen von 84, 95, 112 und 532 Jahren entwickelt,[19] die zumeist regionale Festtraditionen charakterisierten. Von der Umsetzung der Forderung, daß alle Christen Ostern am selben Tag begehen sollten, war man bis in das späte 8. Jh. noch

---

[16] Zur Nicäaproblematik Englisch (1994), 301–4; Springsfeld (2002), besonders 170–1.

[17] Neulicht, das *novilunium,* meint hier gemäß dem mittelalterlichen Verständnis *luna* 1, also das Sichtbarwerden der ersten feinen Sichel nach Neumond; siehe hierzu z.B. Mayr (1955), 307.

[18] Der Vorteil des erstmals von Dionysius Exiguus im 6. Jh. angewandten 532jährigen Zyklus liegt darin, daß nach Ablauf dieser Zeitspanne der Lauf der Sonne, der 19jährige Zyklus des Mondes, die Epakten des Mondes, die Zahl der Konkurrenten der Sonne, der 14. Mond und der Tag des Ostersonntages wieder in geordneter Form zu ihrem Beginn zurückkehren, das Modell also wahrhaft zyklisch ist; siehe hierzu Dionysius Exiguus, *Ciclus ab incarnatione domini*, in Krusch (1938), 69–81. Populär wurde diese Berechnung aber erst durch Beda Venerabilis, der als erster einen vollständigen Osterfestzyklus erstellte. Zur Osterfestberechnung des Beda siehe Jones (1943); Wallis (1999); sowie Englisch (1994), 280–396.

[19] Siehe hierzu Strobel (1977); Strobel (1984); Huber (1969); Krusch (1880); Krusch (1938); Schmid (1907); Walsh und Ó Cróinín (1988).

weit entfernt. Die ältesten Berechnungen, die auf der Einheit von 532 Jahren beruhten, formulierte Victorius von Aquitanien im 5. Jh., die nach dem Konzil von Orléans von 541 für das Frankenreich bis zur Zeit Karls des Großen verbindlich wurden.[20] Ganz allgemein ist aber festzuhalten, daß dort, wo verschiedene Berechnungen aufeinandertrafen, es oft zu massiven Auseinandersetzungen kam, die zu Zerreißproben für die Einheit des Christentums wurden.

So wurde beispielsweise der Hl. Columban aus Luxeuil vertrieben, weil er nämlich den iroschottischen Modus seiner Heimat in seine gallischen Klostergründungen übertrug. Dabei stieß er aber auf den erbitterten Widerstand der burgundischen Bischöfe, die Ostern nach der Ostertafel des Victorius begingen. Dies bewirkte für einige Osterfeste zwischen 590 und 610 (Aufenthalt Columbans in Gallien) eine Diskrepanz von bis zu drei Wochen zwischen dem Ostertermin des Klosters Luxeuil und dem des Umlandes.[21] Die rigorose Haltung beider Parteien[22] ist zu verstehen, wenn man bedenkt, daß es die Erlösungshoffnung war, die mit der Festlegung von Ostern auf dem Spiel stand. Jede Abweichung von der göttlichen Zeitordnung bedeutete nicht nur ein irdisches Gericht, sondern vor allen Dingen auch eine Bestrafung am Ende aller Tage.

Auch für das Frankenreich und insbesondere für Karl den Großen war die einheitliche und gesicherte Terminierung des Osterfestes von grundlegender Bedeutung. Seine Regierung war innenpolitisch vom ersten Tage an dadurch gekennzeichnet, dem Reich durch konkrete Normsetzungen ein einheitliches und funktionales Gepräge zu verleihen. Dies betraf insbesondere die Bewahrung der allgemeinen Wohlfahrt, die Ausbreitung und Stabilisierung des christlichen Glaubens und die Reorganisation der Priesterschaft, wie er es in einer Vielzahl von Erlassen, den sogenannten Kapitularien, formulierte.[23] Die berühmteste dieser Anweisungen ist die sogenannte *Admonitio generalis*,[24] die 'Allgemeine

---

[20] In den Akten des Konzils von Orléans §1 heißt es (CCSL 148A, 132): *Placuit itaque Deo propitio, ut sanctam pascha secundum laterculum Uictori ab omnibus sacerdotibus uno tempore celebretur.* Siehe hierzu Krusch (1938), 11 und ferner Jones (1934).

[21] Die Auseinandersetzung um das Osterfest prägen eine Vielzahl von Columbans Schriften; siehe hierzu z.B. Brunhölzl (1975), 179; Schmid (1904); Schmid (1907), 69–81.

[22] Diese fand auch Ausdruck in mehreren Briefen an Papst Gregor den Großen, der aber auch keine eindeutige Entscheidung finden konnte. Siehe hierzu z.B. den Brief Columbans an Papst Gregor I., der sich dem korrekten Mondalter am Ostertage widmet (Walker (1957), 2–13).

[23] Hägermann (2000), 287–302; Hägermann (2003), 103–5.

[24] Die *Admonitio generalis* sind gedruckt in MGH LL 1, 53–67; siehe dazu Fleckenstein (1953).

Ermahnung', an deren Abfassung vermutlich auch Alkuin beteiligt war.[25] Diese ist aus mehreren Gründen für unsere Fragestellung aufschlußreich. Zum einen verrät der Eingangssatz einiges von der Herrschaftsauffassung Karls des Großen, wenn dieser sich auf den König Josias des Alten Testaments beruft,[26] der das ihm von Gott überantwortete Reich durch 'Bessern und Ermahnen' zum wahren Gottesdienst zurückzurufen versucht habe.[27] Gerade die Bezugnahme auf Josias ist in diesem Zusammenhang auffällig, da Karl ansonsten, in den abendlichen Runden der Hofgesellschaft, unter dem Pseudonym David firmierte.[28] So war Josias (2 Kgs 22–23) nicht nur derjenige Herrscher, der auf die getreue Einhaltung aller Gesetze drang und die Priesterschaft zur Ordnung mahnte, sondern auch der Herrscher, der dezidiert die korrekte Feier des Passahfestes anordnete, also des jüdischen Festes, dessen bestimmende Parameter auch die Festlegung des Osterfestes prägten.[29]

Exakt dieser Anspruch läßt sich auch in der *Admonitio* nachweisen, welche von der Intention geprägt ist, 'eine gottgefällige Ordnung herzustellen'.[30] Dabei nimmt neben der allgemeinen Bildung in Gestalt des Ausbildungskanons der 7 *Artes liberales*, der sieben freien Künste, vornehmlich der Komputus eine besondere Position ein; es wird sogar die Forderung aufgestellt, jeder Priester müsse in der Lage sein, das Osterfest zu berechnen.[31] Damit wird dem Osterfest im Reich Karls des Großen eine zentrale Position für ein funktionierendes Gemeinwesen zuerkannt. Vermutlich war es Alkuin, der Karl für die Bedeutung des Osterfestes sensibilisierte,[32] d.h., daß Karl seinen Anspruch, im Einklang mit Gottes Gesetzen zu regieren, u.a. nur erfüllen konnte, wenn er über die korrekte Festlegung des Osterfestes wachte. Ein Reich, in dem das höchste Christenfest umstritten war, konnte weder nach innen noch nach außen den Anspruch erheben, von einem starken Herrscher geeint zu sein.

---

[25] Fleckenstein (1953), 37.

[26] *Admonitio generalis, praefatio* (MGH LL 1, 54).

[27] Hägermann (2003), 103–4; Borst (1998), 235.

[28] Fleckenstein (1953), 43.

[29] Auf die Bedeutung der Anspielung auf Josias machte Borst (1998), 235 aufmerksam, wenn er den Bericht auch vom 2 Kgs 22:1–23:30 in ein viertes Buch verlegt.

[30] Hägermann (2000), 288, 656.

[31] So z.B. in *Admonitio generalis* 71 (MGH LL 1, 65); siehe hierzu Fleckenstein (1953), besonders 49–54, 75–85.

[32] Hierauf deutet z.B. der Briefwechsel zwischen Karl und Alkuin hin; siehe zu diesem Themenkomplex auch Springsfeld (2002), 33–61.

Umso seltsamer erscheint die Weisung in der *Admonitio generalis*, daß *jeder* Priester den Komputus beherrschen müsse. Geht man davon aus, daß nach dem Befund der erzählenden Quellen zu Zeiten Karls keine Auseinandersetzung über das Osterfest stattfand und die Forschung allgemein von einer raschen Verbreitung des für die Osterrechnung normsetzenden Werkes des Beda Venerabilis, *De temporum ratione*, aus dem Jahre 725 ausgeht,[33] macht beständiges Rechnen im Grunde wenig Sinn. Beda hatte in seiner permanenten Ostertafel bereits alle Festdaten auf der Grundlage eines 532jährigen Zyklus bis zum Jahr 1064 berechnet;[34] danach brauchte man nur wieder von vorne anzufangen. Dennoch ergeht nicht nur die jeden Geistlichen in die Pflicht nehmende Anweisung zur Kalkulation, sondern in den folgenden Jahren beschäftigten sich auch fast alle 'Intellektuellen' des Karlskreises mit der Komputistik, sei es Alkuins Freund Arn von Salzburg,[35] sei es Karls Vetter Adalhard von Corbie[36] und auch Alkuin selbst,[37] um nur einige zu nennen. Daneben entstehen eine Vielzahl anonymer Sammelhandschriften,[38] die diverse Materialien zur Osterrechnung zusammentragen, zumeist derart, daß die bestehenden Traktate kommentiert, nachgerechnet, gegenübergestellt oder auch nur aneinandergereiht werden.[39] Diese Vielfalt ist, wenn wir davon ausgehen, daß diese Materie aufgrund ihrer theologischen, astronomischen wie auch arithmetischen

---

[33] Bergmann (1985), 16; Pedersen (1978), 307. Borst (1998), 184–5 macht auf das Faktum aufmerksam, daß lediglich acht Fragmente des Werkes aus dem 8. Jh. überlebten, während die meisten Handschriften nach 800 auf dem Kontinent entstanden seien.

[34] Beda Venerabilis' *De temporum ratione* ist herausgegeben von Jones (1943) und in CCSL 123B; es ist ins Englische übersetzt von Wallis (1999).

[35] Der Erzbischof von Salzburg gab u.a. den Anstoß zur Erstellung des sogenannten 3-Bücher-Computus; siehe dazu Springsfeld (2002), 31.

[36] Kasten (1986).

[37] Siehe hierzu Springsfeld (2002), besonders 62–90; auch die von Borst postulierte Autorschaft Alkuins für einen 'karolingischen Reichskalender' wäre in diesem Zusammenhang zu erwähnen.

[38] Borst (1994), 126–84; Borst (1998), 177–89.

[39] Ein anschauliches Beispiel hierfür ist die Handschrift Köln, Diözesan- und Dombibliothek, 83² aus dem frühen 9. Jh., die alle zur damaligen Zeit wesentlichen oder als relevant erachteten Werke in sich vereint. Siehe hierzu van Euw (1998), 141–55, welcher die Mängel in den älteren Studien, die z.T. nur sehr pauschale Identifikationen der einzelnen Quellentexte lieferten, absorbiert. Zu den irofränkischen *Lectiones computi* von 760 siehe ausführlich Borst (1994), 116–8, wo in Anm. 92 der Kölner Codex als beste Handschrift der *Lectiones computi* ausgegeben wird; sie sind nun ediert in Borst (2006), 544–659. Zu den Komponenten der Osterrechnung nach Dionysius Exiguus und Victorius von Aquitanien, wie sie z.B. in MS Köln, Diözesan- und Dombibliothek, 83², 59v behandelt werden, siehe Walsh und Ó Cróinín (1988), 87 (Anm. zu Zeile 216), 197 (Anm. zu Zeile 5). Zu dieser HS siehe auch unten, Anm. 63.

Komplexität nicht zur zeitfüllenden Gedankenspielerei taugt, mehr als erstaunlich. All das gibt nur dann einen Sinn, wenn zu Zeiten Karls ein Problem in der Osterrechnung existierte, das so bedeutend war, daß es nicht nur alle Großen des Reiches beschäftigte, sondern möglicherweise auch dafür verantwortlich war, daß Karl *die* Autorität für Zeitfragen seiner Epoche, nämlich Alkuin, an seinen Hof holte.

Welcher Art diese Schwierigkeiten waren, macht eine Übersicht über Alkuins komputistische Schriften deutlich. Diese sind insgesamt von einem Kernproblem bestimmt, das sich durch all seine Schriften zieht: die Frage nach der korrekten Berechnung der Zyklen der Sonne und des Mondes. Diesen Sachverhalten suchte sich Alkuin vornehmlich durch die Auswertung der Aussagen der großen komputistischen Autorität Beda zu nähern, dessen Werk in Angelsachsen das Standardlehrbuch war. Dieser hatte sich zu all den mit der Osterrechnung verbunden Fragen geäußert, insbesondere auch zu den notwendigen Schaltungen in bezug auf Sonnen- und Mondjahr. Diese umfassen zunächst den Schalttag des Sonnenjahres oder den *bissextus*, ein zusätzlicher Tag, der alle 4 Jahre eingeschoben wurde, um der Tatsache Rechnung zu tragen, daß das tropische Jahr ca. 365 1/4 Tage umfaßt, wie es schon Julius Cäsar festgelegt hatte. Der andere Schalttag war der des Mondes, der sogenannte *saltus lunae*, der auf den Sachverhalt zurückzuführen war, daß in einem 19jährigen Mondzyklus ein Tag mehr verteilt wurde, als man zur Verfügung hatte (6940 statt 6939). Diesen ließ man dann am Ende eines solchen Zyklus einfach fort, eine seit der Antike verbreitete Praxis.[40] Korrekt angewandt, wie es Beda seiner Ostertafel zu Grunde gelegt hatte, ergab sich ein stimmiges und funktionales System.

Mehr als auffällig erscheint es jedoch, daß sich genau an diesem Punkt für Alkuin und seine Zeitgenossen ein Problem offenbarte. Es muß für ihn, wie auch für den Intellektuellenkreis um Karl den Großen einen Grund gegeben haben, sich immer wieder explizit mit der Mondalterberechnung, dem Lauf des Mondes durch den Tierkreis, der Darstellung des Mondlaufs, der Länge eines Sonnenjahres und schlußendlich dem Ostertermin zu beschäftigen, welche Springsfeld als die diesbezüglich fünf zentralen Fragen der Karlszeit ermittelt hat.[41] Besonders für Alkuin

---

[40] Grotefend (1982), 2.

[41] Springsfeld (2002), 294–7. Sie nennt hier als Gebiete der Komputistik, auf die Alkuin Einfluß nahm, die Mondalterberechnung im 19jährigen Zyklus, die Abweichung von berechnetem und zu beobachtendem Mondalter, die Ermittlung des Ostertermins, den Lauf der Sonne durch den Tierkreis zur Bestimmung des Schalttagszuwachses pro Jahr und den Lauf des Mondes durch den Tierkreis.

sind diese in seinen Schriften zum Mondalter (*Ratio de luna*), zum *saltus lunae* und zur Ermittlung des Schalttages (*De bissexto*), aber auch in seinem Briefwechsel mit Karl dem Großen immer wieder ein Thema. Dabei muß Alkuin offenkundig stets seine ganze Überzeugungskraft aufbringen, um Karl und dem Hofkreis die diesbezüglichen Sachverhalte näherzubringen, was sich teilweise als aufwendiges Unterfangen darstellt.

Einen guten Eindruck von der Natur dieser vermutlich von Karl geforderten Darlegungen Alkuins vermittelt sein Werk über die Berechnung des Mondlaufs, *Ratio de luna*.[42] Alkuin berechnet dort, ausgehend vom Beginn des Frühlingssternzeichens Widder, d.h. vom 18. März, die scheinbare Aufenthaltsdauer des Mondes in jedem Tierkreiszeichen. Dazu ermittelt er zunächst, wie lange der Mond braucht, um ein Zeichen zu durchlaufen, nämlich 54 2/3 Stunden oder 2 Tage (48 Stunden) und 6 2/3 Stunden. Diesen Wert benutzt er dann als Basis, zu der er für das nächste Sternzeichen wiederum 54 2/3 hinzuaddiert. Das Ergebnis ist also ein wahrer Zahlenwust, der das exakte Mondalter für jede Himmelsposition innerhalb eines Mondmonats angibt, jedoch mit einem nur unbefriedigenden Ergebnis. Aufgrund der nur bedingt reflektierten Kombination von siderischem Monat, also den ca. 27 1/3 Tagen, die der Mond benötigt, um an die ursprüngliche Position im Tierkreis nach einem Umlauf zurückzukehren, und dem 29 1/2 Tage umfassenden synodischen Mondmonat, d.h. die Zeitspanne von Neumond zu Neumond, war das Ergebnis auch unzureichend. Es blieben nämlich ca. zwei Stunden übrig, die Alkuin nur, im Rückgriff auf Plinius, als *coitum solis*, als mystisches Verweilen des Mondes bei der Sonne erklären kann.[43]

Ähnlich problematisch gestaltet sich die Erläuterung des bereits erwähnten *saltus lunae*, also dem rechnerischen Ausgleich am Ende eines 19jährigen Mondzyklus. Auch hier, wie ein Brief an Karl den Großen vom November 797 (*Epistola* 126) belegt,[44] folgt Alkuin Beda, wo er hervorhebt, daß eine Auslassung des *saltus* die korrekte

---

[42] *Ratio de luna* ist ediert in PL 101, 981C–984B.
[43] Zu diesem Sachverhalt ausführlich Springsfeld (2002), 231–3.
[44] In diesem Brief heißt es (MGH Epp. 4, 185–6): *Et videri potest diligenter considerantibus, quantus error et quam perniciosus in luna quarta decima et in die dominico paschae et in aetate lunae eiusdem diei oriri poterit, si quis non curaverit unum diem in aetate lunae ratione saltus in Novembrio mense transilire.* Auf die diesbezüglichen sachlichen Schwierigkeiten macht der den nächsten Absatz einleitende Satz Alkuins aufmerksam: *Est quoque et aliud in huius ratione saltus aliquanto difficilius conputandum, sed necessarie sciendum.*

Osterfestrechnung durcheinanderbringen würde. Auch für dieses Beispiel vollzieht er eine Berechnung für die einzelnen Monate eines Jahres; in der Schrift *De saltu lunae* wird dies in erneut akribischer Manier bis auf die kleinste Zeiteinheit des Atoms,[45] d.h. den 22560 Teil einer Stunde heruntergerechnet.[46]

Insgesamt betrachtet entsteht bei der Durchsicht auch der anderen einschlägigen Schriften der Karlszeit der Eindruck, als manifestiere sich in der Berechnung der Zyklen von Sonne und Mond eine Schwierigkeit, die die Zeitgenossen vor ein solch gewaltiges Problem stellte, daß sie Alkuin zu immer neuen detaillierten Darlegungen über die bedanische Komputistik veranlaßten. Dabei schreckte man allem Anschein nach sogar vor Auseinandersetzungen nicht zurück. So charakterisiert Alkuin im Brief Nr. 145 den 'Palast des berühmten David' mit spürbarer Verärgerung und beißendem Spott als von ägyptischen Knaben bevölkert.[47] Damit greift er den Mißstand auf, daß man während seiner Abwesenheit von den Weisungen Bedas zu den Berechnungen der Ägypter übergegangen sei, wobei der Auslöser des Streites erneut die Ermittlung des korrekten Mondalters und der Umgang mit dem *saltus lunae* war, d.h. faktisch die eindeutige Ermittlung des für das Osterfest relevanten Frühlingsvollmonds.[48]

Dennoch täte man Alkuin wie auch Karl dem Großen und seinem Hof unrecht, wollte man dieses ganze arbeitsreiche Bemühen, die

---

[45] Springsfeld (2002), 37, 149–50.

[46] In ähnlicher Weise demonstriert ein solches Vorgehen für alle zeitlichen Parameter auch Hrabanus Maurus in seiner Schrift *De computo* (ediert von Wesley M. Stevens in CCCM 44, 163–331). Als repräsentatives Beispiel für die in ihrer Form gleichbleibenden Umrechnungen der Zeitangaben sei hier die in Kapitel 17 (p 221) vorgestellte Reihe wiedergegeben: *D. Hora quas subdiuisiones habet? M. [. . .] Habet et minuta decem, partes XV, momenta XL, ostenta LX, atomnos XXII milia DLX.* Die Angaben für Tag, Monat und Jahr werden konform dazu notiert.

[47] MGH Epp. 4, 231–5: 231: *Et ut ad rem veniam ac ignorantiae fomentis caput percussi medicari incipiam, ego inperitus, ego ignarus, nesciens Aegyptiacam scolam in palatio Daviticae versari gloriae, ego abiens.*

[48] Springsfeld (2002), 146 interpretiert dies als eine rein akademische Auseinandersetzung, da es im Grunde gleichgültig sei, ob 'im 19. Jahr des Zyklus alexandrinisch im Juli oder römisch im November 'gesprungen' (werde), da in beiden Fällen der Ausgleich der Mondalterzählung nach dem Ostervollmond des letzten Zyklusjahres am 17. April und vor dem ersten am 5. April' liege. Unberücksicht bleibt dabei aber das Faktum, daß etwaige Finsternisse, die in einer im Grunde 'ungefährlichen' Zeit außerhalb des für das Osterfest relevanten Zeitspanne lagen, zusätzlich durch eine rechnerische Manipulation des Mondalters zu Problemfällen werden konnten.

steten Fragen und ebenso die bis in das arbeitsreiche Detail gehenden Antworten allein als akademische Gedankenspielerei eines möglicherweise gelangweilten Intellektuellenzirkels abtun. Für solche abendlichen Diskussionsrunden waren sicherlich Alkuins heitere Scherz- und Rechenaufgaben in den *Propositiones*[49] ein geeignetes Mittel, nicht aber das Osterfest, bei dem neben der Einheit der Christenheit auch die eigene Erlösung auf dem Spiel stand. Daher muß ein konkreter Grund für all diese Produkte intellektueller Aktivität angenommen werden.

Tatsächlich gab es ein Problemfeld, daß in der Struktur des Kalenders, d.h. der rechnerischen Erfassung des Sonnenjahres selbst begründet war. So hatte, wie es Mathias Schulz im Spiegel (27.12.1999) formulierte, Cäsar den nachfolgenden Jahrhunderten wirklich eine 'Zeitbombe' hinterlassen. So war die Angabe des Jahres mit 365 1/4 Tagen gegenüber der Realität 11 Minuten und 14 Sekunden zu lang. Dieser Fehler addierte sich in 128 Jahren bezüglich des Frühlingspunktes zu einem Tag. Diese Abweichung führte im 16. Jahrhundert zur Gregorianischen Kalenderreform, bei der man die inzwischen aufgelaufenen 10 Tage einfach ausließ. Wiewohl die rigorose Lösung des Problems also zu Lebzeiten Alkuins noch in weiter Ferne lag, muß man dennoch davon ausgehen, daß das Problem selbst bekannt war. So vollzog zunächst einmal das Konzil von Nicäa im Jahre 325 bereits indirekt eine 'kleine Kalenderreform', indem es, in Bezugnahme auf den jüdischen Kalender, den 21. März, und nicht mehr wie zu Cäsars Zeiten den 25. März, zum Frühlingspunkt erklärte. Damit wurde faktisch der in ca. vier Jahrhunderten entstandene Fehler absorbiert. Das Problem innerhalb der Komputistik bestand nun aber darin, daß man diese Werte spätestens seit den Osterschriften des Dionysius Exiguus im 6. Jh.[50] einerseits als unveränderliche Vorgaben interpretierte, und man sich andererseits dem sehr langen 532jährigen Zyklus zuwandte. Dieser garantierte zwar rechnerische Stimmigkeit und permanente Gültigkeit, jedoch zu dem Preis, daß keine Korrekturen des Systems möglich waren. Jede Veränderung am System der Komputistik hätte nämlich genau das konterkariert, was Beda im 8. Jh. zugunsten der Einheit der angelsächsischen Kirche propagiert und konsequent umgesetzt hatte: ein kohärentes, rechnerisch

---

[49] Siehe Anm. 12.
[50] Siehe hierzu den *Libellus de cyclo magno paschae DCCCII annorum* und die *Argumenta de titulis pascalis Aegyptiorum* des Dionysius Exiguus, ediert in Krusch (1938), 63–74, 75–81.

korrektes, auf der Grundlage des Nicäums verankertes System der Osterrechnung.⁵¹

Man sollte dabei allerdings nicht dem Irrtum verfallen, der Kalenderfehler als solcher sei im frühen Mittelalter nicht bekannt gewesen. Bereits Beda war sich des Problems in seiner gesamten Tragweite bewußt.⁵² So widmet er ein ganzes Kapitel von *De temporum ratione* (cap. 43) dem Umstand, daß das berechnete Mondalter vom sichtbaren Mond um zwei Tage differiere.⁵³ Als Gründe hierfür nennt er zunächst die allmählich anwachsende Differenz des *saltus lunae* wie auch die unterschiedlichen Möglichkeiten, einen Tag zu beginnen, also genau die Argumentationsgänge, an die auch Alkuin mit seinen Schriften anknüpft.⁵⁴ Dennoch enden zumindest Bedas Ausführungen bewußt nicht mit einer Detailrechnung, sondern prägnanterweise zunächst mit einem etwas diffusen Verweis auf die

---

⁵¹ Verweise auf das Nicäum ziehen sich wie ein roter Faden durch die gesamten Ausführungen Bedas; sie finden sich z.B. in *De temporum ratione* 6 (Jones (1943), 191): *Neque enim alia servandi paschae regula est quam ut aequinoctium vernale plenilunio succedente perficiatur*; *De temporum ratione* 45 (Jones (1943), 262): *Cuius exordium mensis hac regula debet observari, ut non umquam luna paschae xiiii vernum praecedat aequinoctium, sed vel in ipso aequinoctio.*

⁵² Englisch (1994), 358–63.

⁵³ Beda, *De temporum ratione* 43 (Jones (1943), 257): *XLIII. Quare luna aliquoties maior quam computatur pareat.*

⁵⁴ Beda, *De temporum ratione* 43 (Jones (1943), 257–9): *Notandum sane quod huius ratio saltus lunaris longa sui facit exundantia crementi lunam aliquoties maiorem quam putatur videri, adeo ut etiam de tricesima vesperascente illam non gracilem in caelo apparere contingat; et quanto circuli decemnovenalis terminus amplius instat tanto hoc crebrius patiatur causa existente perspicua, quod saltus ille de quo loquimur iam maxima tunc sit ex parte perfectus.* [. . .] *Verum quia sicut scriptum est a luna signum diei festi et quomodo prima lunae lux a vespera mundum irradiavit, ita omnis dies festus in lege a vespera initiari in vesperam perfici debere praecipitur. Congruentius aetas lunaris a vespertina hora quam aliunde nova computabitur, eandemque aetatem quam vespere inchoat sequentem usque servabit ad vesperam. Et siquidem eam paulo ante vesperam accendi a sole contigerit, mox sole occidente primam computari et esse necesse est, quia videlicet illam temporis horam, qua primum terris fulgere coepit, adiit. Sin autem post occubitum solis accendatur, non tamen prima priusquam vesperam viderit, sed tricesimam potius oportet aestimari. Etiam si xxiii horas post occasum solis accensa suppleverit, illam tamen quam occidente sole habuerat, ne primae conditionis ordo turbetur, usque ad alium eius occasum retinere debebit aetatem.* [. . .] *Quod si qui gravius huic insistens quaestioni, dixerit se novam lunam eo anno quo saltus inserendus est, hoc est ultimo circuli decemnovenalis, biduo priusquam prima caneretur multis cum testibus vidisse, id est quarto nonarum aprilium die, cum eiusdem anni xiiii luna paschalis in circulo memorato xv kal. maiarum sit adnotata, ideoque non nisi pridie nonas apriles esse prima valeat, rationemque a nobis huius causae exegerit.* Siehe hierzu auch die Darstellung von Kaltenbrunner (1876), 5.

Autorität des Konzils von Nicäa, von deren Vorgaben man zwar nicht abweichen könne,⁵⁵ dann mit Kirchenväterzitaten (Cyrill von Alexandria, Paschasinus von Lilybaeum), die das Alter der Diskussion um das Mondalter belegen.⁵⁶ Das Kapitel endet mit einem Verweis Bedas auf die Weisheiten der Altvorderen.⁵⁷ Mit dieser autoritativen Begründung kehrt Beda – vermutlich wider besseren Wissens – das Problem gewissermaßen unter den Teppich, eben um seine vordringliche Zielsetzung nicht zu gefährden: Britannien eine einheitliche Zeitrechnung zu geben und jeden Osterstreit für die Zukunft zu vermeiden.

Doch auch im Frankenreich hatte der Fehler des Julianischen Kalenders zu Unruhe geführt. So findet sich bereits in den Fortsetzungen der Chronik des Fredegar für das Jahr 740 ein Beleg dafür, daß das 'allerheiligste Osterfest' in diesem Jahr umstritten gewesen sei.⁵⁸ Die Gründe hierfür lagen sicherlich nicht allein darin, daß die zu dieser Zeit im Frankenreich allgemein geltende Osterrechnung des Victorius von Aquitanien hier zwei alternative Daten verzeichnete, denn dies war in den Jahren 736 und 743 ebenso der Fall, für die die Chronik keine vergleichbaren Hinweise gibt. Es findet sich jedoch im Nachsatz ein Indiz, welches bei genauerer Prüfung die Kalenderabweichung als Ursache für die komputistischen Differenzen offen legt. So seien, vermerkt der Chronist, an Mond, Sonne

---

⁵⁵ Beda, *De temporum ratione* 43 (Jones (1943), 259): *Hic nostra pusillitas, ne sui fragilitate deficiat, ad paternae immo divinae auctoritatis auxilium concurrat. Paternae etenim auctoritas subsidio fulcimur dum nicaenae synodi scita sectamur, quae quartas decimas festi paschalis lunas tam firma stabilitate praefixit ut decemnovenalis eorum circuitus nusquam vacillare, numquam fallere possit. [. . .] Quid enim? Numquid credendum est quia illam quam nos iiii nonas apriles novam vidimus lunam, nemo viderit de illis cccxviii pontificibus qui in nicaeno concilio residebant? [. . .] et non potius intelligendum quia, cum lunam anni illius paschalem a pridie nonas apr. incipere signabant, aliud maius periculum per hoc declinaverint ne videlicet, si aliter decernerent, indissolubilis ille communium annorum et embolismorum status solveretur, quem inviolabiliter observandum divinae legis auctoritate hebraeis tradentibus agnoverant?*

⁵⁶ Beda, *De temporum ratione* 43 (Jones (1943), 259): *Sed et specialibus divinae auctoritatis indiciis observantiam lunarem quam tenemus defendimus.*

⁵⁷ Beda, *De temporum ratione* 43 (Jones (1943), 260): *Neque enim puntandum est nos vel acutius antiquis diversitatem lunaris discursus deprehendere vel salubrius posse dignoscere, quae in eadem diversitate potissimum sit via sequenda.*

⁵⁸ *Continuationes Chronicarum Fredegarii* 24 (MGH SS rer. Merov. 2, 179): *Interim, quod dici dolor et meror est, sollicitatur ruina, in sole et luna et stellis nova signa apparuerunt, seu et paschales ordo sacratissimus turbatus fuit.* Daß der Chronist das Problem nicht deutlicher bezeichnete, mag darauf zurückzuführen sein, daß keine eindeutige Lösung des Problems vorlag.

und Sternen Zeichen erschienen, was mit dem Faktum korrespondiert, daß sich auf dem Gebiet des Frankenreiches am 18. März eine Mondfinsternis und am 1. April eine Sonnenfinsternis ereignet hatte.[59] Eine solche Verdunkelung der Sonne konnte gleichwohl, und dies war aus der enzyklopädischen Literatur eindeutig bekannt,[60] nur zu Neumond *(luna 30)*, eine Mondfinsternis indes nur zu Vollmond *(luna 15)* stattfinden. Die Nähe dieser Ereignisse zum Osterfest machte es folglich offensichtlich und für jeden einfach nachzuvollziehen, an welchem Datum der korrekte Ostervollmond lag. Brisanterweise stimmte dieser durch die Sonnenfinsternis nachgewiesene Wert von *luna* 30 am 1. April jedoch nicht mit dem durch die Ostertafel des Victorius belegten Datum des 2. April überein.[61] Ein Blick in die Ostertafeln dürfte also für dieses Jahr gegenüber diesem, ergänzend ja auch noch durch eine Mondfinsternis belegten und so nicht wegzudiskutierenden Neumondtermin, genau die Verwirrung hervorgerufen haben, von der der Chronist berichtet. In Reaktion hierauf initiierte die Synode von Soissons 744 eine erste Erörterung dieses bislang ungeklärten

---

[59] Oppolzer (1887), Tafel 94.

[60] Als Beispiel für diese Werke der etablierten Handbuchliteratur sei Martianus Capella, *De nuptiis Philologiae et Mercurii* VIII (Willis (1983), 329, 869–70) angeführt. Ebenfalls in diesem Bereich müssen die *Etymologiae* des Isidor von Sevilla genannt werden. Die immer noch maßgebliche Edition von Lindsay (1911) wurde unlängst in einer spanischen Ausgabe erneut abgedruckt und mit einer spanischen Übersetzung versehen (Oroz Reta, Marcos Casquero und Diaz y Diaz (1993–4)). Diese Ausgabe absorbiert durch einen kritischen Apparat einige der Mängel, die dem dringend einer erneuten kritischen Edition bedürfenden Quellentext anhaften. In Buch III, Kapitel 58 dieses Textes (Oroz Reta, Marcos Casquero und Diaz y Diaz (1993–4), i 468) erwähnt Isidor, daß eine Sonnenfinsternis entstehe, wenn der Mond vor die Sonne tritt. Desweiteren fixierte er diese Himmelserscheinung, konform mit der Sequenz bei Martianus Capella, auf den 30. Tag des Mondzyklus.

[61] Victorius, *Cyclos* (Krusch (1938), 35) legt das Osterfest 740 auf die 15. Kl. Mai. (17. April), *luna* 15. Hieraus resultiert, daß *luna* 1, das erste Licht des neuen Mondes, auf den 3. April fiel und somit *luna* 30 der vorhergehenden Lunation auf den 2. April, also einen Tag später als durch die Sonnenfinsternis angezeigt. Sollte die Osterrechnung des Dionysius resp. des Beda zu diesem Zeitpunkt bereits auf dem Kontinent bekannt gewesen sein, hätte sich dadurch das Problem noch verstärkt, denn aufgrund eines anderen Schaltmodus innerhalb des 19jährigen Mondzyklus war dort das in den Ostertafeln angegebene Mondalter (Epakte, d.h. Mondalter am 22. März, 18, daher *luna* 1 am 4. April, *luna* 30 der vorherigen Lunation am 3. April) um noch einen Tag mehr gegenüber der Realität 'versetzt'; siehe dazu auch Englisch (2002), 83. Der dort benutzte Terminus Neumond meint *luna* 1.

Sachverhaltes, in deren Zuge Mitte des 8. Jahrhunderts vermutlich auf die Initiative des Bonifatius hin die Schriften des Beda Venerabilis erstmals auf den Kontinent gelangten.[62] Dennoch reichte ihre Lektüre offenkundig nicht aus, um das Problem aus der Welt zu schaffen, denn sein Werk wurde den anderen Autoren, vornehmlich Victorius und Dionysius Exiguus, nur beigeordnet, wie ein Sammelkomputus belegt, der sich als Abschrift aus dem Jahre 805 in der Kölner Dombibliothek erhalten hat.[63] Dieses Manuskript enthält zudem auch das älteste Zeugnis des charakteristisch-mittelalterlichen Kalenders, der zum Zwecke der Visualisierung dieses Problems

---

[62] So suchte Bonifatius 745/6, also unmittelbar im Anschluss an die Behandlung des Osterproblems auf der Synode von Soissons, mag er selbst an ihr teilgenommen haben oder nicht, beim Erzbischof von York um die Übersendung der Traktate des Beda Venerabilis nach; siehe hierzu Bonifatius, *Epistola* 75 (MGH Epp. sel. 1, 158): *Preterea obsecro, ut mihi de opusculis Bedan lectoris aliquos tractatus conscribere et dirigere digneris, quem nuper, ut audivimus, divina gratia spiritali intellectu ditavit et in vestra provincia fulgere concessit, et ut candela, quam vobis Dominus largitus est, nos quoque fruamur.* Bonifatius, *Epistola* 76 (MGH Epp. sel. 1, 159): *Interea rogamus, ut aliqua de opusculis sagacissimi investigatoris scripturarum Bedan monachi, quem nuper in domo Dei apud vos vice candellę ecclesiastice scientia scripturarum fulsisse audivimus, conscripta nobis transmittere dignemini.* Vgl. Schieffer (1954), 239.

[63] MS Köln, Diözesan- und Dombibliothek, 83². Diese Handschrift (zu ihr siehe auch oben Anm. 39) enthält beispielsweise ohne Hinweise auf eine bestimmte Präferenz Pseudo-Hieronymus, *Disputatio de sollemnitatibus paschae* (fols 201r–203r); Isidor von Sevilla, *De Ponderibus = Etymologiae XVI* 25–26 (fols 215v–217v), *Liber rotarum = De natura rerum* (fols 126r–143r); Anatolius, *De ratione paschali* (fols 188r–191v); Beda Venerabilis, *De temporum ratione* (fols 86r–125v); sowie verschiedene Briefe von Cyrillus von Alexandrien (fols 173v–175r); Pascasinus (fols 175r–176r); Proterius (fols 178r–181v); Dionysius Exiguus (fols 181v–184r); eines Mönches Leo an einen Archidiakon Sesuldus (fols 184r–185v); sowie Briefe von Papst Leo I. (fols 197r–197v), Papst Hilarus (fols 197v–198r) und Victorius von Aquitanien (fols 198r–201r). Ferner finden sich verschiedene, den etablierten komputistischen Schriftenkanon erweiternde Passagen wie das bis vor kurzem noch nicht identifizierte chronologische Sammelwerk *Ars computi quomodo inventa est* (fols 45r–55v) oder auch die *Lectiones computi* (fols 59r–69r), welche nun ediert sind in Borst (2006), 891–950 und 544–659. Hinzu tritt neben dem besagten Kalendarium auch eine Ostertafel für die Jahre 798–911 (fols 76r–79r).

Dennoch darf man diese Manuskripte nicht allein als unreflektierte Sammelwerke betrachten. So zeigt z.B. die Adaption von Bedas komputistischem Werk *De temporum ratione*, daß nicht alle Kapitel Berücksichtigung fanden. Wurden die meisten Elemente der chronologischen und kosmographischen Tradition wie auch die Kernbestandteile der Osterfestberechnung aufgenommen, erfuhren eine Reihe von *Argumenta*, aber auch einige von Bedas Überlegungen zur Bestimmung des Mondlaufes, also genau des in der Karolingerzeit problematischen Materials, keine Beachtung. Siehe hierzu auch Englisch (2002), 9–10.

vermutlich in St-Médard de Soissons vor 757 angefertigt wurde.[64] Interessanterweise weist auch dieser komputistische Parameter auf, die eindeutig der Osterrechnung des Victorius erwuchsen und noch nicht ausschließlich von Bedas Wissen geprägt waren, wie eine spätere Überarbeitung aus Lorsch aus dem Jahre 789 (840).[65] Wir müssen also insgesamt davon ausgehen, daß Bedas Werke vor Alkuins Wirken zwar bekannt waren,[66] er jedoch nicht als allgemeinverbindliche Autorität angesehen wurde.

Genau dies war aber die Situation in bezug auf die Osterfestberechnung, die Karl der Große bei seinem Regierungsantritt vorfand. Das allerheiligste Osterfest war, um noch einmal die Formulierung des Fredegar-Fortsetzers aufzugreifen, nach wie vor umstritten. Die Brisanz des Fehlers hatte sich gewissermaßen sogar noch verstärkt, da er seit den Lebzeiten Bedas weiter angewachsen und damit noch deutlicher war. Zudem stand Karl nicht einmal mehr der Ausweg zur Verfügung, von dem Gregor von Tours in seinen zehn Büchern fränkischer Geschichte berichtete, daß man wie im Jahr 577 bei einem umstrittenen Datum Hilfe durch ein Wunder erhielt, nämlich durch Quellen in Spanien, die sich zum richtigen Termin mit Wasser füllten.[67] Doch Spanien war spätestens seit 725 in den Händen der Araber, die sicherlich nicht die adäquaten Adressaten für eine Frage nach dem richtigen Osterfest waren.

---

[64] Englisch (2002), besonders 80–96. Vgl. dazu Borst (2004), 70–89, der die besondere Betonung eines einzigen Kalenderwerkes bemängelte. Dabei entging ihm offenkundig das Faktum, daß er selbst ein einziges Kalenderwerk, nämlich den in der Salvator-Abtei Prüm um 840 entstandenen Kalender (siehe Anm. 53), zur Abschrift des Lorscher Prototypen der von ihm propagierten karolingischen Kalenderreform erklärt, auf die dann alle europaweit in der Zeit bis 1200 entstehenden Kalender zurückgeführt werden. Grund für diese Präferenz ist insbesondere die Verzeichnung des Geburtsdatum Karls des Großen am 2. April 742; siehe dazu Borst (1998), 245–6.

[65] Hierbei handelt es sich um den Kalender der Handschrift Berlin, Staatsbibliothek, Phillipps 1869 (131), 1r–11v; dieser ist ediert von Borst (1998), 254–98.

[66] Dies belegt auch *De ratione computi liber*, die im Jahr 771 in Mainz entstandene Bearbeitung von *De temporum ratione* in Dialogform (PL 90, 579–600); siehe hierzu Borst (1998), 95.

[67] Gregor von Tours, *Historiarum libri decem* V 17 (MGH SS rer. Merov. I,1, 215): *Eo anno dubietas paschae fuit. In Galliis vero nos cum multis civitatibus quarto decimo Kalendas Maias sanctum paschae celebravimus. Alii vero cum Spanis duodecimo Kalendas Aprilis solemnitatem hanc tenuerunt; tamen, ut ferunt, fontes illi, qui in Spaniis nutu Dei conplentur, in nostrum pascha repleti sunt. Cainone vero Toronicum vicum, dum ipso glorioso resurrectionis dominicae die missae caelebrarentur, eclesia contremuit, populusque conterritus a pavore, unam vocem dedit, dicens, quod eclesia caderet, cunctique ab ea, etiam effractis ostiis, per fugam lapsi sunt. Magna post haec lues populum devastavit.*

Damit stand für Karl – parallel zur korrekten Ermittlung des Osterfestes – ein zentraler Anspruch seiner herrscherlichen Macht auf dem Spiel: ein Reich zu führen, daß im Einklang mit der göttlichen Ordnung war. Und genau dies dürfte die Fragestellung gewesen sein, die die Intellektuellen in Karls Umkreis zu so vielfältigen komputistischen Aktivitäten herausforderte. Dabei war die Lebens- und Regierungszeit Karls gerade im Hinblick auf das Osterproblem durch zwei unglückliche Gegebenheiten belastet: zum einen fiel gleich an sechs Terminen (22. März 764; 21. März 775; 22. März 783;[68] 21. März 794; 22. März 802; 21. März 813) der Ostervollmond nach Bedas Berechnungen auf den 21. oder 22. März.[69] Damit hing das Osterfest an Daten, die durch die sichtbare Realität dahingehend konterkariert wurden, als man hier an einem nach jedem Zyklus unzulässigen Ostertermin feierte, da der damit bezeichnete Vollmond faktisch vor dem Frühlingsbeginn am 21. März eintrat. Dies dürfte allein noch kein Problem gewesen sein, als diese Abweichung mit dem bloßen Auge bei ungeübten Beobachtern sicherlich nicht so ohne weiteres bemerkt wurde, es sei denn es traten Ereignisse ein, die ganz präzise anzeigten, wann Vollmond und wann Neumond war: eine Sonnen- oder Mondfinsternis.[70]

Diese waren insofern für die Zeitgenossen Alkuins ein gewaltiger Krisenpunkt, als sie im wahrsten Sinne des Wortes unberechenbar waren: es stand keine Möglichkeit zur Verfügung, diese im Voraus zu bestimmen. Damit schwebte über Karls Bestreben nach Ordnung des Reiches gewissermaßen immer das Damoklesschwert, im Hinblick auf die korrekte Definition des Auferstehungsfestes durch die kosmischen, überall sichtbaren und höchst auffälligen Himmelszeichen ad absurdum geführt zu werden. Hiervon gab es zudem gerade zur Herrschaft Karls eine ganze Menge, die – auch dies könnte auf die Bedeutung der Finsternisse hinweisen – in vielen Kalendern, Geschichtswerken und komputistischen Schriften genauestens verzeichnet wurden. So

---

[68] Sicherlich nicht zufällig zeigt die nachträgliche Notiz (*XV regn. Caroli*, was 783 entspricht) im zweitältesten Codex der Victoriusüberlieferung (MS Bern, Burgerbibliothek, 645; ursprüngliche Handschrift von 696) am rechten oberen Rand von f 41v, daß man in diesem Jahr bei Victorius Rat suchte; siehe hierzu auch Krusch (1938), 7, der darauf hinweist, daß weitere Schreibernotizen die Benutzung der Ostertafel des Victorius bis zum 10. Jh. belegen. Zumindest eines der so hervorgehobenen Daten, nämlich 870, ist ebenfalls ein problematisches Datum mit *luna* 14 am 21. März.

[69] Siehe hierzu Bedas Ostertafel, am besten zugänglich bei Wallis (1999), 396–7.

[70] Vgl. hierzu Borst (2004), 85–6.

ereigneten sich im Zeitraum von 773–810 11 Mondfinsternisse[71] und von 753–812 neun Sonnenfinsternisse,[72] die aufgezeichnet, d.h. auch als solche erkannt und bewertet wurden.[73] Der einzige Glücksfall war, daß keines der Jahre betroffen war, welches die oben erwähnten

[71] Mondfinsternisse: 3.12.773 (Irland); 22./23.11.774 (Italien); 26.2.788 (nördl. Frankenreich); 9.4.795 (Frankenreich); 3.10.795 (Frankenreich); 2.11.803 (südl. Frankenreich); 1.9.806 (Frankenreich); 26.2.807 (Frankenreich); 21.8.807 (Frankenreich); 25.12.809 (Frankrenreich, Wales); 20.6.810 (Frankenreich); siehe hierzu Oppolzer (1887), 355; Schove (1984), 153–67, 265; Newton (1972), 143–563.

[72] Sonnenfinsternisse: 9.11.753 (Zentraleuropa); 15.8.760 (Zentraleuropa, Konstantinopel); 4.6.764 (Nordwesteuropa, Deutschland, Irland); 3.4.768 (Südöstliches Mittelmeergebiet); 16.8.779 (Spanien); 16.9.787 (Rheinland, Südwesteuropa, Italien?); 11.2.807 (Frankenreich, Italien, Wales); 16.7.809 (England, Nordosten des Frankenreichs); 14.5.812 (Nordafrika, Mittelmeergebiet); siehe hierzu Oppolzer (1887), Tafel 95–7; Schove (1984), 153–67, 265–9; Newton (1972), 143–563. Der geographische Radius dieser Ereignisse wurde angesichts der 'Internationalität' des Reiches Karls des Großen, aber auch der Sichtbarkeit solcher Ereignisse von anderen Standpunkten, lediglich mit der Einschränkung, daß es sich z.b. um eine partielle Verdunkelung handelt, wahrgenommen.

[73] Ein gutes Zeugnis hierfür sind auch die mittelalterlichen astronomischen Kalendarien, die in den oben genannten Studien keine Berücksichtigung fanden; siehe hierzu z.B. als aussagekräftige Quellen den Kalender der schon erwähnten Handschrift Berlin, Staatsbibliothek, Phillipps 1869 (131), 1r–11v (der von Borst sogenannte Karolingische Reichskalender; siehe dazu Borst (1998), 265, 268–9, 284, 288), welche die Sonnenfinsternisse vom 6.4.760 (*Anno DCCLX. indictione XIII. eclypsin solis*) und vom 16.9.787 (*Anno incarnationis Domini DCCLXXXVII. indictione X. XVI. kl. Octb. luna XXVIIII. ab hora diei prima usque ad tertiam eclipsin solis facta est die dominica*), sowie die Verfinsterungen des Mondes vom 15.2.770 (*Aeclypsis lunae facta est prima noctis vigilia, permanens trium horarium spatio. Ita luna obscurata est, ita ut nec ullam quidem vestigiam globi lunaris inspici potuisset, etiam caelo sereno stellis una filgentibus*), vom 22.9.777 (*Anno DCCLXXVII. eclypsis lunae facta est principio noctis X. kl. Octb. luna XIIII., ut raro, indictione XV*) und vom 23.3.778 (*luna XIIII. facta est eclypsis lunae*), 2. Nov. 784 (*Anno DCCLXXIIII. IIII Non. Nov. aeclypsis lunae facta est quasi VIII hora noctis lunae XIIIImae*) verzeichnet. Insbesondere bei der Finsternis von 784 wird durch den der Ereignisschilderung beigefügten Nachsatz die Unsicherheit auch dieses Autors bei besagtem Ereignis deutlich (Borst (1998), 268–9): *Cui vero placeat scire, quomodo potuisset luna XIIII. aeclypsin pati, dum calculatores veteres non posse esse nisi in XV. dicant, computet horam lunae que fit in mense lunari.* Dieser Erklärungsansatz gipfelt in einer komputistisch-rechnerischen Erklärung, die dieses Abweichen auf die Differenz zwischen synodischem Mondmonat und Sonnenmonat zurückführt, jedoch, wie bereits Borst (1998), 269 anmerkte, sowohl Beda, *De temporum ratione* 11, mißversteht, als auch von logischen Fehlern durchsetzt ist: *Tricies XXIIII faciunt DCCXX, tolle VIIII, remanent DCCXII, partire per medium, medietas eius est CCCLVI. Et ex eo colligere potest certam horam incensionis lunaris usque in horam eclypsis, id est octavam noctis quartarum Nonarum. Et invenies ab hora incensionis usque in horam eclypsis horas CCCLVI. Sed in partiones signorum caelo conveniunt eidem rationi. Nam in secunda parte Scorpionis accensa est luna, et iterum in III. parte Tauri eclypsis est facta, quod sunt VI signa plena peracta, remanentibus aliis VI. Et ex hac ratione curiosus quisque potest horam incensionis scire per singulas lunas quae XII horas anticipat in omni luna.* Erneut entsteht der Eindruck, daß hier für ein im Grunde unerklärliches Phänomen nach irgendeiner plausiblen Erklärung gesucht wurde. In die gleiche Richtung deutet der Nachsatz bei der Mondfinsternis von 777.

Eckdaten aufwies. Aber dies war ja, wie oben erwähnt, zunächst einmal nicht abzusehen.

Dabei muß man den Gelehrten an Karls Hof, allen voran Alkuin, sicherlich das Verdienst zubilligen, daß sie allesamt die Wurzel des Übels erkannt hatten. Besonders die Analyse von Alkuins Werken verdeutlichte, wie oben gezeigt, daß er eine Inkommensurabilität an den Stellen vermutete, in denen die Zyklen ineinandergriffen, der *saltus lunae* am Ende des 19jährigen Mondzyklus und die Einfügung eines Schalttages alle 4 Jahre. Genau an diesem Punkt setzte auch die Gregorianische Kalenderreform *c.*800 Jahre später an, indem sie verordnete, daß zur zukünftigen Vermeidung des Fehlers in den Säkularjahren, also 1700, 1800 und 1900, die Schalttage ausfallen sollten; lediglich die darüber hinaus noch durch 4 teilbaren Jahre, also 1600 und 2000, sollten wie gewöhnlich Schaltjahre sein.

Die unüberwindbare Hürde stellte aber für Alkuin und seine Zeitgenossen der Tatbestand dar, daß die Werte für Jahr und Lunation durch alle Überlieferungen nicht nur traditionelles Bildungsgut darstellten, sondern theologisch auch vielfach gedeutet waren. Es bestanden damit genau in dieser Hinsicht Barrieren, die nach der Mentalität der damals lebenden Menschen nicht überwunden werden konnten. Ausgangspunkt aller Überlegungen war, daß Gott die Welt nach Maß, Zahl und Gewicht geordnet hatte.[74] Der Nachweis, daß irgendwo in diesem Gebäude keine harmonische Ordnung herrschte, daß z.B. ein Jahr nicht 365 1/4 Tage lang war, war für Alkuin nicht nur jenseits alles Vorstellbaren, er wäre damit auch der Zielsetzung entgegengelaufen, der man mit der korrekten Osterfestrechnung dienen wollte: sich alljährlich des Einklanges mit Gottes Ordnung zu versichern.

Alkuin griff daher, da eine inhaltliche Lösung dieses Dilemmas nicht erreichbar war, auf einen pragmatischen Weg zurück: er suchte die Schrift seines Lehrers Beda, deren Durchsetzung in Angelsachsen alle Streitigkeiten absorbiert hatte, auch auf dem Kontinent verbindlich zu machen. Damit sind es Bedas Werke, die für ihn die allein legitime Antwort auf alle komputistischen Fragen bereitstellten; auf diese Autorität griff Alkuin stets zurück und suchte ihre Stimmigkeit durch Differenzierung und Erläuterung der Hofgesellschaft klar zu machen. Als große Leistung Alkuins ist es anzusehen, daß er mit diesem sicherlich gewagten Unternehmen Erfolg

---

[74] Isidor von Sevilla, *Etymologiae* III 4, 1 (Oroz Reta, Marcos Casquero und Diaz y Diaz (1993–4), i 426): *Omnia in mensura et numero et pondere fecisti.*

hatte. Zwar gab es hin und wieder einigen Widerspruch, wie die Episode um die ägyptischen Knaben 'verdeutlicht', doch schaffte es Alkuin, daß – wie die Handschriftenüberlieferung beweist – die Schriften und die Ostertafel Bedas im gesamten Einflußbereich Karls nicht nur verbreitet, sondern auch rezipiert wurden. Selbst der Kalender, die geordnete Visualisierung des Jahreslaufs, die der Verzeichnung von Heiligendaten, komputistischen Notierungen und astro-meteorologischen Erscheinungen Raum bietet, erscheint in einer Handschrift aus Lorsch aus dem Jahr 789 in einer ganz durch Beda geprägten Form.[75] Er wird so zwar vom aktiven Hilfsmittel komputistischer Berechnung zu einem Bestandteil enzyklopädischer Gelehrsamkeit, doch alle Facetten der Osterfestberechnung tragen nun ein einheitliches Gepräge. Zwar wurden auch nach Alkuins Rückzug nach Tours und besonders auch nach seinem Tod noch mehrfach komputistische Schriften verfaßt, wie der sogenannte 3-Bücher-Komputus und der 7-Bücher-Komputus,[76] die in z.T. immensem Ausmaß und eklatanter Detailverliebtheit alle Bestandteile des Osterfestes erneut aufrollen. Das Bemerkenswerteste ist aber, daß sie in ihrer Wirkung gewissermaßen ins Leere laufen. Zukunftsträchtig und zukunftsbestimmend bis in die Gegenwart ist allein die bewußt gewählte Norm, die Alkuin hinsichtlich der Zeit für Europa setzte.

[75] Borst (1998), 254–98.
[76] Zu diesen zuletzt Springsfeld (2002), 105–26.

DAVID HOWLETT

COMPUTUS IN HIBERNO-LATIN LITERATURE

Abstract

The essay begins with an Introduction to the history of the Latin language, computus, and related disciplines in Antiquity before knowledge of the subjects among the Irish; it proceeds with Part I, three Hiberno-Latin computistic texts, a note about the introduction of computus among the Irish, analysis of the beginning of Cummian's Letter of 633 to Ségéne and Béccán, and an edition, translation, and analysis of the preliminaries and dating clause of the Oxford computus of 658; it proceeds with Part II, a survey of Computistic Phenomena in Hiberno-Latin Literature under twenty-three headings, considering texts from the fifth century to the twelfth; it ends with a Conclusion.

Keywords

Adomnanus, Aileranus Sapiens, Anatolius, Anonymus ad Cuimnanum, Augustinus Hibernicus, Auxilius, Béccán, Bede, Brianus Molosi Belli, Pope Celestine, Cogitosus, Columbanus, Conchubranus, Cormac, Cú Chuimne Sapiens, Cummianus Longus, Dicuill, Eusebius, Faustus Reiensis, Gildas Sapiens, Gregory the Great, Gregory of Tours, Isidore, Isserninus, Jerome, Joseph Scottus, Laidcenn mac Baíth, Martianus Capella, Mo Sinu maccu Min, Muirchú moccu Macthéni, Palladius, Patricius, Pelagius, Prosper of Aquitaine, Ruben of Dairinis, Secundinus, Sedulius Scottus, Ségéne of Iona, Theodore of Canterbury, Tírechán; *Ailerani Interpretatio Mystica, Canon Euangeliorum,* Annals of Ulster, *Auraicept na n-éces, Cantemus in omni die,* Collectanea Pseudo-Bede, Collectanea Tirechani, Collectio Canonum Hibernensis, Columbani Carmen de mundi transitu, Precamur Patrem, Cummiani Celebra Iuda, Comm. in Marcum, De figuris apostolorum, De ratione conputandi, Penitentiale, De locis sanctis, De mirabilibus sacrae scripturae, De ratione paschali, Josephi Scotti Carmen figuratum, Liber *de mensura orbis terrae, Laidcenni Egloga, Lorica, Memoria abbatum nostrorum, Patricii Epistola, Confessio,* Saint Sechnall's Hymn, *Synodus episcoporum, Versiculi familiae Benchuir, Versus de annis a principio, VV.SS. Brigitae, Maedoci, Monenne, Patricii*; *triuium, quadruuium,* gematria, Hiberno-Latin grammar, computus, and literature, infixing of values of cardinal, ordinal, and calendrical numbers.

Introduction

When military administrators and the legions left Britain early in the fifth century the official language was, as elsewhere in the Western empire, high-register Latin, as illustrated in the literary works of the Romano-Britons Pelagius (†*c.*430) and Faustus, fourth son of Vortigern, abbot of Lérins, and bishop of Riez (†*c.*490),[1] and in the epigraphic works of official inscriptions of *The Roman Inscriptions of Britain*.[2] Personal inscriptions and graffiti show that Latin was widely spoken and written in a demotic form that might have evolved into something like the Old French of Gaul, but that language gave way to Old Welsh and Old English, leaving only the high-register literary Latin taught in the schools run by the civilian and ecclesiastical administrators who remained in Britain. Post-Roman Britons maintained the correctness of their Latin partly because, uniquely in the West, they learned and wrote it by the book, without interference from a vernacular Romance language. Speakers of Vulgar Latin, proto-French in Merovingian Gaul, proto-Italian in Ostrogothic Italy, and proto-Spanish in Visigothic Spain, had to cope with a written language as far removed from their spoken language as modern English is from the Middle English of Chaucer, Gower, Langland, and the Gawain poet. They faced difficulties comparable with those we would experience if we spoke as we do, but tried to write newspapers, parliamentary reports, wills, and funeral monuments in Middle English. The Briton Gildas Sapiens (496 – *c.*570) continued to use Roman units of measurement, *himina Romana* and *sextarius Romanus,* in his Penitential, and

---

[1] For bibliography of their works see Lapidge and Sharpe (1985), 2–24. For analysis see Howlett (1995a), 56–65; Howlett (1998a), 6–15.

[2] Collingwood et al. (1965).

in *De excidio Brittanniae*, published in 540, he wrote of Latin as *lingua nostra*, distinct from *lingua eorum*, the Old English of the Saxons, of which he preserves the earliest written word, *cyula*, 'keel', 'boat'.³ In matters of orthography, grammar, syntax, and style the Latin of Gildas compares favourably with that of the native speaker Gregory of Tours.

During the fifth century the Irish were evangelized by two missions, one of the Romano-Gaul Palladius, a papal deacon sent by Pope Celestine in 431,⁴ the other of the Romano-Briton Patricius, traditionally from 432.⁵ Both missions introduced converts to the literary Latin of the Bible and the liturgy. The Irish, like the Britons, learned and wrote Latin by the book without interference from their vernacular Goidelic Old Irish. Unlike the Britons, however, they had no historical experience of the Christian Roman Empire. As they were first outwith the empire to learn Latin as a completely foreign language, they needed forms of help the Britons had not needed. Recent studies by Vivien Law have shown that they had access to a wide range of Late Latin grammatical texts, sometimes in forms superior to those that have descended directly to us.⁶ From these, devising a new form of Latin grammatical treatise that inculcated many things a native speaker would not need to be told, they created an extensive library of Insular Latin grammars,⁷ as well as new forms of script, half-uncial and minuscule, and either devised or developed many of the features we now take for granted on a printed page – a hierarchy of scripts,⁸ diminuendo, separation of words by space, paragraphs, systems of punctuation, construe-marks, and glossing.⁹ As an index of the quality of Hiberno-Latin literature from its very beginnings, the Latinity of the works of Columban of Bangor compares favourably with that of

---

³ Gildas, *De poenitentia* 1; *De excidio Britanniae* 23,2 (Winterbottom (1978), 26, 97).

⁴ Lapidge and Sharpe (1985), 1244–50, and possibly Lapidge and Sharpe (1985), 599, the *Synodus episcoporum*. Howlett (1998d); Ó Cróinín (1986); Ó Cróinín (2000).

⁵ Lapidge and Sharpe (1985), 25–6. Howlett (1994a); Howlett (2003–4a).

⁶ Law (1982); Law (1995).

⁷ Lapidge and Sharpe (1985), 295–8, 306, 331–4, 337, 663–5, 670, 681–4, 731, 750–1.

⁸ For a perfect example see Faris (1976), text 1–8, facsimile 64–75.

⁹ Stokes and Strachan (1901); for facsimiles of glossed manuscripts see *CLA* 2, 271 and *CLA* 9, 1403 and Stern (1910); Draak (1957), 261–82; Parkes (1987), repr. in Parkes (1991), 1–18; Saenger (1997).

Gregory the Great as the Latinity of Gildas compares favourably with that of Gregory of Tours.[10]

As grammar is the foundation of the human linguistic arts of the *triuium*, the three ways, grammar, rhetoric, and logic, so computus is the foundation of the divine sciences of the *quadruuium*, the four ways, arithmetic, music, geometry, and astronomy, the first as static number, the second as moving number, the third as measurement of the static earth, the fourth as measurement of the moving heavens, all understood as reflexions of the mind of God expressed in number. Prosecuting the study of computus with the same vigour as in their study of grammar, the Irish created an extensive library of computistical treatises.[11]

They began with what we used to call Pseudo-Anatolius, until a happy discovery by Dáibhí Ó Cróinín issued in publication by Dan Mc Carthy and Aidan Breen of a Late Latin translation, made before 402, of the real Anatolius.[12] For a reprint with minor changes of this text, arranged *per cola et commata* 'by clauses and phrases', with parts and lines numbered, see *Peritia* 20.[13] Given the constraints of the technical subject under discussion, the Latin is competent, pellucid, and packed with phenomena that implicitly corroborate explicit statements. The translator embedded words for cardinal and ordinal numbers in positions that corroborate from within the structures of single lines. For example, in line 125, the twentieth line of part III, there are twenty letters and spaces between words before | *uigesimam* and twenty letters and spaces between words from | *uigesimam* to | *uigesimam*. Again, in line 297 there are from the beginning to the first *undecim* | and from the second | *undecim* to the end eleven words. The twentieth syllable from the beginning is the last of *uigesimam*, and the twenty-first syllable from the end is *uigesimam* | *primam*. There are more than one hundred examples of comparable infixing. The translator embedded phenomena that corroborate from within structures larger than single lines and single sentences in more than one hundred places. In line 25, for example, *In quo adnuntians in die paschae non solum lunae cursum et aequinoctii transitum* there are eighty letters and

---

[10] Lapidge and Sharpe (1985), 532, 639–42, 819. Howlett (1993); Howlett (1994b), 1–10; Howlett (1995a), 82–91, 156–77.

[11] Lapidge and Sharpe (1985), 288–9, 315, 317–9, 321–4, 720, 732, 752.

[12] Lapidge and Sharpe (1985), 320. Mc Carthy and Breen (2003).

[13] Howlett (2006a).

spaces between words, coincident with the number of days from 1 January to 21 March, the day of the vernal equinox. Between *solis ascensum descensumque relinquentes* | in line 2 and | *et solis transcensum* in line 26 there are 172 words, coincident with the number of days from 1 January to 21 June, the day of the summer solstice.[14] Within these bounds there are eighty-four words before | *nonnulli octoginta quattuor* line 13, in which the thirtieth letter is the last of *triginta*.

In connecting part IIII with part V, if one begins counting at | *Conputa* in line 169 the seventieth word brings one to | *septuaginta* 178.

In part VI, if one begins counting at | *primi* in line 202, the one-hundredth word brings one to *centessimam* | 217.

In part VII sentence ii the sixteenth word is *sextam* | *decimam* 227, and there are seventeen words before | *et septimam decimam* 227. After *septimam decimam* | there are seventeen words to the end of the sentence.

In part VIII sentence i the nineteenth word is *decem* | *et* | *nouem*, *nouem* being the nineteenth word from the end of the sentence.

In part X sentence i line 292, conspicuous for play on 9 and 19, there are nineteen words from *decem et nouem* | to the end of the sentence.

In part XI sentence ii there are from *octo* | 316 to *duodecim* | 317 twelve words, and from | *id est prima pars* 318 to *duodecim* | 319 twelve words. At the end of part XI in sentences x and xi there are after *octo* | 353 eight (8×1) words to the end of the chapter, between *in octo* | 351 and | *in octo* 353 sixteen (8×2) words, from | *in octo* 346 to | *in octo* 351 thirty-two (8×4) words. In the same passage there are from | *duodecim* 352 to *duodecim* | 353 twelve (12×1) words, from | *duodecim* 345 to *duodecim* | 351 twenty-four (12×2) words, from | *duodecim* 349 to *duodecim* | 352 twenty-four (12×2) words, and from | *duodecim* 345 to *duodecim* | 349 thirty-six (12×3) words.

The Introduction to this twelve-part text consists of the title and twelve sentences. Anatolius states the need to incorporate solar reckoning into calculation of the date of Easter, and the translator supports the calendrical subject of his author's narrative with a title that contains thirty letters, for days of a solar month. The twelve sentences represent months of a solar year, and the 360 words the number of degrees in a circle. Including the title the 365 words represent the number of days of an ordinary solar year.

---

[14] See below p 311.

The Introduction both announces the subject and fixes the structure of the entire composition, the twelve introductory sentences prefiguring the twelve parts, and the reference to the eighty-four-year cycle prefiguring the eighty-four sentences, and the 365 words prefiguring the 365 lines of the whole.

After the Introduction parts I and XII, the beginning and the end, confirm again the structure of the entire composition. Part I fixes the beginning of the first month four ways, as reckoned by the Egyptians, by the Macedonians, and by the Romans in two ways, forward from the beginning of March and backward from the beginning of April. The chiastic pair to part I is part XII, which contains twelve lines, the subject being the four seasons. The number of lines in the Introduction (48) equals the number of lines in part I (36) added to the number of lines in part XII (12).

The forms of corroboration are related but independent: first, confirmation of the numerical values of numerical words by their placement within discrete lines of text; second, confirmation of the numerical values of numerical words by their placement within sentences longer than discrete lines and between adjacent sentences; third, coincidence of the numerical values of numerical words with the numbers of the lines in which they occur, both within discrete chapters and in the entire composition; fourth, coincidence of the numerical values of numerical words with the numbers of the sentences in which they occur, both within discrete chapters and in the entire composition; fifth, play with calendrical numbers of cycles, years, months, days, hours, and moments, and with the numbers of equinoxes and solstices; sixth, incremental infixing of relations among numbers of progressively larger units, letters, syllables, words, sentences, lines, parts or chapters, and the entire composition.

This is a text with which the Irish began, from which they learned much and forgot nothing. In addition to the calendrical reckoning learned from this fundamental text, another feature that involves counting recurs throughout Hiberno-Latin literature – gematria, Hebrew גִּימַטְרִיָּא, perhaps borrowed from Greek γεωμετρια, the calculation of numerical values of personal names, place-names, and words. As every letter of the alphabet, in Hebrew, Greek, and Latin alike, bears numerical value as well as phonetic value, א = 1, ב = 2, ג = 3, A = 1, B = 2, Γ = 3, Λ = 1, B = 2, C = 3, every word exhibits a number as well as a semantic meaning.[15] The text of the Hebrew Old Testament is filled with this artifice, a notable example occurring in Jds 3:7–11, in which the name כלב *Caleb*, of which the alphanumeric value is 52, is the fifty-second word.[16] The most famous example in the Greek New Testament occurs in Apoc 13:18, in which the number of the beast, εξακοσιοι εξηκοντα εξ, 666, exhibits descending value of the Roman numerals DCLXVI. This number is the sum of the letters of the Latin name NERO CAESAR, spelled in Hebrew letters נרון קסר, 50+200+6+50+100+60+200 or 666. The example best known in a Latin literary text is Martianus Capella's *De Nuptiis Philologiae et Mercurii*, in which at the beginning of book II Philologia reckons multilingual gematria on her name and that of Mercury to determine their compatibility. For a 'smoking gun', consider a fourteenth-century Anglo-Norman poem that refers to gematria during the practice of it:[17]

Dieux soit ou vous *quant* vous ali*etz*	a	7	8	30
Et de vos malx vous doynge p*ard*o*nne*	b	7	8	29
Pensez de moy *quant* meux porr*etz*	a	6	8	27
Qe 78 [*i.e.* setante huit] hauetz *a* n*onne*.	b	6	8	25
	2	26	32	111

'May God be where you [be] when you go,
and for your sins may He grant you pardon.
Think of me when best you can,
you who have for a name [the number] 78.'

In the right margin against these lines, of which the sense is clear, the metre perfect, and the rhymes pure, is a crown over the number 78. In the twenty-three-letter Latin alphabet

E D V A R D I I I = 5 + 4 + 20 + 1 + 17 + 4 + 9 + 9 + 9 or 78.

---

[15] For a tabulation of the Hebrew system see Gesenius, Kautzsch and Cowley (1910), 26 (§5.2). For evidence of knowledge of the shapes and names of letters of the Hebrew alphabet see the seventh-century Old-Irish text *Auraicept na n-éces* (Calder (1917), 86–87, 229–30). For early Insular explications of the Greek system see Bede, *De temporum ratione* I (CCSL 123B, 272–3), and the seventh-century *Auraicept na n-éces* (Calder (1917), 230–1). For actual calculation of Greek values see the eighth-century Hiberno-Latin MS Milano, Biblioteca Ambrosiana, F 60 sup, 61r.

[16] Howlett (1995a), 34–7.

[17] Howlett (2006d), 90–2.

Part I: Computistic Texts

On Mo Sinu maccu Min

One of the oldest extant Latin texts written in Ireland is a note, now pasted into MS Würzburg, Universitätsbibliothek, M.p.th. f. 61, about the introduction of computistic learning into Ireland during the second half of the sixth century,[18] that offers valuable evidence of various forms.

> Mo Sinu maccu Min scríba et àbbas Bénncuir [space]
> primus Hibernensium computum a Graeco quodam sapiente memoráliter dídicit
> Deinde Mo Cuoroc máccu Net Sémon
> quem Romani doctorem totius múndi nominábant [space]
> alumnúsque praefáti scríbae —
> in insula quae dicitur Crannach Duin Lethglaisse hanc scientiam líteris fíxit
> ne memória làberétur.

In this text the scribe has separated the words with spaces, marking four of the seven commas specially, two with extraordinarily long spaces, one with a dash, and the last with a punctuation point. The writer has simultaneously fulfilled the requirements of two systems of elegant prose composition, both quantitative metrical *clausulae* and stressed rhythmic *cursus*, making patterns of the forms. The number of letters, 276, coincides with the number of days from the Annunciation, celebrated on 25 March, to the Nativity, celebrated on 25 December.[19]

Two cultures that exhibit long traditions of transmission in literary form, one Sanskrit, another Hebrew, also exhibit long traditions of transmission in oral form. This note bears eloquent witness to immersion in both forms of transmission from the very beginnings of the Hiberno-Latin tradition: *primus memoraliter didicit, deinde scientiam literis fixit ne memoria laberetur.*

Epistola Cummiani

I have tried on earlier occasions to understand the oldest extant computistic text from Ireland, Cummian's Letter to Ségéne abbot of Iona

---

[18] Lapidge and Sharpe (1985), 288. Ó Cróinín (1982a), 283; Howlett (1995b), 1–3; Howlett (1997b), 59–60.

[19] For other examples see Howlett (1997a), 294 n 59; Howlett (2000a), 84, 86–97, 157–60; Howlett (2007b).

and Béccán the hermit about the Paschal controversy,[20] a text that attests the author's knowledge of ten different computational cycles, a text filled with varied forms of computistic artifice. In the Letter, that can be dated internally to 633, Cummian wrote to precede and follow his fifteen-part argument an introduction and a conclusion, chiastically arranged ABC–C'B'A', that contain 633 words. Here follow some phenomena from the beginning of the text, the invocation, salutation, and first two sentences, not noted earlier.

IN NOMINE DIVINO DEI SUMMI CONFIDO

Dominis sanctis et in Xpisto uenerandis Segiéno abbáti
Columbae sancti et ceterorum sanctórum successóri
Beccanoque solitario caro carne et spiritu fratri cum suis sápiéntibus
Cummianus supplex peccator magnis minimus apologiticam in Xpísto salútem

Verba excusationis mee in faciem sanctitatis uestre proferre procáciter non aúdeo
sed excusatum me habere uos ut pátres cúpio
testem Deum inuocans in ánimam méam
quod non contemptus uestri gratia nec fastu moralis sapientie cum ceterórum despéctu
sollemnitatem festi paschalis cum ceteris sapiéntibus suscépi
Ego enim primo anno quo cyclus quingentorum triginta duorum annorum a nostris celebrari orsus est non suscépi sed sílui [. . .].

From *Segieno abbati* | to *Beccanoque solitario* | there are seventy-one letters and spaces between words, coincident with the alphanumeric value of SEGIENO, $18 + 5 + 7 + 9 + 5 + 13 + 14$ or 71. From | *Dominis* to *Columbae* | there are sixty-three letters and spaces between words, coincident with the alphanumeric value of COLUMBA, $3 + 14 + 11 + 20 + 12 + 2 + 1$ or 63. Between *Columbae* | and | *Beccano* there are forty-one letters and spaces between words, coincident with the alphanumeric value of BECCANO, $2 + 5 + 3 + 3 + 1 + 13 + 14$ or 41. One notes that the value of Cummian's father's name, FIACHNAE, $6 + 9 + 1 + 3 + 8 + 13 + 1 + 5$ or 46, added to the value of his mother's name, MUGAIN, $12 + 20 + 7 + 1 + 9 + 13$ or 62, equals that of his

---

[20] Lapidge and Sharpe (1985), 289. This text is discussed, edited, and translated by Walsh and Ó Cróinín (1988), 1–97. Howlett (1994b), 10–7; Howlett (1995a), 91–102; Howlett (1995b), 3–6.

own name, CUMMIANUS, 3 + 20 + 12 + 12 + 9 + 1 + 13 + 20 + 18 or 108.[21] From *Cummianus* | to | *Cummiani* there are 108 letters and spaces between words. One notes further that from | *Dominis sanctis et in Xpisto uenerandis* to *Ego enim primo anno quo cyclus quingentorum triginta duorum* | *annorum* there are 532 letters.[22]

Cummian wrote his *Epistola* three years before the death of Isidore of Seville in 636, seven years before Braulio's publication of Isidore's *Etymologiae* in 640. By the time someone in the circle of Cummian, probably Cummian himself, wrote *De ratione conputandi*, texts of Jerome's *Chronicon*, Isidore's *Etymologiae*, and Martianus Capella's *De nuptiis Mercurii et Philologiae* had arrived in Ireland. All are cited explicitly in the following text.

Computus Oxoniensis

MS Oxford, Bodleian Library, Bodley 309, 62r–v, a manuscript of the eleventh century from Vendôme, contains a seventh-century Hiberno-Latin computus.[23] A version of the text was printed incompletely during the nineteenth century in *Patrologia Latina* XC 647–52, and the prologue and chapter headings were printed once during the twentieth century.[24] In the new edition of the preliminaries that follows, to the left of the text Roman numerals mark numbers of sentences. Within the text *litterae notabiliores* and punctuation marks in boldface represent features of the manuscript; acute and grave accents mark the rhythms of the cursus (*planus, maiórum collécta; tardus, pártes diuíditur; uelox, paúca dictúri súmus; medius, Iánuárii; trispondiacus, númerus procéssit; dispondeus dactylicus, créscunt de minóribus*); and italics suggest possible rhymes. To the right of the text letters represent a possible rhyme scheme, and Arabic numerals mark numbers of lines.

i	De numero igitur fratres dilectissimi Deo adiuuante	
	paúca dictúri súmus. secundum modulum	
	ingenioli nostri sed tamen ex auctoritate	
	maiórum coll*écta*.	a

---

[21] For play on the identical values of the names of Patrick's father *Calfarnio* and his mother *Concessa*, each worth 75, coincident with the number of syllables from | *Calfarnio* to *Concessa* |, see Howlett (2006c), 140.

[22] See below p 286.

[23] Lapidge and Sharpe (1985), 323.

[24] Jones (1943), 393–4.

ii	Primum nobis interrogandum est unde primum; hec ars que numerus uel computus dicitur ínitiáu*it*.	b	
iii	Deinde postea scire debemus ex qua radice sapientie númerus procéss*it*.	b	
iiii	Scimus enim quod omnis sapientia siue diuina siue humana philosóphia *nùncupátur*.	c	
v	Et illa philosophia in tres pártes diuíd*itur*.	c	5
	et pars philosophie que dicitur phisica in quatuor partes diuísa uidé*tur*;	c	
vi	Sic etiam et ethica secunda pars philosophie quatuor diuísiónes hábe*t*.	b	
vii	Tertia autem logica dúas diuìsiónes; ex qua ergo diuisione hec ars númeri procéd*it*. demonstrári necésse es*t*.	b	
viii	Et postea etiam scire conuenit quis primus inuenit numerum apud Hebreos et Chaldeos et Persos et Egiptios et apud *Grécos et Latínos*;	d	
viiii	Ac deinde inuestigari oportet quo modo numerus in linguis principálibus *nùncupátur*.	c	10
	hoc est apud Hebreos et Chaldeos et Persos et Egiptios et Macedones et *Grécos et Latínos*;	d	
x	Et illud nomen quod dícitur númeru*s*	d	
	si simplex ést an conpós*itum*.	e	
	si primitiuum sit an dériuat*íuum*.	e	
	et quo modo definitur scíre nos cónuen*it*.	b	15
xi	Dehinc etiam interrogare debemus quo modo numeri nominantur apud Grecos ab uno usque ad mílle et mỳriáde*s*.	d	
	et que note significant illos números àpud Gréco*s*.	d	
xii	Nec non etiam scíre nos opórtet. que note significant istos numeros ab uno usque ad mille et myriades ápud *Latínos*.	d	
xiii	Deinde etiam oportunum interrogandum est quot sunt diuisiónes témpor*is*.	d	
	et ille diuisiones maiores créscunt de minóribu*s*.	d	20
xiiii	De athomis étiam tractándum es*t*.	b	
xv	De momen*tis*.	d	
xvi	De minu*tis*.	d	
xvii	De punc*tis*.	d	
xviii	De hor*is*.	d	25
xviiii	De quadránte naturáli.	f	
xx	De quadrante ártificiáli.	f	
xxi	De die*bus*.	d	

xxii	De ébdomád*ibus*.	d	
xxiii	De ménsibus lun*áribus*.	d	30
xxiiii	De ménsibus sol*áribus*.	d	
xxv	De órdine ménsi*um*.	e	
xxvi	De inuentióne e*órum*.	e	
xxvii	De numero mensium apud antíquos Rom*ános*;	d	
xxviii	Deinde etiam interrogandum est ex quo tempore menses inuenti súnt et nuncup*áti*	f	35
	et in quo numero dierum menses sunt in sole et luna hoc est in anno communi et embolismo et in ánno sol*ári*.	f	
	et quo modo inter se dissentiunt et conuéniunt ìlli *ánni*	f	
	et qui sunt qui primi menses soláres inuen*érunt*.	b	
	et quo modo menses nominantur apud Hebreos et Egiptios et Macedones et *Grécos et Latínos*.	d	
	et quot sunt cause ex quibus menses nómina àccep*érunt*;	b	40
xxviiii	Et postea inuestigandum est de quatuor tempóribus *ánni*;	f	
xxx	Deinde etiam tractandum est de annis et de generibus annorum et de numero annorum ab origine mundi usque ad incarnátionem Xp*ísti*.	f	
xxxi	De numero annorum ab incarnatione Xpisti úsque ad prèsens témp*us*.	d	
xxxii	De bissex*to*.	g	
xxxiii	De sal*tu*.	g	45
xxxiiii	De indictiónibus Ròman*órum*.	e	
xxxv	De cyclo décennouen*áli*.	f	
xxxvi	De cýclo lun*ári*	f	
	et quis inuenit primum cyclum decennouenalem et cýclum lunáre*m* et in quo témpore inuénti *sunt*.	b	50
xxxvii	De cyclo epactarum in .xi. kaléndas Apríli*s*.	d	
xxxviii	De epactis que currunt in *kalendis Ianuárii*.	f	
xxxviiii	De epactis in kalendis .xii. [duódecim] ménsi*um*.	e	
xxxx	De epactis uniuscuiusque diéi per tòtum ánn*um*.	e	
xxxxi	De feria monstranda in *kalendis Ianuárii*.	f	55
xxxxii	De feria in kalendis .xii. [duódecim] ménsi*um*.	e	
xxxxiii	De feria in omni díe quer*énda*.	a	
xxxxiiii	De feria in término pasch*áli*.	f	
xxxxv	De luna in qua hora uel in quo puncto accenditur uel sua etas commutátur omni dí*e*.	h	
xxxxvi	De luna quot horis lucet in únaquaque nócte.	h	60

xxxxvii	De initio quádragésim*i*.	f	
xxxxviii	De terminis quádragesimál*ibus*.	d	
xxxxviiii	De prima lúna primi méns*is*.	d	
l	De términis paschálibus. et de illorum régulár*ibus*.	d	
li	De pasc*h*a.	a	65
lii	De terminis rogation*um*.	e	
liii	De rogátió*nibus*.	d	
liiii	De cóncurrént*ibus*.	d	
lv	De cyclo solari et lunari per dígitos dèmonstránd*is*.	d	
lvi	De concurrentibus demonstrandis *per cýclum solárem*.	e	70
lvii	De bissexto monstrando *per cýclum solárem*.	e	
lviii	De zodiaco circulo et de xii signis celi et eórum nom*ínibus*;	d	
lviiii	De cursu solis et lune per xii [duódecim] sígn*a*.	a	
lx	De vii stéllis errántibu*s*;	d	
lxi	De ascensu solis et descensu hoc est quo modo crescit dies uel no*x*.	d	75
lxii	De eclypsi sólis et lúne.	h	
lxiii	De numero momentorum et minutorum et punctorum et horárum totìus *anni*.	f	
lxiiii	De epacta monstranda et de feria querenda a presente qualibet die: úsque ad cèntum án*nos*.	d	
lxv	De computo Grecorum et Latinorum per litteras et dígitos dèmonstránd*o*.	g	
lxvi	De epístolis Grec*órum*.	e	80
lxvii	De Anatholio ét Macrób*io*.	g	
lxviii	De Victorio et Díonís*io*.	g	
lxviiii	Dé Boét*io*.	g	
lxx	De calcul*o*.	g	84

1 sumus deest. 2 compotus. iniciauit. 6 fisica. quattuor. 7 a&hica. quattuor. 8 loica. 9 et persos desunt. 13 deriuatiuum *with* i *written in a later hand above* e. 15 diffinitur. 16 miriades. 19 quot *with* t *written above erased* d. diuisionis *with fourth* i *changed to* e. 27 astificiali. 36 embolismi. 37 desentiunt *with* is *written above erased first* e. 38 salares *with* o written *above first subpuncted* a. 40 mensis *with* i *corrected to* e. 41 quattuor. 59 a&as. 72 zoziaco. 78 annis *with* o *written above subpuncted* i. 79 compoto. gregorum. 80 gregorum.

i   About number, therefore, most beloved brothers, with God helping, we are about to say a few things according to the little measure of our little intellect, but nevertheless gathered together from the authority of greater men.

ii	First it must be asked by us whence first this art, which is called number or computus [lit. 'reckoning together'], began.
iii	Then after these things we ought to know from what root of wisdom number proceeded.
iiii	For we know that all wisdom, whether divine or human, is designated philosophy.
v	And that philosophy is divided into three parts, and the part of philosophy that is called physics is seen divided into four parts.
vi	And so also ethics, the second part of philosophy, has four divisions.
vii	The third, however, logic, two divisions, from which division, therefore, this art of number proceeds needs to be demonstrated.
viii	And after these things it is convenient [lit. 'it comes together'] to know who first discovered [lit. 'invented', 'came upon'] number among Hebrews and Chaldeans and Persians and Egyptians and among Greeks and Latins.
viiii	And then it is opportune that it be investigated in what manner number is designated in the principal tongues, this is among Hebrews and Chaldeans and Persians and Egyptians and Macedonians and Greeks and Latins.
x	And that word [lit. 'name'] that is called number, whether it is simple or composite, whether it may be primitive or derivative, and in what manner it is defined it is convenient for us to know.
xi	Then also we ought to ask in what manner numbers are named among the Greeks from one up to a thousand and ten thousands, and what notes signify those numbers among the Greeks.
xii	And also it is opportune for us to know what notes signify those numbers from one up to a thousand and ten thousands among the Latins.
xiii	Then also it is opportune for it to be asked how many are the divisions of time, and in what manner those greater divisions grow from the lesser.
xiiii	About atoms is also to be treated.
xv	About moments.
xvi	About minutes.
xvii	About points.
xviii	About hours.
xviiii	About the natural quadrant.
xx	About the artificial quadrant.

xxi	About days.
xxii	About weeks [lit. 'seven-day periods'].
xxiii	About lunar months.
xxiiii	About solar months.
xxv	About the order of months.
xxvi	About their discovery [lit. 'invention', 'coming upon'].
xxvii	About the number of months among the ancient Romans.
xxviii	Then also it is to be asked from what time months were discovered [lit. 'invented', 'come upon'] and designated, and in what number of days months consist in the sun and the moon, this is in a common year and an embolismic [year] and in a solar year, and in what manner those years disagree [lit. 'sense or experience apart'] and agree [lit. 'come together'] among themselves, and who they are who first discovered [lit. 'invented', 'came upon'] solar months, and in what manner the months are named among Hebrews and Egyptians and Macedonians and Greeks and Latins, and how many are the causes from which months received names.
xxviiii	And after these things is to be investigated about the four times of the year.
xxx	Then also is to be treated about years and about kinds of years and about the number of years from the origin [lit. 'rising'] of the world up to the Incarnation of Christ.
xxxi	About the number of years from the Incarnation of Christ up to the present time.
xxxii	About the bissextile [lit. 'twice-sixth' (day), 24 February and the following intercalary day inserted into leap years of the Julian calendar].
xxxiii	About the leap [year].
xxxiiii	About the indictions of the Romans.
xxxv	About the nineteen-year cycle.
xxxvi	About the lunar cycle, and who discovered [lit. 'invented', 'came upon'] the first nineteen-year cycle and the lunar cycle, and in what time they were discovered [lit. 'invented', 'come upon'].
xxxvii	About the cycle of epacts on the eleventh of the kalends of April.
xxxviii	About the epacts that run on the kalends of January.
xxxviiii	About the epacts on the kalends of the twelve months.
xxxx	About the epacts of every single day through the whole year.

xxxxi	About the holy day to be shown on the kalends of January.
xxxxii	About the holy day on the kalends of the twelve months.
xxxxiii	About the holy day to be sought on every day.
xxxxiiii	About the holy day in the term [limit] of Easter.
xxxxv	About the moon, in which hour or in which point it is illuminated or its own age is changed on every day.
xxxxvi	About the moon, in how many hours it shines on every single night.
xxxxvii	About the beginning of Lent [lit. 'the forty-day period'].
xxxxviii	About Lenten terms [limits].
xxxxviiii	About the first moon of the first month.
l	About Lenten terms [limits] and their rules.
li	About Easter.
lii	About the terms [limits] of Rogation [days].
liii	About Rogation [days].
liiii	About concurrent [numbers, lit. 'that run together', *e.g.* concurrence between 1 January and 24 March, the former occurring on Sunday entailing the latter occurring on Friday].
lv	About the solar cycle and the lunar to be demonstrated on the fingers.
lvi	About concurrents to be demonstrated through the solar cycle.
lvii	About the bissextile to be shown through the solar cycle.
lviii	About the circle of the zodiac and about the twelve signs of heaven and their names.
lviiii	About the course of the sun and the moon through the twelve signs.
lx	About the seven wandering stars.
lxi	About the ascent and descent of the sun, this is in what manner the day or night increases [lit. 'grows'].
lxii	About the eclipse of the sun and the moon.
lxiii	About the number of moments and minutes and points and hours of the whole year.
lxiiii	About the epact to be shown and about the holy day to be sought from a certain present day up to a hundred years.
lxv	About the computus [lit. 'reckoning together'] of the Greeks and Latins to be demonstrated through letters and fingers.
lxvi	About the epistles of the Greeks.
lxvii	About Anatolius and Macrobius.
lxviii	About Victorius and Dionysius.
lxviiii	About Boethius.
lxx	About calculus.

The short list of rejected manuscript readings suggests that this is a good copy of the original text. Three features suggest the influence of a Continental Francophone scribe, who for the author's *e*, as in *hec* 1, sometimes wrote *e caudata*, as in *quę* 2, *sapientię* 3, *philosophię quę* 6, *hęc* 8, *hebręos* 9, *quę notę* 17, 18, *illę* 20, *quę* 52, *cęli* 72, *lunę* 73, 76, restored here to *e*, and *iniciauit* 2 for original *initiauit*, and *loica* 8 for *logica*.[25] To the author's original *quatuor* 6 the scribe added a superscript *t* and wrote *quattuor* subsequently. Two features suggest that an exemplar was Insular. In line 27 *astificiali* implies a misreading of Insular long *r* as long *s*. In lines 79 and 80 *gregorum* illustrates the identity of voiced *g* with unvoiced *c* among speakers of Celtic languages. If one reckons critically, the scribe miswrote sixteen characters per thousand or, omitting consideration of *e caudata*, eleven characters per thousand.

Our computist's prose is elegantly rhythmical. Of lines containing more than four syllables all but two (0.0238 of the whole) exhibit faultless cursus rhythms, and with simple reversal of word order both of those would exhibit correct rhythms, *rogatiónum términis* 66 and *créscit nox uel díes* 75.[26]

Our computist made his words for cardinal and ordinal numbers exhibit their values by their placement within discrete lines.

The seven words of line 5 divide by *epitritus* or sesquitertian ratio 4:3 at | *tres*.

In line 6 the fourth word from the end is *quatuor*.

The ten words of line 7 divide into two parts at *secunda* | *pars*. They divide into quarters in *quatuor*. From | *secunda* to *quatuor* | there are four words.

In line 8 the five words *tertia autem logica duas diuisiones* divide by *hemiolus* or sesquialter ratio 3:2 at | *tertia* and | *duas*.

In line 41 the eight words divide into quarters at *quatuor* |.

In line 53 the twelfth syllable is the last of *duodecim*, and the twelfth letter from the end is *duo* | *decim mensium*.

In line 56 the twelfth syllable is the last of *duodecim*, and the twelfth letter from the end is *duo* | *decim mensium*.

In line 72 there are twelve syllables after *xii* |.

---

[25] For other examples of the influences of Francophone scribes on early Insular texts see Howlett and Thomas (2003), 29; Howlett (2006c), 154.

[26] An additional argument for reversal of these words is the priority of night to day in Gen 1:5: *factumque est uespere et mane dies unus*.

Employing another form of artifice our computist wrote thirty-three letters before *Deo* 1, coincident with the thirty-three years of the life of Jesus.

Our computist also made his words for cardinal and ordinal numbers exhibit their values by their placement in units that extend beyond discrete lines, as in *pars philosophie* 6, *secunda pars philosophie* 7, and *tertia autem* 8.

From | *unde primum hec ars* 2 to *usque* | *ad mille* 18 there are 1000 letters.

From *numerus* | 3 to *ab uno usque* | *ad mille* 16 there are 1000 letters and spaces between words.

There are from | *quadrante* 26 to *quadrante* | 27 four words.

In lines 47–49 there are from *cyclo* | *decennouenali* to | *cyclum decennouenalem* nineteen syllables.

In lines 73–74 there are from | *xii* [*duodecim*] to | *vii* seven syllables.

From | *duodecim* 73 to *centum* | 78 there are one hundred syllables.

From *centum* | *annos* 78 to *De epistolis Grecorum* | 80 there are one hundred letters and spaces between words.

Our computist has arranged words and ideas in a comprehensive chiastic pattern.

A	1	De numero
B	1	ex auctoritate maiorum
C1	2	computus
C2a	7	diuisiones
C2b	9	scire conuenit
C2c	9	numerum
C2d	9	apud Hebreos et Chaldeos et Persos et Egiptios et apud Grecos et Latinos
C2e	10	quo modo numerus in linguis principalibus nuncupatur
C2d'	11	apud Hebreos et Chaldeos et Persos et Egiptios et apud Grecos et Latinos
C2c'	12	numerus
C2b'	15	scire nos conuenit
C2a'	19	diuisiones
D1	20	crescunt
D2	22–5	De momentis, minutis, punctis, horis
E	45	De bissexto
F	49	cyclum decennouenalem et cyclum lunarem
G	51–4	De epactis

H	55–8	De feria
I₁	59–60	De luna
I₂	61–2	De quadragesima
I'₁	63	De prima luna primi mensis
I'₂	64–5	De pascha
H'	66–7	De rogationibus
G'	68	De concurrentibus
F'	69	De cyclo solari et lunari
E'	71	De bissexto
D'₁	75	crescit
D'₂	77	De numero momentorum et minutorum et punctorum et horarum
C'₁	79	computo
C'₂	79	Grecorum et Latinorum
B'	81–3	De Anatholio, Macrobio, Victorio, Dionisio, Boetio
A'	84	De calculo

The crux of the chiasmus shows the focus of the computist's interest, the reckoning of the dates of Lent and Easter.

The text contains 631 words. In his *Epistola*, published AD 633, Cummian wrote 633 words in the Prologue and Epilogue, and in his *Historia ecclesiastica gentis Anglorum*, published AD 731, Bede wrote 731 words in the Prologue.[27] If our computist did something similar, one might infer a date of composition in *Anno Passionis* 631.

Professor Dáibhí Ó Cróinín has drawn attention to the dating clause in the text of the computus that follows in MS Oxford, Bodleian Library, Bodley 309:[28]

> Omnibus annis temporibus diebus ac luna maxime que iuxta Hebreos menses facit rite discussis a mundi principio usque in diem quo filii Israhel paschale mysterium initiauere
>
> anni sunt IIIDCLXXXVIIII precedente primo mense VIII kalendas Aprilis luna XIIII VI feria
>
> Passum esse autem Dominum nostrum Ihesum Xpistum peractis VCCXX et VIII annis ab exortu mundi eadem cronicorum relatione monstratur VIII kalendas Aprilis primo mense luna XIIII VI feria

---

[27] Howlett (1997b), 110–6. For other examples see Howlett (1995a), 138–52, 187–93; Howlett (1997a), 493–7; Howlett (1998a), 42–3, 67–8, 80–3, 149, 152; Howlett (1998c), 216; Howlett (1999a), 13–22, 43–7, 50–5, 87–8, 91–4; Howlett (2000a), 181–4.

[28] Ó Cróinín (1983a), 83, repr. Ó Cróinín (2003), 83–4.

Inter primum pascha in Egipto et passionem Domini anni sunt
ĪDXXXVIIII R,

Ex Domini uero passione usque in pascha quod secutum est Suibini filii Commainni anni sunt DCXXXI

A pascha autem supradicto usque ad tempus prefinitum consummationis mundi id est sex milibus consummatis anni sunt CXLI

The mathematics of the text are straightforward enough: from the Passion to the year of writing is 631 (DCXXXI) years. To reduce a Victorian *anno passionis* date to its AD equivalent, add 27: 631+27=AD 658. The second clause in the dating formula provides confirmation of this: from the time of writing to the end of the world – traditionally set at AM 6000 (as in the poem *Deus a quo facta fuit*)[29] – is 141 (CXLI) years. AM 6000−141=AM 5859. To reduce this Victorian *anno mundi* reckoning to AD, subtract 5201: 5859−5201=AD 658. Both formulae, therefore, give the date of writing as AD 658. But the date is further specified by the addition of a most important detail: the author counted from the Passion *usque ad pascha quod secutum est Suibini filii Commani*, 'the Easter of Suibine mac Commáin which has just ensued'. That is to say, the author was referring to a Victorian Easter-table in the margin of which, beside the data for the current year, was noted the death of one Suibine mac Commáin. And the annal is preserved independently in the Annals of Innisfallen *s.a.* AD 658: *Kl. Dub Tíre ua Maíl Ochtraig 7 Conaing mac Muricáin 7 Suibne mac Commáin.*[30] Most importantly of all, however, the Irish compiler of the computus calculated from the pasch of Suibine mac Commáin, that is the Easter of the year in which he died. Easter fell on 25 March in AD 658, so the computist must have been writing after that date. He was undoubtedly resident in southern Ireland.

As the Prologue contains 631 words, so this passage contains 631 letters. Further confirmation of the date *Anno Passionis* 631 and *Anno Incarnationis* 658 is that in the Prologue from | *Deinde etiam interrogandum est* 35 to *De numero annorum ab incarnatione Xpisti* | *usque ad presens tempus* there are 658 letters and spaces between words.

The text of the Prologue contains seventy sentences and eighty-four lines, appropriate numbers for a computist. We may observe in

---

[29] Lapidge and Sharpe (1985), 315. Howlett (1996a), 1–6. This poem, composed AD 645, may provide both a *terminus ante quem* for the arrival of Jerome's *Chronicon* and Isidore's *Etymologiae* in Ireland and a model for our computist's threefold manner of dating his text. See below pp. 283–4.

[30] Mac Airt (1988), 94–5.

passing that 25 March, regarded as the anniversary of the Creation, also the anniversary of the Annunciation, on which renewal of a fallen creation began, is the eighty-fourth day of a year that begins on 1 January, a number coincident with that of an eighty-four-year cycle. The text of the Prologue also contains exactly 3333 letters, in Roman numerals a twelve-figure sum, MMMCCCXXXIII, that omits the intervening numerals DLV.[31]

Our computist has provided in the prologue to his text from *De numero* to *De calculo* a brisk, efficient, and elegant introduction to the study of number, a composition that has retained its freshness and accessibility through fourteen centuries.

This is the background, a fundamental ancient text about computus, a note about the introduction of computus, oral and written, among the Irish during the second half of the sixth century, and three important Irish computistical texts composed during the central third of the seventh century. With this cursory guide to the first century of computus among the Irish, the period from the foundation of Bangor, about 550, to the publication of the Oxford computus in 658, let us turn to consideration of computistic phenomena in Hiberno-Latin literature.

Part II: Computistic Phenomena in Hiberno-Latin Literature

Synodus episcoporum

MS Cambridge, Corpus Christi College, 279, copied from an Insular exemplar and written at the end of the ninth century or the beginning of the tenth in a scriptorium under the influence of Tours, preserves a copy of what may be the oldest extant Hiberno-Latin text, the *Synodus episcoporum*.[32] Though it is often called the *Synodus prima Sancti Patricii*, it may have been attracted by seventh-century propagandists for Armagh to the Patrician dossier from its original milieu, a dossier of texts associated with the Palladian mission. Assuming that Auxilius, Isserninus, and Secundinus were

---

[31] For comparable play in early Insular texts see Howlett (1996b), 132–5; Howlett (1998b), 63–6; Howlett (2000a), 142, 192; Howlett (2005a), 266.

[32] Lapidge and Sharpe (1985), 599. Howlett (1998d), as above nn. 4, 8.

suffragans to Palladius, not to Patrick, one infers from the dedications of churches to them that they worked in central and southeastern Ireland, while Patrick worked in northern and western Ireland.[33] The acts of the synod might be dated after the arrival of Palladius in 431 and the others in 439, between the death of Secundinus in 447 and the death of Auxilius in 459, perhaps to the year 457, when *Anno Domini.cccc°.l°.uii°. Calcedonensis senodus congregatus est*. The acts issue from a time during which authority over a *clericus qui de Brittannis ad nos uenit* mattered, a time during which *more Romano capilli tonsi* mattered. The text, which is well composed, exhibits both the elegant quantitative metrical rhythms of *clausulae* and the stressed rhythms of the *cursus*. In addition to many other phenomena infixed to guarantee the authenticity and integrity of the text, the word 'half way' through sentence 20, line 39, is *dimidium*, and the twentieth letter from the end of the line is the first of *uiginti*. There are forty sentences before canon xxviii, in which the second line, containing the word *quadragesimum*, contains forty letters. Canon xxxiv contains thirty-four syllables. The Incipit, which contains fifty-three letters, is the key to the preliminaries, which contain fifty-three words, and to the complete text, which contains fifty-three sentences, the penultimate sentence containing fifty-three letters and spaces between words. The fifty-three sentences divide by extreme and mean ratio at 33 and 20, 33 dividing by the same ratio at 20 and 13. In the twentieth sentence, as noted above, the twentieth letter from the end is the first of *uiginti*, and after twenty intervening sentences the thirteenth sentence from the end, the line that contains the word *quadragesimum*, contains forty letters. From the beginning of the *Salutatio* to the end of the *Captatio beneuolentiae*, from | *Gratias agimus* to *Melius est arguere quam irasci* | there are thirty-six words. The preliminaries end *Exempla definitionis nostrae inferius conscripta sunt. et sic inchoant.*, two phrases which contain seventy-two letters and spaces and punctuation points (2×36), which introduce thirty-six canons. From | *Gratias agimus* to *salutem.;* | there are 143 letters and spaces and punctuation marks, the key to 143 syllables in the preliminaries. From | *satius* to *irasci* | there are 115 letters and spaces and punctuation marks, the key to 115 lines of text.

---

[33] Thomas (1981), 305 (fig. 57).

Patricii Epistola et Confessio

Two other fifth-century texs that certainly belong in the Patrician dossier are Patrick's own *Epistola ad milites Corotici* and *Confessio*,[34] the texts of which he fixed in various ways.

The *Epistola* is exactly CCXXII or 222 lines long, the number of lines in part I, 39, relating to that in part II, 39, by symmetry, 1:1, that in part III, 89, relating to that in part IIII, 55, by extreme and mean ratio, 144 dividing by the golden section at 89 and 55. One can be certain that the number of lines is correct by comparing the thirty-sixth line of part I, **lupi rapaces deuorantes plebem Domini**, with the thirty-sixth line of part III, **lupi rapaces deglutierunt gregem Domini**, and with the the thirty-sixth line from the end of part III, *tu potius interficis et uendis illos genti exterae ignoranti Deum quasi in **lupa**nar tradis membra Xpisti*. Patrick refers to God thirty-three times.

In the *Confessio* Patrick validates his words for numbers by their placement. In chapter i (1), relating his age at the time of his capture, he states *Annorum eram tunc fere sedecim*, with fifteen letters before | *fere sedecim*.

In chapter viii (17): *Et intermisi hominem cum quo fueram sex annis*, with six words before *sex*, the last letter of which is sixth from the end of the line.

Still in chapter viii (19):

> Et post triduum terram cepimus
> et uiginti octo dies per desertum iter fecimus

in which the third word from the beginning and from the end of the first line is *triduum*. The ten syllables divide into thirds at *tri|duum*, and the twenty-six letters divide into thirds at *tri|duum*. There are twenty-eight letters before | *uiginti octo*. From the next line to the end of the account of Patrick's supplying of food for his starving companions there are twenty-eight lines, followed immediately by the beginning of the account of Patrick's dream of trial by Satan and his dream of release from subsequent capture, in which the twenty-eighth line is *uiginti et octo dies per desertum iter fecimus*.

From the beginning of the prophecy | *Duobus mensibus eris cum illis* | to *nocte illa sexagesima* | there are sixty-one letters, representing two months, one of thirty-one days and one of thirty days. Because of the reckoning of Gen 1:5: *factumque est uespere et mane dies unus*,

---

[34] Lapidge and Sharpe (1985), 25–6; Howlett (1994a); Howlett (2003–4a). Chapter numbers in Roman numerals refer to those of my edition; those in Arabic numerals refer to the traditional chapter numbers.

the day beginning at sundown, Patrick writes punctiliously here, the two months of sixty-one days being fulfilled on the evening of the sixtieth.

At the beginning of chapter viiii (26–28), writing of trial by his elders, Patrick states *Occasionem post annos triginta inuenerunt me aduersus*, with thirty letters and spaces between words from | *occasionem* to *triginta* |.

In chapter xvii (35):

> Breuiter dicam qualiter piissimus Deus de seruitute saepe liberauit et de periculis duodecim qua periclitata est anima mea

with twelve words before | *duodecim* and twelve syllables after *duodecim* |. In chapter xx (52–53): *Et quartodecimo die absoluit me Dominus de potestate eorum*, in which the fourteenth letter is the last of *quartodecimo*. Six lines later:

> Censeo enim non minimum quam pretium quindecim hominum distribui illis ita ut me fruamini
> et ego uobis semper fruar in Deum

in which the fifteenth syllable is the middle of *quindecim*, and the fifteenth word from the end is *quindecim*.

In chapter iii (4) Patrick writes a Creed of thirty-three lines and 144 words and in the chiastic pair to that, chapter xxiiii (57–60), a Doxology of forty-four lines and 265 words. The thirty-first line of the former mentions *filii Dei et coheredes Xpisti*, and the thirty-first line of the latter mentions *filii Dei et coheredes Xpisti*, both echoing Rom 8:16–17, a text quoted only here.

The six chapters that begin part I of the *Confessio*, i–vi (1–8), contain ninety-six lines of text, and the six chapters that end part V, xxi–xxvi (54–62) also contain ninety-six lines of text.

The crux of the chiasmus of the central part III is also the crux of the entire text, the internally chiastic chapter xiiii, which recounts Patrick's elevation to the episcopate. Its seventeen lines divide by extreme and mean ratio at 11 and 6, and its ninety words divide by the same ratio at 56 and 34. The eleventh line is *Ecce dandus es* | *tu* | *ad gradum episcopatus*, and the fifty-sixth word is the central word *tu*. The central 2285th word of the entire *Confessio* is *tu*.

Audite omnes amantes Deum

The 'Hymn of Saint Secundinus' or 'Saint Sechnall's Hymn' *Audite omnes amantes Deum* in praise of Patrick is a ninety-two-lined, pentadecasyllabic composition in twenty-three stanzas, of which the first letters follow the order of the Latin alphabet from A to Z.[35] As the text must be earlier than the seventh-century Hiberno-Latin texts which quote from it and allude to it, the 577 words of the hymn may represent the year of its composition, AD 577.

Columbani Carmen de Mundi Transitu et Precamur Patrem

The earliest Hiberno-Latin author whose works have descended to us under his own name, Columban of Bangor, founder of Luxeuil, Annegray, and Bobbio, composed *Epistolae*, *Sermones*, a *Poenitentiale*, a *Regula Coenobialis*, and two poems before his death in 615.[36] One, about the transitory world, *Carmen de Mundi Transitu*, is calendrical, containing seven syllables in each line, for days of a week, four lines in each stanza, for weeks in a month, thirty stanzas, for days in a month, 120 lines, for a decade of months, and 365 words, for days in a year. The other, an Easter hymn, *Precamur Patrem*, contains twelve syllables in each line, six lines each in the Prologue and Epilogue, together twelve, two twelve-lined halves in part I, together twenty-four, and two twenty-four-lined halves in part II, together forty-eight. The eighty-four-lined hymn thus exhibits repeated diminution in duple ratio 2:1.

Versus de annis a principio

An occasional poem, *Versus de annis a principio, Deus a quo facta fuit huius mundi machina*, contains, like *Audite omnes amantes Deum*, fifteen syllables in each line.[37] It considers the six ages of the world, first in the Biblical tradition and second in the Classical tradition. At the end of the Biblical passage the poet resumes the six ages from Abraham to Christ, and at the beginning and the end of the Classical passage he synchronizes pagan with Biblical history. Then third he considers the Irish tradition, dating his own time three ways, from the baptism of Christ to the death of Domnall king of the Irish

---

[35] Lapidge and Sharpe (1985), 573. Howlett (1995a), 138–52.

[36] Lapidge and Sharpe (1985), 532, 639–42, 819. Howlett (1995a), 82–91, 156–77; Howlett (1997b), 70–5.

[37] Lapidge and Sharpe (1985), 315. Howlett (1996a), 1–6. See above p 282.

(8+10+600+27) or AD 645, second from that date forward to the end of the period of his reckoning (6000−155=5845), and third backwards to the Creation, with which his poem began (5845−5200=645). The number of the ages of the world (6) squared yields the number of lines in the poem (36), which squared yields the number of letters in the poem (1296).

Aileranus Sapientis Interpretatio Mystica et Canon Euangeliorum

Aileranus Sapiens, lector of Clonard (†c.655), wrote two works that have survived to modern times. One is prose, an exposition of the list of forty-two ancestors of Jesus in Matthew's Gospel,[38] extant in three manuscripts of the ninth, tenth, and fourteenth centuries, and in two manuscripts of a recension by Sedulius Scottus, both of the ninth century. The title

AILERANI SCOTTI INTERPRETATIO MYSTICA PRO-GENITORUM DOMINI NOSTRI IESU CHRISTI IN NATIVITATE SANCTAE GENITRICIS IPSIUS LEGENDA

contains CXI or 111 letters, which precede the Introduction of 111 words and 333 syllables.

The other work is verse, a poem, an exposition of the 650 numbers in the Eusebian Canons that allow a reader to coordinate similar passages in varying combinations of the four Gospels.[39] This most widely disseminated Hiberno-Latin poem survived into the modern period in fifteen manuscripts. Composed in rhyming hendecasyllables it contains ten stanzas, 1–6 and 8–9 four lines long, 7 two lines long, and 10 eight lines long, together forty-two lines, coincident with the number of ancestors in the *Interpretatio Mystica*. The stanzaic structure and rhyme scheme coincide exactly with the sentence structure, except in stanza 10, which exhibits a single sentence but two rhymes.

The subject is numbers, but Aileran has presented the text as a form of mythology, in which a man, a lion, an ox, and an eagle, the symbols of the Four Evangelists, talk with each other in ten varying combinations. The number of canons in the lists of both canonists is identical,

---

[38] Lapidge and Sharpe (1985), 299. Howlett (1996a), 6–11.
[39] Lapidge and Sharpe (1985), 300. Howlett (1996a), 11–20; Howlett (2001b), 22–6.

650, though Eusebius and Ailerán have disposed the canons differently. The number may represent, as well as the number of canons, the year of composition, AD 650, about five years before the poet's death. The number of letters in the title, 34, AILERANI SAPIENTIS CANON EVANGELIORUM, an octosyllabic couplet, and in the poem, 1077, is MCXI or IIII, with which one may compare the III letters of the title and III words of introduction in Ailerán's other work.

Augustinus Hibernicus De mirabilibus sacrae scripturae

An Irishman who named himself as Augustinus son of Eusebius wrote for the clerics of the church of Carthage, *Carthaginensis*, a treatise in three books *De Mirabilibus Sacrae Scripturae*.[40] Here follows part of Book II chapter iv, about the halting of sun and moon in their courses in Jos 12–14, which allows us to fix the date.

1   Ut enim hoc manifestis approbationibus pateat cyclorum etiam ab initio conditi orbis recursus in se bréuiter dìger*émus*
    quos semper post **quingentos triginta duos** annos sole ut in principio et luna per omnia conuenientibus nullis subuenientibus impedimentis in id unde coeperant redíre ostend*émus*
    Quinto namque cyclo a mundi principio anno centesimo quarto decimo generale totius mundi diluuium súb N*oe* u*énit*
    qui post diluuium quadringentesimo decimo octáu*o* defécit
5   Et inde alius incipiens id est **sextus** in octauo aetatis Abrahae ánno fin*ítur*
    Et nono eius anno **septimus** incipiens tricesimo quinto anno egressionis filiorum Israhel de Aegypto quinquennio ante mortem Móysi conclúd*itur*
    Post quem **octauus** in quo etiam istud signum in sole et luna factum tricesimo sexto anno egressionis Israhel de Aegypto incipiens in tricesimum primum annum Asa regis Iúda incéd*it*
    cuius tricesimo secundo anno **nonus** exordium capiens in quo etiam aliud signum in sole Ezechiae regis tempore de quo paulisper dicemus fáctum lég*itur*
    centesimo octauo anno post templi restaurationem quae sub Dario facta est sui cursus spátium cònsummá*uit*

---

[40] Lapidge and Sharpe (1985), 291.

10   donec **decimus** inde oriens nonagesimo secundo anno post
     passionem Saluatoris Auiola et Pansa consulibus cúrsibus
     cònsummátur
11   Post quem **undecimus** a consulatu Paterni et Torquati ad
     nostra usque tempora decurrens extremo anno
     Hiberniensium moriente Manchiano inter ceteros sapiéntes
     perági*tur*
12   Et **duodecimus** nunc tertium annum agens ad futurorum
     scientiam se praestans a nobis qualem finem sit habitúrus
     ignorá*tur*
     quorum unusquisque uniformi statu peractis **quingentis
     triginta duobus** annis ín semetípsum
     id est in sequentis initium reuóluitur complét*is*
15   uidelicet in unoque solaribus octouicenis nónodécie*s*
16   et in lunaribus decemnouenalibus uicies ócties círcul*is*

Our author composed, despite the technical nature of his subject, elegantly rhythmical rhyming prose. After his account of the *quintus cyclus* he wrote six words before | *sextus*, then seven syllables before | *septimus*, then eight letters before | *octauus*. In the account of the eighth cycle the thirty-second word is *tricesimo secundo* |, and from | *tricesimo secundo anno* there are nine syllables before | *nonus*. In the account of the eleventh cycle the words *extremo anno Hiberniensium moriente Manchiano* refer to the death of Manchianus of Min Droichit in AD 652. In the account of the twelfth cycle in the twelfth line the twelfth letter is the last of *duodecimus* |, the third year of that cycle, *nunc tertium annum agens*, being the year of composition, AD 655. From the second | *quingentis triginta duobus* the nineteenth word is *nonodecies*. In the penultimate line the nineteenth syllable is *nono* | *decies*. After *nonodecies* | the nineteenth letter is the first *n* of *decemnouenalibus*. After *octies* | there are eight letters to the end of the passage. From | *post quingentos triginta duos annos* to *peractis quingentis triginta duobus annis* | there are exactly 532 syllables.

Laidcenni Egloga et Lorica

The *Moralia in Job*, written by Gregory the Great between 579 and 600 or 602, was abbreviated by Laidcenn mac Baíth Bannaig of Clúain Ferta Mo Lua, who died in 661.[41] The work survives in nine manuscripts, seven written during the eighth and ninth centuries, one

---

[41] Lapidge and Sharpe (1985), 293–94. Howlett, (1995b), 6–18; Howlett (1996a), 68–9.

during the eleventh, and one during the thirteenth, and references to it in medieval library catalogues attest its wider diffusion. Laidcenn also composed a *Lorica* that survives in seven manuscripts written from the eighth century to the sixteenth. Part I invokes the powers of heaven to protect the poet, and part II invokes their protection of 144 parts of his body. Each line contains, like Ailerán's *Canon Euangeliorum*, eleven syllables. Each of the two parts is introduced by a two-lined *Inuocatio* of twenty-two syllables. There are twenty-two quatrains, each containing two couplets of twenty-two syllables. The twenty-second stanza contains twenty-two words. The word that gives the poem its title occurs twice, with twenty-two words from | *lorica* 49 to *lorica* | 53. The entire poem contains MMCCCCXXXXII or 2442 letters (111×22).

Laidcenn refers to himself fifteen times at places determined by epogdous or sesquioctave ratio, 9:8 and 1/9 and 8/9.

The central word of a central line 47, *capitali centro cartilagini*, is *centro*, and the line is an anagram with the same rhyme and the same rhythm as the verse in which it is embedded, *ti[bi] practica Lorica Laitgenni* 'for you a practical *Lorica* of Laidcenn'.

The first six lines may suggest the date.

Suffragare Trinitatis Unitas	Undertake support, Unity of Trinity,
Unitatis miserere Trinitas	be merciful, Trinity of Unity;
Suffragare queso mihi posito	undertake support, I ask, for me placed
Maris magni uelut in periculo	as if in the peril of a great sea,
Ut non secum trahat me mortalitas	so that the mortality of this year may not
Huius anni neque mundi uanitas.	draw me with it nor the vanity of the world.

The last syllable of *huius anni* is the fifty-ninth of the poem, representing perhaps the year of the century. The sixth line might represent the month of the year. The twenty-third word might represent the day of the month. In the year AD 659 23 June was the second Sunday after Trinity, the twenty-second day from Whitsun inclusive, which occurred after the twenty-second week of that year, two years before the Annals of Ulster record the death of *Laidggnen sapiens mac Baith Bannaigh*.[42]

---

[42] Mac Airt and Mac Niocaill (1983), 132–3.

Cummiani Longi Opera

About the time of the supposed departure of Columban from Bangor, about 590, the Annals of Ulster record the birth *s.a.* 592 *Natiuitas Cummeni longi*, known in Irish as *Cuimmíne Fota*, in Latin as *Cummianus Longus*.[43] In the forty-second year after that Cummian published the *Epistola* considered above, and shortly after that *De ratione conputandi*. Kenney described a *Penitentiale* ascribed in the manuscripts to Cummean as 'the most important of all',[44] and Bieler described the same work as 'the most comprehensive of the Irish [Latin] penitentials'.[45] The Prologue contains twelve headings, for months of a year, thirty lines, for days of a month, and 365 words, for days of a year. The Epilogue contains thirty-one lines, for days of a month, and half as many words, 182, ninety-one words in lines 1–16 and ninety-one words in lines 17–31.[46]

Citing a reference in 'the Carolingian MS. Angers 44(48) [. . .] fol. 7r [. . .] *novellum auctorem in Marcum nomine Comiano*', Bernhard Bischoff ascribed a Commentary on the Gospel according to Mark that survives in nearly ninety manuscripts, one of which contains Old Irish glosses, to Cummian.[47] The Prologue contains thirty sentences.

The hymn *Celebra Iuda* survives in two copies of the Irish *Liber Hymnorum*, both written during the eleventh century, bearing ascription to *Cumaini Fota* or *Cumineus Longus*.[48] The metre is, like that of Columban's *Precamur Patrem*, dodecasyllabic, appropriate for the subject of the Twelve Apostles. Like Laidcenn Cummian refers to himself at 1/9 and 8/9, the 197 words of *Celebra Iuda* dividing at 22 and 175. Twenty-two words before the *Gloria Patri* we read *pro | nobis*.

From nine manuscripts Bischoff edited a little mnemonic text *De figuris Apostolorum*, ascribed in one manuscript to *Comianus Longus*.[49] The text contains twelve lines, for months of a year, thirty words (eighteen lemmata and twelve names, 30 dividing by extreme and

---

[43] Mac Airt and Mac Niocaill (1983), 94–5.
[44] Lapidge and Sharpe (1985), 601; Kenney (1966), 241 (nr. 73).
[45] Bieler (1975), 5.
[46] Howlett (1996a), 33–6.
[47] Bischoff (1954), 257–9 (nr. 27); Lapidge and Sharpe (1985), 345; Howlett (1996a), 36–40; CCSL 82.
[48] Lapidge and Sharpe (1985), 582. Howlett (1996a), 40–6.
[49] Bischoff (1932), repr. in Bischoff (1967), 167–8; Lapidge and Sharpe (1985), 292. Howlett (1996a), 46–7.

mean ratio at 18 and 12), for days of a month, and 365 letters, for days of a year.

The grammatical tract *Anonymus ad Cuimnanum*,[50] described by Vivien Law as 'one of the most interesting products of seventh- and eighth-century grammatical scholarship to survive',[51] begins with a *Dedicatio operis* that contains twelve sentences, thirty lines, and 365 words.

The Annals of Ulster record *s.a.* 662 that *Cummeni Longus.lxx.ii⁰.anno etatis sue quieuit*.[52] The names CVMMENE and LADCEN survive on the same grave slab.[53]

Tírecháni Episcopi Collectanea

Bishop Tírechán may have been a grandson of Amolngid, a descendant of Conall son of Énde, and may have come from Tirawley. To advance the claim of the church at Armagh to metropolitan status he composed the *Collectanea*, which he introduced with one statement in the third person, *Tirechan episcopus haec scripsit ex ore uel libro Ultani episcopi cuius ipse alumpnus uel discipulus fuit*, and one statement in the first person, *Inueni quattuor nomina in libro scripta Patricio apud Ultanum episcopum Conchuburnensium*.[54] As his tutor Bishop Ultán died in AD 657, Tírechán is often assumed to have written the *Collectanea* some time later, about AD 670. The second paragraph states that

> Patricius sexto anno babtitzatus est, uigesimo captus est, quindecim seruiuit, quadraginta legit, sexaginta unum docuit, tota uero aetas centum undecim.
> Haec Constans in Gallis inuenit.

Tírechán's reckoning differs from one that supposes Patrick, born about AD 390, to have died according to one obit in the Annals of Ulster *s.a.* 461 in about his seventy-second year, or according to another obit in the same text *s.a.* 492 when he was about 101.[55] It differs also from the

---

[50] Lapidge and Sharpe (1985), 331. Howlett (1996a), 47–8.
[51] Law (1982), 90.
[52] Mac Airt and Mac Niocaill (1983), 132–3.
[53] Lionard (1961), 104–5. Moloney (1964).
[54] Bieler (1979), 124.
[55] Mac Airt and Mac Niocaill (1983), 46–7, 54–5.

120-year life with which Muirchú moccu Macthéni credits Patrick, like Moses.⁵⁶ It differs more particularly from the total of years that Tírechán himself supplies. One might infer that Tírechán took the numbers of years in distinct parts of Patrick's career from his acknowledged source, but then behaved as an independent computist, reckoning for himself the gematria value of the name PATRICIVS as 15+1+19+17+9+3+9+20+18 or 111 and attributing to Patrick a life of 111 years.

Memoria abbatum nostrorum

One poem in the Antiphonary of Bangor, entitled *Memoria abbatum nostrorum*, beginning *Sancta sanctorum opera*, is composed in rhyming octosyllables.⁵⁷ Both diction and rhyme of the eight-lined Prologue are echoed in the eight-lined Epilogue, which with a two-lined refrain surround four six-lined stanzas, the initial letters of twenty-three of the twenty-four lines following, as in *Audite omnes amantes Deum*, the order of the Latin alphabet. Prologue, Epilogue, and refrains occupy twenty-four lines and seventy-five words and 192 syllables; the four stanzas of alphabetic verse occupy twenty-four lines and seventy-five words and 192 syllables. Prologue, alphabetic stanzas, refrains, Epilogue, and final couplet together comprise 153 words, 153 being the number of fishes in the net in Jn 21:11 and the triangular number 1–17. There are 152 letters in the Prologue and 153 letters in the Epilogue, exactly 1000 letters in the entire poem, of which the 500th is the first letter of the central twenty-sixth line.

The reference to the most recent abbot Crónán in the present tense subjunctive mood and future tense indicative mood are usually assumed to indicate composition during the period of his régime, AD 680–691. The number of letters in the Prologue, the four alphabetic stanzas, and the refrain in the future tense about Crónán, 152+120+124+121+125+41, total 683, which may suggest the year of composition of the poem.

Versiculi Familiae Benchuir

Another poem in the Antiphonary of Bangor, entitled *Versiculi familiae Benchuir*, beginning *Benchuir bona regula*, is composed in rhyming

---

⁵⁶ Howlett (2006c), 120.
⁵⁷ Lapidge and Sharpe (1985), 576. Howlett (1995a), 187–9.

heptasyllables, arranged in ten stanzas of four lines each.[58] The subject is the monastic family of Bangor, to which the poet refers in lines 1 and 5. The 119 words of the poem divide by 1/9 and 8/9 at 13 and 106. Between *Benchuir* | 1 and | *Benchuir* 5 there are thirteen words. The number of letters in the complete poem, 686, may suggest the year of composition.

Adomnanus De Locis Sanctis

After a journey to the Holy Land, perhaps during the third quarter of the seventh century, a Gaulish bishop named Arculf, driven off course on a sea voyage, landed at Iona, where he dictated a description of holy places he had seen to Adomnán, ninth abbot of Iona, AD 679–704, biographer of the founder, Saint Columba, and author of some scholia on Vergil's Eclogues. Adomnán's *Liber de Locis Sanctis* enjoyed a wide circulation during the Middle Ages.[59] About AD 686 the author presented a copy to Aldfrith king of the Northumbrians, who had further copies made. Before AD 735 the Venerable Bede revised and rewrote the book. Sedulius Scottus quoted it. More than twenty manuscripts and fragments survive to this day.

In the text that follows, the first chapter of the first book, Adomnán's account of the walls of Jerusalem, letters and punctuation marks in boldface represent features of MS Wien, Österreichische Nationalbibliothek, 458 (*olim* Salisburgensis 174), 2r–v, written for Baldo, a teacher at the cathedral school of Salzburg about the middle of the ninth century.

DE SITU HIERUSALEM

De situ Hierusalem nunc quedam scribénda sunt paúca
ex his que mihi sanctus dictáuit Arcúlfus.
Ea uero que in aliorum libris de eiusdem ciuitatis positione répperiúntur.
a nobis prétermitténda sunt.

i	In cuius magno murorum ambitu.idem Arculfus.LXXXIIII númerauit túrres.	10	29	60
	et portas bis ternas quarum per circuitum ciuitatis órdo sic pónitur.	11	23	58
ii	Porta Dauid ad occidentalem montis Sion partem. príma numerátur	9	22	54
iii	Secunda porta Vílle Fullónis	4	10	25
iiii	Tertia porta Sáncti Stéphani	4	10	25

---

[58] Lapidge and Sharpe (1985), 574. Howlett (1995a), 189–93.
[59] Lapidge and Sharpe (1985), 304–5.

v	Quarta pórta Béniamin	10	3	7	19
vi	Quinta portula. hoc est páruula pórta		6	12	31
vii	Ab hac per gradus ad uallem Iósafat descénditur		8	15	40
viii	Sexta pórta Tecúitis		3	8	18
viiii	Hic itaque ordo per earundem portarum et turrium Íntercapédines.		9	24	55
	a porta Dauid supramemorata per circuitum septémtrionem uérsus.	15	8	23	55
	et exinde ad oriéntem dirígitur		5	13	27
x	Sed quamlibet sex porte in múris numeréntur.		7	14	37
	Celebriores tamen ex eis portarum intróitus frèquentántur.		7	21	51
	Unus ab óccidentáli.		3	8	17
	alter a septémtrionáli.	20	3	9	20
	tertius ab órientàli párte,		4	11	23
	Ea uero pars murorum cum interpósitis túrribus.		7	17	40
	que a supra descrípta Dauid pórta.		6	11	28
	per aquilonale montis Sion súpercílium.		5	15	34
	quod a meridie superéminet cìuitáti.	25	5	15	31
	usque ad eam eiusdem montis fróntem dirígitur.		7	16	39
	que prerupta rupe orientalem réspicit plágam.		6	16	39
	Nullas habere pórtas conprobátur	28	4	11	29
			144	360	855

1 quędam. 2 quae. 3 quae. 4 praetermittenda. 5 *from* quarum *to* ponitur *in margin with a signe de renvoi.* 8 uillae. 14 Intercapidines. 17 quaelibet. portę. numerantur. 23 quae. 27 quae praerupta.

ABOUT THE SITE OF JERUSALEM

About the site of Jerusalem some few things are now to be written
from those which holy Arculf dictated to me;
those things, however, which are found in others' books
about the position of the same city
are to be omitted by us.

i   In the great ambit of its walls the same Arculf
      numbered eighty-four towers   5

	and twice-three gates, of which through a circuit	
	of the city the order is placed thus.	
ii	The Gate of David on the western part of	
	Mount Zion is numbered first.	
iii	Second the Gate of the Villa of the Fuller.	
iiii	Third the Gate of Holy Stephen.	
v	Fourth the Gate of Benjamin.	10
vi	Fifth a small gate, that is the Little Gate.	
vii	From this by steps one is brought down to the	
	Valley of Josaphat.	
viii	Sixth the Gate of Tekoa.	
viiii	And so this order is directed through the spaces	
	between the same gates and towers	
	from the abovementioned Gate of David through a	
	circuit toward the north	15
	and thence toward the east.	
x	But although six gates are numbered in the walls,	
	from among them, however, the more celebrated	
	entrances of gates are used frequently,	
	one from the western,	
	a second from the northern,	20
	a third from the eastern part,	
	for that part of the walls with interposed towers	
	which from the Gate of David described above	
	through the northern ridge of Mount Zion,	
	which towers over the city from the south,	25
	until it is directed as far as the side of the same	
	mountain,	
	which with an abrupt cliff looks round on the	
	eastern expanse	
	is proved to have no gates.	28

The passage begins with an introductory sentence of thirty words, which divide by symmetry at 15 and 15, at the name of Adomnán's informant, *Arculfus* |. The thirty words divide by extreme and mean ratio at 19 and 11, at Adomnán's reference to himself, | *mihi*, and his reference to other writers, | *aliorum*. There is another golden section in the nineteen words between *Hierusalem* | and | *eiusdem ciuitatis*. Adomnán refers to his subject and himself at one-ninth and eight-ninths of the sentence, at 3 and 27, *Hierusalem* | and | *nobis*.

The 144 words divide by symmetry at 72 and 72, at *per* | *circuitum*. They divide by 1/9 and 8/9 at 16 and 128, at *per* | *circuitum*. There are seventy-two words from | *portas bis ternas* to *sex* | *porte*. There are seventy-two words from | *occidentalem* to | *occidentali*. There are twenty-five words from *septemtrionem* | to | *septemtrionali*. There are twenty-five words from | *orientem* to *orientali* |.

The 144 words divide by duple ratio 2:1 at 96 and 48. There are ninety-six words from | *montis Sion* to *montis Sion* |. There are also ninety-six words from | *porta Dauid* to *Dauid porta* |, which divide by symmetry at 48 and 48, at *porta* | *Dauid*.

As Adomnán knew from Apoc 21:17 that the walls of the heavenly Jerusalem contain 144 cubits, his account of the walls of the earthly Jerusalem occupies 144 words. As Adomnán knew from Apoc 21:12 that the heavenly Jerusalem has twelve gates, he writes, after stating that the earthly Jerusalem has only *portas bis ternas*, the word *porta* twelve times. As Adomnán states twice that he is taking us *per circuitum* 'through a circuit', his account occupies as many syllables as there are degrees in a circle, 360. The 144 words of the account divide by extreme and mean ratio at 89 and 55. Between *circuitum* | and | *circuitum* there are fifty-five words. From | *dirigitur* to *dirigitur* | inclusive there are fifty-five words. On the circuit of the walls are eighty-four towers, appropriate for an Irish computist who reckoned Easter on a cycle of eighty-four years. The use of space to separate words is a convention devised in these islands not later than the sixth century, exhibited above in the note about Mo Sinu maccu Min and easily demonstrable in a wide range of Cambro-, Hiberno-, and Anglo-Latin texts of the seventh century. In Adomnán's prose we observe that the number of letters, 855, added to the number of spaces between words and before the first word and after the last word is 1000, so that the number of letters and spaces is the cube of the number of sentences, perfection.

The number of lines in the passage is a perfect number, 28. All but three lines end in good clausular rhythms, two, 11 and 17, exhibiting hexameter endings, and one, 22, exhibiting a *cursus tardus*.

Cogitosus of Kildare Vita Sanctae Brigitae

To advance the claim of the church at Kildare to metropolitan status Cogitosus published in AD 693 a *Vita Sanctae Brigitae* in thirty-two chapters. Here follows the beginning and the end of the thirty-second chapter.

Part I

1 XXX II [i.e. triginta duo]
2 Nec et de miraculo in reparatione eclesiae fácto tacéndum es*t* a
3 in qua gloriosa **amborum** hoc est archiepiscopi Conleath
    et huius uirginis florentissimae **Brígitae** córpor*a* b
4 a dextris et a sinistris altaris decorati in monuméntis pósit*a* b
5 ornatis uario cultu auri et argenti et gemmarum et pretiosi
    lapidis atque coronis aureis et argenteis
    dé super pendént*ibus* c
6 ac diuersis imagin*ibus* cum celaturis uariis et colór*ibus*
    rèquiéscun*t* cca 64

Part VI

73 Veniam peto a fratribus ac <u>lec</u>toribus haéc <u>leg</u>én*tibus*
74 <u>im</u>mo <u>é</u>mendán*tibus*
75 qui <u>cau</u>sa obed<u>ién</u>tiae <u>co</u>ác*tus*
76 nulla praerogatiua <u>sci</u>én*tiae* <u>s</u>ubfúl*tus*
77 pelagus inmensum uirtutum <u>b</u>eátae **Brígitae**
78 et <u>uir</u>is perití<u>ssimis</u> <u>fo</u>rmidándum
79 his paucis rustico sérmone díc*tis*
80 uirtutibus de maximis et innúmerabíl*ibus*
81 parua líntre cucúrri
82 Orate pro mé | **Cogitóso** 693
83 nepote culpábili **Aédo**
84 et ut audaciae méae indùlgeá*tis*
85 atque orationum uestrarum clipe*o* Domino me
    commendétis exór*o*
86 Et **Deus** uos pacem euangelicam sectántes exaúdiat 67
87 Amen
88 Explicit Vita Sanctae Brígitae uírginis 6
    73
    723

Part I

1 XXX II [Thirty-two]
2 Nor, also, is it fit for silence to be kept about a miracle
    performed in the repair of the church
3 in which the glorious bodies of both, this is of
    Archbishop Conlaed and of this most flourishing
    Brigit, rest
4 placed in monuments on the right and on the left [sides]
    of a decorated altar

5   with ornaments hanging from above with varied
       adornment of gold and silver and gems and precious stone
       and crowns gilded and silvered
6   and with diverse images with varied carvings
       [or 'paintings', 'panellings', 'canopies'] and colours.

Part VI

73 Pardon I seek from brothers and readers reading these things
74 indeed emending
75 who compelled for the sake of obedience
76 supported by no prerogative of learning
77 the immense sea of the virtues of blessed Brigit
78 to be feared by the most learned men
79 with these few sayings in rustic speech
80 about very great and innumerable virtues
81 with a small boat I have run on to.
82 Pray for me, Cogitosus,
83 reprehensible nephew to Áed
84 and that you may be indulgent to my audacity
85 and that you may commend me to the Lord with the
       shield of your prayers I pray earnestly.
86 And may God hear you out pursuing evangelical peace.
87 Amen.
88 The Life of holy Brigit the virgin ends.

Reckoning one line and two words for the chapter number XXX II, *triginta duo*, and five lines and sixty-two words of narrative, there are six lines and sixty-four words. The six lines prefigure the six parts of the narrative. The sixty-four words represent the alphanumeric value of the name BRIGITA, 2+17+9+7+9+19+1 or 64, which divide by extreme and mean ratio at 40 and 24, at *florentissimae | Brigitae*. In line 3 the thirty-second syllable is the central syllable of *Brigitae*, coincident with her commemoration on 1 February, the thirty-second day of the year, the thirty-second line of a calendar that begins at 1 January. The entire paragraph contains 365 letters, one for each day of an ordinary solar year.

In Part VI coincident with the alphanumeric value of the name BRIGITA, there are sixty-four syllables from the beginning to | *Brigitae*, and sixty-four syllables from *Brigitae* | to *Cogitoso nepote culpabili Aedo* |. Coincident with the alphanumeric value of the nominative form of the name COGITOSVS, 3+14+7+9+19+14+18+20+18 or 122, there are 122 syllables from the beginning to *Cogitoso* |. Coincident with the alphanumeric value of the ablative form COGITOSO,

3+14+7+9+19+14+18+14 or 98, there are ninety-eight letters and spaces between words from the space before | *pro me Cogitoso* to the space after *me* |. Coincident with the alphanumeric value of the name AEDO, 1+5+4+14 or 24, the twenty-fourth word after *Brigitae* | is *Aedo* |. Coincident with the alphanumeric value of DEVS, 4+5+20+18 or 47, there are from the space before | *Deus* to the space after *exaudiat* | forty-seven letters and spaces between words.

Let us recapitulate. Part I of the concluding chapter suggests by its number, 32, the day of the year on which Brigit is commemorated. It suggests by the number of its words, 64, the alphanumeric value of Brigit's name, which relates to the former number by duple ratio 2:1, 64:32. It suggests by the number of its lines the number of parts in the chapter, 6, and by the number of its letters the number of days in an ordinary solar year, 365. It describes three artefacts, one altar flanked by two tombs, one of Conlaed and one of Brigit.

The last lines of part VI contain six words, which recall the six lines of part I. The seventy-three words of part VI confirm the line number 73, with which part VI begins.

From the beginning of chapter XXXII to *Orate pro me* | *Cogitoso* there are 693 words, suggesting perhaps that Cogitosus published his *Vita Sanctae Brigitae* in AD 693, about thirty years after publication of Ailerán's *Vita Sanctae Brigitae*, six or seven years before Muirchú published his *Vita Sancti Patricii* not later than the year 700. The year 693 was slightly more than 300 years after the birth of Patrick, supposing that to have occurred about 390, slightly more than 250 years after the birth of Brigit, supposing that to have occurred about 439.

Muirchú moccu Mactheni, Vita Sancti Patricii

To supersede the claim of the church at Kildare and to advance the claim of the church at Armagh to metropolitan status Muirchú moccu Mactheni published not later than AD 700 his *Vita Sancti Patricii*, of which chapter X follows.[60]

> In illis autem diebus quibus haec gesta sunt
> in praedictis regionibus fuit rex quidam magnus ferox gentilisque
> inperator barbarorum regnans in Temoria
> quae erat caput Scotorum.
> Loiguire nomine filius Neill   5

---

[60] Lapidge and Sharpe (1985), 303. Howlett (2006c), 58–61, 146–8.

origo stirpis regiae huius paene insulae
Hic autem sciuos. et magos. et aruspices.
et incantatores et omnis malae artis inuentores habuerat.
qui poterant omnia scire et prouidere ex more gentilitatis et
    idolatriae antequam essent.
e quibus duo prae caeteris praeferebantur   10
quorum nomina haec sunt
Lothroch qui et Lochru. et Lucetmail qui et Ronal.
et hii duo ex sua arte magica crebrius prophetabant morem quendam
    exterum futurum
in modum regni cum ignota quadam doctrina molesta
longinquo trans maria aduectum   15
a paucis dictatum.
a multis susceptum.
ab omnibus honorandum.
regna subuersurum.
resistentes reges occisurum.   20
turbas seducturum.
omnes eorum deos destructurum.
et eiectis omnibus illorum artis operibus. in
    saecula regnaturum.
Portantem quoque suadentemque hunc
    morem signauerunt
et prophetauerunt hiis uerbis   25
quasi in modum uersiculi crebro ab hiisdem dictis.
maxime in antecedentibus aduentum Patricii duobus
    aut tribus annis.
Haec autem sunt uersiculi uerba *propter linguae idioma*
    non tam manifesta :.
"Adueniet ascicaput cum suo ligno curuicapite
Ex sua domu capite perforata incantabit nefas   30
A sua tabula ex anteriore parte domus suae.
respondebit ei sua familia tota '.*Fiat Fiat.*'"
Quod nostris uerbis potest manifestius exprimi.
Quando ergo haec omnia fiant.
regnum nostrum quod est gentile non stabit   35
quod sic postea euenerat.
euersis enim in aduentu Patricii idolorum culturis.
fides Xpisti catholica nostra repleuit omnia.
De hi*is ista sufficiant.*
Redeamus ad propositum.   40

In those days, however, in which these things were accomplished
in the foresaid regions there was a certain king great, fierce,
    and gentile

an emperor of the barbarians reigning in Tara
which was the capital [lit. 'head'] of the Scots [*i.e.* Irish]
Loiguire by name, son of Neill   5
the origin [lit. 'rising'] of the royal lineage of almost
    all this island.
This man, however, had men possessing knowledge and
    magi [or 'druids'] and diviners
and enchanters and finders of every evil art
who could know all things and foresee by the custom of
    gentile belief and idolatry before they would be
from among whom two before the others were preferred,   10
whose names are these,
Lothroch, who also [was named] Lochru, and Lucetmail,
    who also [was named] Ronal
and these two by their own magic [or 'druidic'] art used
    rather frequently to prophesy that a certain alien custom
    was coming
in the manner of a realm with a certain unknown
    distressing teaching
borne across the seas from a long way away,   15
preached by few,
undertaken by many,
to be honoured by all,
about to subvert realms,
about to kill resisting kings,   20
about to seduce throngs,
about to destroy all their gods,
and with all the works of the art of those men hurled away
    about to reign for ages.
Also the one bearing and persuading to this custom they signaled
and prophesied in these words,   25
as if in the manner of a little verse frequently in these same words
especially in the two or three years leading up to the coming
    here of Patrick.
These, however, are the words of the little verse, not very manifest
    on account of the idiom of the language.
'Adze-head will come here with his own curve-headed wood
from his own house with the head perforated he will sing
    something unutterable   30
from his own table from the forward part of his own house
all his own family will respond to him "let it be done, let it be done"'
which can be expressed more manifestly in our words:
when therefore all these things may be done
our realm, because it is pagan, will not stand   35
which came about thus afterward

> for with the cults of idols overturned at the coming here of Patrick
> our catholic faith of Christ refilled all things.
> About these matters those things suffice.
> Let us return to the proposed matter.   40

Chapter X is remarkable for many reasons. One cannot read this passage aloud without noting Muirchú's close attention to rhythm and rhyme and many other structural phenomena. In lines 2–6 there are from | *fuit* to *Neill* | forty-six syllables and from | *Neill* to *insulae* | forty-six letters and spaces between words, coincident with the alphanumeric value in the eighteen-letter Irish system of NEILL, 12+5+9+10+10 or 46. The twenty-eighth line of the chapter introduces four verses that contain twenty-eight words, the verses before the caesuras containing twenty-eight syllables.[61] The number 28 is a perfect number and also a triangular number 1+2+3+4+5+6+7. The verses are noteworthy octodecasyllables. The verses in the first couplet share an identical number of syllables arranged on each side of the caesura (8–10) and an identical number of letters (39). The verses in the second couplet share an identical number of syllables arranged on each side of the caesura (6–12) and an identical number of letters (35). The verses end with the word FIAT, which bears a numerical value of 6+9+1+20 or 36. As the word occurs twice, its value of 72 coincides with the number of syllables in the verses it ends. The number 72 is half of the number of cubits in the wall of the heavenly Jerusalem, 144.

The entire passage contains 245 words, which divide into 2/3 and 1/3 and 163 and 82, at *duobus aut tribus* |. From | *Et hii duo* [...] *prophetabant* to *fides Xpisti* | there are exactly 888 units, coincident with the value in the Greek alphanumeric system of the name IHCOYC 10+8+200+70+400+200 or 888.[62] From | *e quibus duo* to *hii duo* | inclusive there are twenty-two words.

The most securely fixed date in Patrician scholarship is AD 431, the year in which Prosper of Aquitaine wrote in his Chronicle *ad Scotos in Xpistum credentes ordinatus a Papa Caelestino Palladius primus episcopus mittitur.*[63] Under the year 432 the Annals of Ulster record that *Patricius peruenit ad Hiberniam.*[64] In this chapter there are from *prophetabant* |

---

[61] For other examples of this see Howlett (1996a), 28–32; Howlett (2001a), 16.

[62] For other examples of this see Howlett (1995b), 30; Howlett (1995a), 151; Howlett (1996b), 131–2; Howlett (1997a), 119–21, 539–40; Howlett (1998a), 82; Howlett (1999a), 89–91; Howlett (2000a), 1–9, 159–60.

[63] MGH Auct. ant. 9, 473.

[64] Mac Airt and Mac Niocaill (1983), 38–9.

13 to *aduentum* | *Patricii* exactly 428 letters, coincident with the year AD 428, between which and AD 432 there are three years. Muirchú has infixed a letter count as a means of confirming his statement that the druids prophesied Patrick's advent during three years before his arrival, and set that within a larger pattern coincident with the alphanumeric value of the name of Jesus.

Collectanea Pseudo-Bedae

The text now known as *Collectanea Pseudo-Bedae* first took shape in Ireland during the seventh century.[65] Two passages suffice to illustrate the computistic nature of this remarkable text. First, some questions about Adam.

16 Dic mihi quis primus prophetauit?
    Adam quando dixit 'Hoc nunc os ex ossibus meis et
     caro de carne mea'.
17 Omnis homo qui in dolore positus est memor
     sit illius sententiae 'Ne quid nimis'.
18 Vidi filium cum matre manducantem cuius pellis
     pendebat in pariete.
19 Sedeo super equum non natum cuius matrem in manu teneo.
20 Quaero barbarum quem inuenire non possum.
    In aquilonali parte ciuitatis ubi aqua attingit parietem tolle saxum
     quadratum ibi inuenies barbarum.
21 Dic mihi quae est illa res quae cum augetur minor erit et dum
     minuitur augmentum accipit?
22 Nemo in ecclesia amplius nocet quam qui nomen et opinionem
     sanctitatis habet.
23 Cui plus creditur plus ab eo exigitur.
24 Potentes potenter tormenta patiuntur.
25 Sic ut per prophetam dicitur 'Ducunt in bonis dies suos et in
     puncto ad inferna descendunt'.
26 Quattuor claues sunt sapientia uel industria legendi assiduitas
     interrogandi honor doctoris contemptio facultatum.
27 Dic quot annos uixit primus parens Adam? Noningentos triginta.
28 Qui sunt tres amici et inimici sine quibus uiuere nemo potest?
     Ignis aqua et ferrum.

In §16 between *Adam* | and | *mea* 'mine', that is 'of Adam', there are forty-six letters, coincident with the alphanumeric value in the Greek system of AΔAM, 1+4+1+40 or 46. In §17 the sixty-first letter is the first of *nimis*, coincident with the alphanumeric value in the Latin system of NIMIS,

---

[65] Lapidge and Sharpe (1985), 1257. Bayliss and Lapidge (1998); Howlett (2006b).

13+9+12+9+18 or 61. In §20 there are twenty words. There are forty syllables before | *quadratum* and four words from | *quadratum* to the end. In §21 the sixteen words divide by extreme and mean ratio at 10 and 6, so that the minor part of the golden section begins at *minor* |. The six words of the minor part divide by the same ratio at 4 and 2, so that the minor part of the minor part begins at *minuitur* |. From the space before | *Dic mihi quis primus prophetauit? Adam* §16 to the *N* of *Noningentos triginta* §27 there are 930 letters and spaces between words. In §28 the third word is *tres*. From the space before | *qui* to *ferrum* | there are eighty-three letters and spaces between words, coincident with the alphanumeric value of FERRUM, 6+5+17+17+20+12 or 83.

Second, some observations about the Three Wise Men, whose names appear here for the first time in a Latin text.

52   Magi sunt qui munera Domino dederunt.
     Primus fuisse Melchior senex et canus barba prolix et capillis
     tunica hyacinthina sagoque mileno
     et calceamentis hyacinthino et albo mixto opere pro mitrario
        uariae compositionis indutus
     aurum obtulit Regi Domino.
53   Secundus nomine Caspar iuuenis imberbis
     rubicundus milenica tunica sago rubeo
     calceamentis hyacinthinis uestitus
     thure quasi Deo oblatione digna Deum honorabat.
54   Tertius fuscus integre barbatus Balthasar nomine
     habens tunicam rubeam albo uario sago
     calceamentis milenicis amictus
     per myrrham filius hominis moriturum professus est.
55   Omnia autem uestimenta eorum Syriaca sunt.

This passage about the three *magi* is introduced by a sentence of six words and concluded by a sentence of six words, together twelve words, which surround twelve lines of description, in three parts, introduced *Primus*, *Secundus*, and *Tertius*. The names of the *magi* bear alphanumeric values, MELCHIOR 12+5+11+3+8+9+14+17 or 79, CASPAR 3+1+18+15+1+17 or 55, BALTHASAR 2+1+11+19+8+1+18+1+17 or 78, together 212. Their gifts bear alphanumeric values, AURUM 1+20+17+20+12 or 70, THUS 19+8+20+18 or 65, MYRRHA 12+22+17+17+8+1 or 77, together 212.[66] The three *magi* represent the

---

[66] Compare the play in Rhygyfarch ap Sulien's *Vita Sancti Dauid* §2 (Howlett (2007a), 253–8, 263–8) on the identical alphanumeric value of DAVID and AQUA as 38, and on the values of the antenatal gifts, FAVUS, PISCIS, and CERVUS.

peoples of the three continents, Europe, Asia, and Africa, and the account of them occupies seventy-two words, coincident with the number of peoples and languages after the confusion of tongues. From the beginning of the account of Melchior to the end there are seventy-nine words, coincident with the value of the name Melchior. The account of Melchior contains 212 letters and spaces between words, coincident with the value of the names of the three magi and the value of their three gifts.

Cú Chuimne Sapiens

One of the editors of the *Collectio Canonum Hibernensis*, Ruben of Dairinis, died in 725, leaving the other, Cú Chuimne Sapiens abbot of Iona, to complete the work before his death in 747.[67] In the text of the Prologue that follows capital letters and punctuation marks in boldface represent features of MS Oxford, Bodleian Library, Hatton 42, 1r, copied in Brittany during the ninth century, formerly at Glastonbury and Canterbury. Those in square brackets represent features of the other manuscripts.

INCIPIT IN NOMINE PATRIS ET FILII ET SPIRITVS SANCTI AMEN;			10	21	48
I					
1	Sinodorum exemplarium. in numerositátem con*spíciens* [.]	a	5	20	46
2	ac plurimorum ex ipsis obscuritatem rudibus minus útilem *próuidens* [:]	a	9	24	58
3	[n]ec non ceterorum diuersitatem inconsonam. [d]estruentemque magis quam edificántem [. **p**]*rospíciens* :	a	10	32	82
4	[b]reuem planamque ac consonam [.] silua de ingenti scriptorum in unius uoluminis textum. expositiónem degéss*i*.	b	14	37	90
5	plúra ádd*ens* [.]	a	2	4	11
6	[**p**]lúra mínu*ens* [.]	a	2	5	12
7	[**p**]lura eodem trámite dírig*ens*.	a	4	11	25
8	[plura sensu ad sensum **neglecto** uerborum trámite áds*erens*.,.]	a	8	19	49
9	**h**oc ergo solum pre ómnibus conténd*ens* [.]	a	6	12	32

---

[67] Lapidge and Sharpe (1985), 612–3. See now the Oxford D.Phil. thesis of Roy Flechner (2006). Howlett (2003–4b).

10	ne meo iudicio qué uideb*á*n*tur*	c	5	12	25
11	uelut commendaticia déscriber*é*n*tur* [.]	c	3	13	32
II					
12	[S]ingulorum nomina [.] singulis testimoniis prescrípta pósu*i* :	b	6	21	50
13	[n]e uelut incertum [. q]uisque quod dicat [. m]ínus lúce*at* :	d	8	16	42
III					
14	Sed hoc lectórem nón fall*at*.	d	5	8	23
15	ut [. c]um [.] ad generales titulos quos necessario preposúimus recúrr*at* [.]	d	9	24	56
16	numeros diligénter obs*é*r*uet*.	e	3	10	25
17	quibus obseruatis [:] questiónem quam uolú*erit*	e	5	15	38
18	sine ulla cunctatióne rep*ériet* ;	e	4	14	27
	FINIT PROLOGUS.		2	5	13
excluding line 8			112	304	735
including line 8			120	323	784

IT BEGINS IN THE NAME OF THE FATHER AND OF THE SON AND OF THE HOLY SPIRIT AMEN.

I
1. Gazing together on the numerousness of the model judgements of synods,
2. and foreseeing dark confusion, less useful, from the very crude states of very many of them,
3. and also gazing with an eye to the future on the diversity of others, unharmonious and destructive rather than constructive,
4. I have set in order an exposition, brief and clear and harmonious, into a text of one volume, from an immense forest of writers,
5. adding to many [of the judgements],
6. reducing many,
7. arranging in order many in the same course,
8. [setting in a row many, from sense to sense, with the course of the words not followed,]
9. striving therefore toward this alone before all things,
10. so that things might be written down as commendatory,
11. which were not seen [so] by my judgement [alone].

II
12. I have placed the forewritten names of individual men in individual testimonies,

13	so that what anyone would say might shine as less uncertain.
III	
14	But this should not deceive the reader,
15	as when he would recur to the general titles that we have necessarily placed before,
16	he should diligently observe the numbers,
17	which observed, any question that he will have wished,
18	without any delay he will find.
	THE PROLOGUE ENDS.

The two most recent authorities named in the *Collectio* are Theodore archbishop of Canterbury, who died AD 690, and Adomnán abbot of Iona, who died AD 704. Thomas Charles-Edwards has suggested[68]

> 'the A recension is where the work had got to when one of the compilers, Ruben, died in 725 [. . .]. It may have been his collaborator, Cú Chuimne, who revised and expanded the text [. . .]. The prologue in two B recension MSS and one A recension copy appears to have been written by a single person. [. . .] Since the work on the B recension is a completion of the unfinished labours on the A, it is likely that the B recension was compiled in Ireland and in a centre associated with the production of the A recension.'

In the twenty-three-letter alphabetic system the name CU CUIMNE IAE bears a numerical value of 3+20 and 3+20+9+12+13+5 and 9+1+5 or 23+62+15 or 100. The numerical value of his complete name *Cú Cuimne Iae* coincides with the number of words in the Prologue proper, 100. From | *Incipit* to *Finit prologus* | there are 735 letters, which may represent the year in which Cú Chuimne published the text, ten years after the death of Ruben and twelve years before his own death.

The only significant variant is line 8, absent from the text of three manuscripts of recension B, but present in two manuscripts of recension A. The additional text raises the number of words in the Prologue proper to 108, which coincides with the numerical value of the name CU CHUIMNE IAE, 3+20 and 3+8+20+9+12+13+5 and 9+1+5 or 23+70+15 or 108. In this expanded form of the Prologue there are from | *Sinodorum* 1 to *Finit Prologus* | inclusive 736 letters.

Both forms of the Prologue may have issued from the mind of a single orderly legist, who described accurately what he and his late colleague had done. The text extant in three manuscripts of recension B

---

[68] Charles-Edwards (1998), 237.

may have been published by Cú Chuimne in AD 735. The text extant in two manuscripts of recension A may have been revised by the same man in AD 736.

Cú Chuimne's hymn *Cantemus in omni die*, the oldest extant Latin hymn in praise of the Virgin Mary, survives in four manuscripts.[69]

> Cantemus in omni die concinnantes uarie
> conclamantes Deo dignum ymnum Sancte Marie.
> Bis per chorum hinc et inde conlaudemus Mariam
> ut uox pulset omnem aurem per laudem uicariam.
> Maria de tribu Iuda Summi mater Domini   5
> oportunam dedit curam egrotanti homini.
> Gabriel aduexit Verbum sinu prius Paterno
> quod conceptum et susceptum in utero materno.
> Hec est summa, hec est sancta uirgo uenerabilis,
> que ex fide non recessit sed exstetit stabilis.   10
> Hüic matri nec inuenta ante nec post similis
> nec de prole fuit plane humane originis.
> Per mulierem et lignum mundus prius periit;
> per mulieris uirtutem ad salutem rediit.
> Maria mater miranda Patrem suum edidit   15
> per quem aqua late lotus totus mundus credidit.
> Hec concepit margaritam — non sunt uana somnia —
> pro qua sani Christiani uendunt sua omnia.
> Tunicam per totum textam Christi mater fecerat
> que peracta Christi morte sorte statim steterat.   20
> Induamus arma lucis loricam et galiam
> ut simus Deo perfecti suscepti per Mariam.
> Amen, Amen, adiuramus merita puerpere
> ut non possit flamma pire nos dire decerpere.
> Christi nomen inuocamus angelis sub testibus   25
> ut fruamur et scribamur litteris celestibus.
>   Sancte Marie meritum
>   imploramus dignissimum
>   ut mereamur solium
>   habitare altissimum.   30
>   Amen.   31

Let us sing on every day, putting together variously,
shouting together to God a hymn worthy of Saint Mary.
Twice in a chorus on this side and that let us together praise Mary,
so that the sound may strike every ear in successive praise.

[69] Lapidge and Sharpe (1985), 581. Howlett (1995b), 19–30.

Mary from the tribe of Judah, mother of the Highest Lord,
gave an advantageous cure to ailing man.
Gabriel conveyed first from the Father's bosom the Word,
Which [was] conceived and received in the mother's womb.
This is the highest, this is the holy venerable virgin,
who has not drawn back from the faith, but stood out stable.
To this mother a like has been found neither before nor since
nor after her offspring of fully human birth.
By a woman and a tree the world first perished;
by the virtue of a woman it came back to salvation.
Mary the wondrous mother brought forth her own Father,
through Whom the whole world far and wide washed by water [in baptism] has believed.
This woman conceived a pearl — these are not empty dreams —
for which sane Christians sell all their own possessions.
The mother of Christ had made a garment woven throughout,
which, when Christ's death was brought about, had remained constant in its order.
Let us put on the arms of light, breastplate, and helmet,
that we may be perfect for God, received through Mary.
Amen. Amen. We affirm on oath the merits of the child-bearer,
so that the flame of the dreadful pyre cannot snatch us away.
We invoke the name of Christ with angels as witnesses,
so that we may enjoy and be written in celestial letters.
> We beseech the most worthy
> merit of Saint Mary
> that we may be worthy
> to inhabit the loftiest throne.
>   Amen.

Cú Chuimne has embedded within the text of the first quatrain instructions for performance, singing by a group (*cantemus, concinnantes, conclamantes, conlaudemus*), divided into an antiphonal pair (*per chorum hinc et inde*) to produce successive responsorial praise (*laudem uicariam*). He has stated within the last quatrain of long lines what the choir do and why, then restated this in the last quatrain of short lines. Enclosed within these four lines of introduction and eight lines of conclusion are eighteen lines in praise of the Virgin Mary.

The hymn's long lines echo the pentadecasyllabic verses of 'Saint Sechnall's Hymn' *Audite Omnes Amantes Deum* in praise of Saint Patrick. Its short lines echo the rhyming and alliterating octosyllabic couplets and quatrains invented by Irish poets, probably during the fourth quarter of the sixth century. The hymn consists of twenty-six pentadecasyllabic lines, four octosyllabic lines, and a concluding *Amen*, thirty or thirty-one lines, one for every day of the month. After the end of the twenty-sixth line manuscripts C and D read *Cantemus*, implying, as the second couplet states explicitly, that the hymn is to be sung *bis* 'twice'. Those who do that sing fifty-two pentadecasyllabic lines, one for every week of the year. The pentadecasyllabic lines sung and repeated and the octosyllabic quatrain and the concluding *Amen* sung once contain 365 words, one for every day of the year, appropriate for those who sing *Cantemus in omni die*.

Note particularly the couplet *Gabriel aduexit Verbum sinu prius Paterno, quod conceptum et susceptum in utero materno*. In MS A this is preceded by the punctuation marks //., and followed by..,. In MS B the central word of the first line of the couplet is pointed *Verbum*.. The event referred to is the Annunciation, celebrated on 25 March, the eighty-fourth day, 23% of the way through a 365-day year. The seventh line is 23% of the way through a thirty-one-lined poem. The word *Verbum* is the forty-fifth of 188 words, 23% of the poem. The *m* of *Verbum* is the 234th of 1021 letters, 23% of the poem.

From *Amen Amen adiuramus merita puerpere* 23 to the last mention of *Sancte Marie* in line 27 there are twenty-seven words, of which the central fourteenth word is *Christi* 25. From *Amen Amen* to the M of *Marie* inclusive there are 152 letters. In Greek notation the value of the name MAPIA is 40+1+100+10 +1 or 152. Eight-ninths of the way through this passage the poet refers to himself at *scribamur litteris | celestibus*. In the immediately preceding line he writes *Christi nomen inuocamus*, of which the last *s* is the 888th letter of the hymn. In Greek notation the value of the name IHCOYC is 10+8+200+70+400+200 or 888.

Joseph Scotti Carmen figuratum

Joseph Scottus was an Irishman at the court of Charlemagne, of whose career we learn a few facts from the correspondence of Alcuin during the years 790–796. Joseph wrote an *Abbreuiatio commentarii Hieronimi in Isaiam* and six *Carmina*, one of which we may note here.[70] The poem contains thirty-five lines, each of which contains thirty-five letters. By arranging the letters vertically one notes a further aspect of Joseph's art. Embedded in this *carmen figuratum* are four further hexameter verses:

---

[70] Lapidge and Sharpe (1985), 648, 649. Howlett (1995a), 116–20, 213–6.

Ille pater priscus elidit edendo nepotes.
Mortis imago fuit mulier per poma suasrix.
Iessus item nobis ieiunans norma salutis.
Mors fugit uitae ueniens ex uirgine radix.

The first and third of these verses form a parallelogram descending from upper left to lower right. The second and fourth of these verses form a parallelogram descending from upper right to lower left. At the very centre of the poem three words as half of a hexameter in the shape of a cross read *Lege feliciter Carle*.

```
P R I M V S A V V S  VI  V E N S E N N O S I  N MO R T E  R E D E G I  T
H E V S  I  C E T MV LI  E R P R A E B E N D  O P OM A  P E R Y D R V M
F E C I  T N O S P  L A G A S I  V S T E P E R C V R R E R E M V L T A S.
E N N O S  T R A E M O R T I S  S E M E N T A F V I S  S E V I  D E M V S.
S C O R P I  O P V L S A N D O V A L V I  T N A M P E R F I  D E F A R I :
E V A P A R A T  A P A R E T I  V S I  S H E V V A L D E S  V P E R B I S.
V I  N C I  T T E M V L I  E R P O M I P V L C H E R R I  M A G R A N D I
V I  S  P E C I  E S P E R I  S A M MO T A T V N C V I  R G I  N I  T A T E.
H O R R V I  T A V E R S V S P  A R I  T E R S E V C O N I  V G E T A L I
D E P R A V A T V S A V V S   G V S T V S  M I S E R A B I  L I S A  V S  V :
E R G O V I  R E X L I  G N O M A N D E N S  N O S  N O X A P E R E M I T.
I  N C L Y T A V I  R G O F I  D E S E D R A R  O V O T A S A C R A V I T
I  S T A M A R I  A N O V A M V I  T A M V E S V  B I  R E S V P E R N A M
E T C A R N I  S P R I  O R A S C V L T E T S I  B I  I  V R A T E N E R E
A R D V A  T V N C T O T O C V P I E N S C A S T I  S  S I  M A M V N D O.
T A L I  S  E N I  M M E R V I  T R E  G E M Q V E P A R I  R E D E V M V E
Q V I  P I  E S V B V E R T I  T F R E G I  T E T F E R R E A N O S T R I
I  V R A M A L I. O V E R E F E L I  C I T E R  H I N C P R I  O R I  B A T
L E G E S I  B I  P L A C I  T A P R A V O R V M F E R R E T V T I  R A S.
I  A M D E V S A  D L I  G N V M S  E R V I  L I  N O M I N E P E N D E N S
P O R R O S V I  M I  S E R V M S O L V E N S O P E S O L I  V S A D A M
E X P I  R A V I  T I  T A E T C E L E R A T D E F E R R E S E R E  N A M
D E  S V M M I S L V C  E M.S V N T H A E C V E XI  L L A R E V E R  S A
R E G N I  Q V A E P R I  D E M C L A V S E R  V N T L I  M I N A D E I  N
I  P S E V I  R E T M V L I  E R A R B O R V I  T A E Q V E R E L A T  O R
P R O C E R V M O S E R P E N S V L T O R V E N A T V S I  L I  D R V M
P R O M V N T H I  C V I  T A M D E L I  G N I  V E C T E R E D E M T A M.
H I  C N A M C V M S  E R V V S  D O M I N V M S E P A R E T H A B E R I
M O R T I  S  E V M I  V S T E S E V  N O S T V N C V L T I  O S V M S I  T.
S E D V E R V S  D O M I N V S S  E S  E R V I V E S T I  T A M I  C T V M
E T R E P A R A T  M V N D V M R E X O P T I M V S I  L L E P I  O R V M.
I  N L I  C I  T I  S A S T E V A C  I  B I  S  O S  C O N T V L I  T A V D A X
S E D M O R T E  M R A P I  D O C O N T A C T V D E T V L I  T O R B I S.
I  N D E M A R I  A V I  R I  E X T E I  V R A R  E C I  D I S  H A B E N D A
H I  N C G E N E T R I  X V E R A E T V S V M I  S  S E M I N A V I T A E.
```

The 237 words of the thirty-five verses and the twenty-nine words of the embedded verses together total 266, which divide by one-ninth and eight-ninths at 29.56 and 236.44, at *uitae*, the last word of the thirty-fifth verse, having arrived at which one begins to read the embedded verses. Of the twenty-nine words in the embedded verses one-ninth and eight-ninths falls at 3 and 26, at the end of the central words addressed to Charlemagne, *lege feliciter Carle*. Joseph exhibits in his verse repeated diminutions, of the number 35, once in lines of the entire poem and again in letters of each line, and of the fraction 1/9, once in the embedded verses as a fraction of the entire poem and again in the words *lege feliciter Carle* as a fraction of the embedded verses.

Dicuilli Liber de Mensura Orbis Terrae

From a period slightly later than Joseph is a monument to the scientific curiosity of an Irishman at the Carolingian court. Dicuill's *Liber de Mensura Orbis Terrae* ends with thirty-one hexameter verses, the first of which which names him, *Dicuil accipiens ego tracta auctoribus ista*, and the twenty-eighth of which dates the work to the year 825, *post octingentos uiginti quinque peractos*.[71]

In Dicuill's three-part account of an eighth-century scientific expedition by Irish clerics to observe the solstice in Iceland the sixty-two words of the first part, about islands, divide by the ratio 1:1, in the 'middle', at *aliaeque | mediae* 3. They divide by duple ratio 2:1 at 41 and 21, at *aliae | minimae* 2. They divide by extreme and mean ratio at 38 and 24, at *| aliae* 4. They divide by sesquitertian ratio at 35 and 27, at *aliae | magnae* 3. They divide by sesquioctave ratio 9:8 at 33 and 29, at *aliae | paruae* 3 and *sunt | aliae* 4. They divide by 1/9 and 8/9 at 7 and 55, at *insulas |* 1 and *habitaui | alias* 5. The second part begins *Plinius Secundus in quarto libro edocet quod Pytheas Massiliensis sex dierum nauigatione in septentrionem a Brittannia Thílen distàntem nárrat*, so that the second word of the second part is *secundus* and the fourth word is *quarto*, after which the sixth word is *sex*. In sentence 7 in *De eadem semper deserta in eodem quartodecimo* the fourteenth syllable is the first of *quartodecimo*. In sentence 19 the twenty-three words divide by the ratio 1:1 at *| una |*. They divide by the ratio 2:1 at 15 and 8, at *duobus |* and *| in duorum*, there being also 8 words between *una |* and *|*

---

[71] Lapidge and Sharpe (1985), 662. This text is edited and translated by Tierney (1967), with contributions by L. Bieler. Howlett (1995a), 124–9; Howlett (1999b), 127–34.

unam. In sentence 20 there are from the space before *Illae* to the space after *centum* inclusive 100 letters and spaces between words. In the third part the first word is *trigesimus* 16. The 188 words divide by extreme and mean ratio at 116 and 72, so that the golden mean is at *in | medio orbis terrae* 26. They divide by the ratio 4:3 at 107 and 81, at *in medio | illius* 24. They divide by the ratio 5:4 at 104 and 84, at *| medium noctis* 24. They divide by the ratio 9:8 at 100 and 88, at *in | medio orbis terrae* 24.

There is a calendrical trick in the repeated statements that *Ultima Thule* and the islands north of Britain were 'always deserted'. From *semper | deserta* 7 to *desertae | semper* 41 inclusive there are 365 words, one for every day of the year, 'always'.

Yet another cluster of calendrical features is that in the words from *a uernali | aequinoctio* 28 to *ad | uernale aequinoctium* 29 there are eighty-one letters, the vernal equinox occurring on 21 March, the eighty-first day of a leap year. From *| aestiuum solstitium* 8 to *in hiemali solstitio |* 25 inclusive there are 172 words, the summer solstice occurring on 21 June, the 172nd day of an ordinary year. The word *solstitii* 31 is the 356th of the passage, the winter solstice occurring on 22 December, the 356th day of an ordinary year.

In sentence 2 from the space after *Hiberniam* inclusive to the space before *Brittanniam* exclusive there are sixty-four letters and spaces between words, coincident with the value of HIBERNIA 8+9+2+5+17+13+9+1 or 64. In sentence 6 from *| Plinius Secundus* to *Brittannia |* inclusive there are exactly 103 letters, coincident with the value of BRITTANNIA, 2+17+9+19+19+1+13+13+9+1 or 103. The name THILE VLTIMA bears a numerical value of 19+8+9+11+5 and 20+11+19+9+12+1 or 52+72 or 124. From *| Plinius* 6 to *de | eadem [insula Thile]* 7 there are fifty-two syllables. From *| Thilen* 6 to *Thile Ultima insula |* 8 inclusive there are 124 letters and spaces between words. From *Ultima |* 8 to *Ultima |* 15 there are seventy-two words. From *Thile Ultima in qua |* 15 to *| in Thile* 25 there are 124 words.

By writing in line 17 that the *clerici* [. . .] *a kalendis Febroarii usque kalendas Augusti in illa insula manserunt* and in line 30 of them *nauigantes in naturali tempore magni frigoris* Dicuill implies that the clerics embarked in January, presumably in a boat not much larger than that in which an *aliquis presbyter religiosus* described himself as *nauigans in duorum nauicula transtrorum* to islands nearer the north of Britain than Iceland.

Dicuill dates his *Liber* to the year AD 825, thirty years from the year AD 795, in which Irish clerics told him of their voyage, not the year of the voyage itself. Dicuill does suggest a date by writing in line 17 that the clerics *in illa insula manserunt* and in line 31 *manentes in ipsa*. 'Remaining' on the island from February to August was what gave authority to the clerics' observations. From | *manserunt* to *manentes* | inclusive there are 792 letters. Between *manserunt* | and | *manentes* there are 775 letters. One might infer that the clerics sailed north either in AD 792 or in AD 775. If the later date is correct Dicuill acquired his information from eyewitnesses within three years of their expedition. If the earlier date is correct Irishmen were in Iceland 95 years before settlement by Icelanders. The punctilious composition of sentence 20 implies that Irishmen were living in the Orkneys and the Shetlands and the Faroes by AD 725.

Briani Molosi Belli Rubisca

A poet who names himself *Brianus Molosi Belli*, a Latin rendering of an Irish name *Brían mac Con Catha* 'Brían son of Hound of Battle' composed *Rubisca*, a poem in rhythmic double adonics in ninety-six lines in twenty-four four-lined stanzas.[72] Stanza 1 is a prologue, stanzas 2–19 address a red bird, stanzas 20–21 describe the poet and his friends, stanzas 22–23 present a doxology in Greek, and stanza 24 is an epilogue. In three-quarters of the poem the poet considers the bird. The remaining quarter is divided into thirds, two stanzas, one each for prologue and epilogue, two stanzas for the poet and his friends, and two stanzas for the doxology.

The twenty-four stanzas may represent the twenty-four letters of the Greek alphabet. Apart from the two stanzas of doxology in Greek there are twenty-four words in the poem derived from Greek but not borrowed into Classical Latin. The first letters of the twenty-three stanzas 2–24 follow the order of the twenty-three letters of the Latin alphabet from A to Z. The twenty-two stanzas which are not entirely in Greek may represent the twenty-two letters of the Hebrew alphabet. Apart from the divine name *Ia* 93 half that number of words in the poem, eleven, are derived from Hebrew. The same number of words, eleven, are used in ways not recorded in standard literary Classical Latin.

[72] Howlett (1996c).

The poet has arranged some words and ideas in a chiasmus.

A	5	amica
B	11	modo-quoquo
C	12	nedulam
D	13	ignaram
E	33	hiulcusque forceps sic aera cogit
	34	hiulcumque qui de- caladum -hiscit
F	40–1	aequiparatis kastis ambobus
E'	42	nigrioribus spectu coruino
	44	ainis uitreo neu nigerrimo
D'	50	gnostici
C'	53	nedulos
B'	53	quibus
A'	60	amica

In the eighteen stanzas about the bird the centre of the stanzas falls between 10 and 11, and the centre of the lines falls between 40 and 41. The two halves of the 301 words are equally arranged 150–1–150 round *aequiparatis kastis ambobus*, 'both' 'made equal'. The 1791 letters are equally arranged so that the first *a* of *aequiparatis* is 890th, and there are 890 letters from *kastis ambobus* to *uoculas* inclusive. The middle letters of the entire poem are equidistant from beginning and end, *par*.

The poet has arranged other words and ideas in a chiasmus that extends from the first word of the poem to the last.

A	1	parce
B	1	domine
C	1	narranti
D	2	peccanti
E	5	bonus
F1	8	rubisca rara estin aduentus
F2	9	ab heri nudiusque-tertius
F3	13–5	cantricem [...] cantus [...] antris
G	18	nexam
H	19	aeque
I	40	aequiparatis
J	41	ambobus
K	51	inter
J'	53	utrae
I'	54	aeque
H'	66	aequas
G'	67	strictas
F'1	69	rubisca rata redi

F'2	69	de mane
F'3	71	cantus in cripta cane
E'	93	bona
D'	96	misero
C'	96	mihi
B'	96	Domine
A'	96	parce

In this pattern the central word of the entire poem, which lies 'between' the two halves, is *inter* at the crux of the chiasmus, the central 206th of 411 words.

The poet has fixed many other numerical features into the text. In the second stanza the second line begins with the disyllabic word *bipes* 6. In the second alphabetic stanza the eighteen words, beginning with the disyllabic word *bifax*, divide by duple ratio 2:1 at 12 and 6, so that the first third of the words of the stanza ends with *nudiustertius* 10. In line 31 the second word is *binis*. In line 38 the second word is *duis*, between which and *septenis* 40 there are seven words. In line 41 the second word is *ambobus*. In line 65 the second word is *bis*. From *aequiparatis* 40 to *aeque* inclusive in line 54 there are fifty-four words. From *aeque* 54 to *aequas* 66 inclusive there are also fifty-four words.

There are seventeen words in stanza 1, the prologue, and seventeen words in stanza 20, about the poet, and seventeen words in stanza 21, about the poet's friends.

The poem contains 411 words. If we remove from consideration the two stanzas of doxology in Greek, the remaining 372 words divide by 1/9 and 8/9 at 41 and 331. The last word of our poet's name, *Belli*, is the forty-first word from the end.

Brían mac Con Catha composed *Rubisca* certainly after the fourth quarter of the seventh century, later than the Irish poet Laidcenn mac Báith and the *Hisperica Famina* and the *Orationes Moucani*, probably after the fourth quarter of the eighth century, later than the rhythmic syllabic adonics composed by the English poet Alcuin and the Irish poet Columbanus of Saint-Trond, and certainly before compilation of the Anglo-Latin and Old English Harley Glossary and composition of Æthelstan's charter of AD 928, both of which quote its diction. He must have written *Rubisca* not later than the first quarter of the tenth century. He may have spent some time at the court of King Æthelstan.[73]

---

[73] Howlett (1995a), 352–3; Howlett (1995b), 30–48.

Vita Sancti Maedoci

The earliest extant form of the *Vita Sancti Maedoci* is represented by MS London, British Library, Cotton Vespasian A XIV (V), written about 1200. The narrative survives thereafter in three recensions of the *Vitae Sanctorum Hiberniae*, the first represented by two sister manuscripts in MS Dublin, Primate Marsh's Library, Z3.1.5 (formerly V.3.4) (M), and MS Dublin, Trinity College, E.3.11 (175) (T), both written during the fifteenth century and copied from a common source, the original of which derived probably from the early thirteenth century; the second represented by a single manuscript, the Codex Salmanticensis, MS Bruxelles, Bibliothèque Royale, 7672–4 (S), written during the fourteenth century; the third represented by two manuscripts in MSS Oxford, Bodleian Library, Rawlinson B.485 (R1) and B.505 (R2), dated variously from the thirteenth century to the fifteenth.[74]

In the text which follows capital letters and punctuation marks in boldface represent features of MS London, British Library, Cotton Vespasian A XIV, f 101. I have arranged the text in lines *per cola et commata* 'by clauses and phrases', marking the rhythms of the cursus and numbering sentences to the left and lines, rhymes, words, syllables, and letters to the right.

1	Alio autem tempore exiit sanctus					
	Aidus ad sanctas uirgines filias Aido					
	fílii Cóhirbri.		a	13	32	73
	secumque arátrum cum bóbus túlit:		b	5	11	28
	ut apud éas aráret.		c	4	8	15
2	Cumque boues ad arándum iúngerent.		d	5	11	29
	ecce quaedam mulier leprósa aduénit	5	b	5	13	31
	rogans ut sibi Aidus bóuem largíret.		c	6	12	30
3	Cui dedit Aidus electum de céteris bóuem.		e	7	15	34
4	Tunc aratóres dixèrunt éi.		a	4	10	22
5	'Quíd faciémus.		f	2	5	12
	et quomodo aráre potérimus?'	10	f	4	11	23
6	Quíbus dixit Aídus.		f	3	6	16
7	'Expectate paulisper bouem ad uos					
	uelóciter uèniéntem.'		e	7	19	46
8	Subito autem ex propinquo mari bouem					
	ad se uenire conspíciunt.		d	10	22	52
	qui suam uocem exaltans tribus uícibus					
	clamáuit.		b	7	16	41

[74] Lapidge and Sharpe (1985), 381, 403, 475. Howlett (2002).

9	Suumque collum in iugum alterius bouis humíliter pósuit:	15	b	8	21	48
	et tribus mensibus ueris apud íllos aráuit.		b	7	15	36
	qui in initio diei uniuscuiusque ad arandum dé mari uéniens.		g	10	25	50
	et ter uócem exáltans :		g	4	7	18
	iterum in fine diei reuertebátur ín mare.	19	h	7	17	34
				118	276	638

MS reading: 5 quedam. 12 uenientem uelociter. 17 inicio.

1   At another time, however, saint Aéd went out to the holy virgins, daughters of Aéd the son of Cohirbri, and with him he took a plow with cows, so that he might plow among them.

2   And when they were yoking the cows for plowing, behold, a certain leprous woman came to [him], asking that Aéd generously give to her a cow.   5

3   To whom Aéd gave a cow chosen from the others.

4   Then the plowmen said to him,

5   'What shall we do, and how shall we be able to plow?'   10

6   To whom Aéd said,

7   'Wait a little while for a cow coming to you quickly.'

8   Suddenly, however, from the nearby sea they stare together at a cow coming to them, which raising its own voice on three occasions called.

9   And it placed its own neck humbly into the yoke of the other cow,   15
and in three months of spring among them it plowed, which at the beginning of each day coming from the sea for plowing, and three times raising the voice, again at the end of the day it used to return into the sea.   19

In the Latin alphabet of twenty-three letters the name AID bears a numerical value of 1+9+4 or 14 and the name AIDUS a numerical value of 1+9+4+20+18 or 52. In the first occurrence of the name the first syllable of *Aidus* 1 is the fourteenth of the narrative. In sentence 2 *Aidus* is the fourteenth word. In the last occurrence of the name there are fourteen letters and spaces between words from the beginning of line 11 to the A | of *Aidus*. After the first occurrence of *Aidus* | 1 *Aidus* 11 is the fifty-second word.

The clearest sign of comprehensive order in this composition is that the author divided his 118 words by duple ratio 2:1 into 79 and 39, so that the last third of the text begins with the word | *tribus* 14. That last third divides by the same ratio at 26 and 13, at *tribus* | 16. The last twenty-six words divide again by the same ratio at 17 and 9, at *ter* | 18. This makes three divisions by successive diminution of text into thirds at the word 'three'.

Wondering what, if anything, the numbers of sentences, lines, words, and syllables suggest, one might associate the nine sentences with the square of the number of plays on 'three', the nineteen lines with the cycle of nineteen years in which solar and lunar calendars come into synchrony, the 118 words with the number of the longest text in the Psalter, the alphabetical Psalm 118, and the 276 syllables with the number of days from the Annunciation on 25 March to the Nativity on 25 December.

Let us consider next the text in MS Dublin, Primate Marsh's Library, Z3.1.5, 54r.

1	Quodam tempore beatus antistes Moedhog exiuit uisitare sanctas uirgines filias Eda fílii Córpri.		a	13	35	83
2	Et duxit secum aratrum eius in elemósinam cum bóbus		b	9	19	43
	ut arárent uirgínibus.		b	3	8	19
3	Cumque statim aratores boues ad arándum iúngerent		c	7	17	43
	uenit múlier leprósa	5	d	3	8	18
	rogans ut uir Dei Moedhog largirétur ei bóuem.		e	8	15	38
4	Cui dedit sánctus eléctum bóuem.		e	5	11	27
5	Aratores dixérunt uiro Déi,		a	4	11	23
6	'Quíd faciémus		b	2	5	12
	quia par numerus bóum non est nóbis?'	10	e	7	12	29
7	Vir sánctus ait éis,		e	4	7	16
8	'Expectate donec ueniat bonus bos missus nóbis á Deo.'		f	9	19	43
9	Subito mirabile dictu de mari bouem conspíciunt ueniéntem		e	8	22	50
	qui exaltans mugítum ter clamáuit		c	5	11	29
	suumque collum aptius in iugum alterius bouis humíliter pósuit.	15	c	9	24	54
10	Et tribus uicibus ueris ibi ípse bos aráuit		c	8	16	36
	qui in initio cuiuscunque diei quo debebat arare de mari ad arándum ueniébat.		c	13	31	64

11	Et ter uocem exaltans iterum in fine diei				
	reuertebátur ín mare.		g	11 24	52
12	Omnes audientes et uidentes hoc				
	miraculum glorificabant Deum et famulum				
	súum sanctum Moédhog.	19	h	13 32	80
				141 327	759

MS readings: 1 moedog. corppri. 2 elymo'. 15 possuit. *All occurrences of the lexeme arare, except arare 17, were penned with initial arr-, the first r being marked with a subscript point for cancellation.*

1   At a certain time the blessed bishop Moedhog went out
       to visit holy virgins, the daughters of Eda son of Corpri.
2   And he led with him his plow for alms with cows,
       that they might plow for the virgins.
3   And when immediately the plowmen were yoking the
       cows for plowing,
       there came a leprous woman,   5
       asking that the man of God Moedhog might generously
       grant to her a cow.
4   To her the holy man gave a select cow.
5   The plowmen said to the man of God,
6   'What shall we do,
       since there is not for us an equal number of cows?'   10
7   The holy man says to them,
8   'Wait until a good cow may come sent to us from God.'
9   Suddenly, wondrous in the saying, they stare together at
       a cow coming from the sea, which called, raising a
       moo three times,
       and it humbly placed its own neck quite snugly into the
       yoke of the other cow.   15
10   And on three occasions in the spring there the same
       cow plowed,
       which at the beginning of each day on which it used to
       plow used to come from the sea for plowing.
11   And three times raising its voice again at the end of the
       day it used to return into the sea.
12   All those hearing and seeing this wonder used to
       glorify God and His own servant, holy Moedhog.   19

The name MAEDOC bears a numerical value of 12+1+5+4+14+3 or 39. The name MOEDHOG bears a numerical value of 12+14+5+4+8+14+7 or 64. The value of these names combined is 39+64 or 103. From | *Quodam* to the space after *Moedhog* | 1 inclusive there are 39 letters and spaces between words. From | *Quodam* to |

Moedhog 6 there are 39 words. From | *Moedhog* | 6 to *Moedhog* | 19 there are 102 words.

The name EDA bears a numerical value of 5+4+1 or 10. Before | *Eda* 1 there are 10 words.

Moedhog is commemorated on the 31ˢᵗ day of the first month. In the first line there are from | *Quodam* to | *Moedhog* 31 letters and spaces between words.

The passage about the day on which the cow, like the sun, used to come from the sea and return to the sea, *qui in initio cuiuscunque diei quo debebat arare de mari ad arandum ueniebat et ter uocem exaltans iterum in fine diei reuertebatur in mare*, contains twenty-four words, one for each hour of a day.

The chapter consists of twelve sentences, one for each month of the year, and nineteen lines for the cycle of nineteen years in which solar and lunar calendars come into synchrony. Sentence 6 contains nine words, of which the central fifth word is *par numerus* | 'equal number'. The twelve sentences of the chapter divide by symmetry 1:1 at 6 and 6 in this central sixth sentence. The nineteen lines divide by symmetry in this central line 10. The 141 words divide by extreme and mean ratio at 87 and 54, at | *quia par numerus*, from the beginning of the chapter to which there are 365 letters and spaces between words, one for each day of an ordinary year.

The 141 words of the chapter divide by hemiolus or sesquialter ratio 3:2 at 85 and 56, at | *ter* 14. The remaining fifty-six words of the chapter divide by the same ratio at 34 and 22, at *ter* | 17. The thirty-four words from | *ter* 14 to *ter* | 17 divide by the same ratio at 20 and 14, at *tribus uicibus* | 16. After three successive diminutions by the ratio 3:2 on the word 'three' these 14 words divide by the same ratio at 8 and 6, so that a fourth diminution by sesquialter ratio falls at *alterius* | 15.

Our author has retained from the first passage the number of lines, 19, but increased the number of sentences from 9 to 12, the number of words from 118 to 141, the number of syllables from 276 to 327, and the number of letters from 638 to 759. He has incorporated multiple indications of the alphanumeric values of two forms of the saint's name. He has incorporated into his text counts of letters, of syllables, of words, of spaces between words, and of lines, that represent 31 January, the day on which Saint Moedhog is commemorated, twenty-four hours in a day, twelve months in a year, and 365 days in a year. Of particular interest is the alteration of his predecessor's successive diminutions in the ratio 2:1, with three plays on 'three', to successive diminutions in the ratio 3:2, with three plays on 'three' and a fourth on the word *alterius* that echoes the name of the new ratio.

Conchubrani Vita Sancte Monenne

In MS London, British Library, Cotton Cleopatra A II, fols 4–61 (olim 1–58) written early in the twelfth century, one poem about Saint Monenna in octosyllabic abecedarian stanzas, beginning *Audite sancta studia*, with a refrain beginning *Deum deorum Dominum*, precedes the *Vita Sancte Monenne*, and another poem about the same saint in dodecasyllabic abecedarian lines follows it.[75] The author of the *Vita* signed himself thus fols 59r-v.

> **Har**um uirtutum lectorem símul et aùditórem. pér Deum téstor.
> ut pro me ualde misero Domini séruo. Conchubráno.
> peccati sarcína opprésso.
> piis orationibus intercédant ad Dóminum.
> ut quod impediente áduersário.
> uiribus meis inplere de Dei praeceptis non uáleo ut débet.
> sororum mearum meritis pro me íntercedéntibus.
> ante mortem meam perfícere póssim.
> ut mortis uinculis absolutus. per eárum suffrágia.
> Xpisto praestante transire merear in sanctorum consortium célicolárum.
> in mansionibus simul perfectórum cum Xpísto
> qui regnat in sécula sèculórum. AmeN.

Our author's name, *Conchubranus*, looks like an Irish name *Conchobhar*, with an Irish diminutive *-an* and the nominative singular inflexion of a Latin noun of the second declension *-us*, meaning 'little Connor'. In the alphanumeric system of the twenty-three letter Latin alphabet the name CONCHUBRANO bears a numerical value of 3+14+13+3+8+20+2+17+1+13+14 or 108. From the space before | *Harum* to the space after *Conchubrano* | inclusive there are 108 letters and spaces between words. The name CONCHUBRANUS bears a numerical value of 3+14+13+3+8+20+2+17+1+13+20+18 or 132, coincident with the number of syllables between *Conchubrano* | and | *Xpisto qui regnat in secula seculorum Amen*. From the space before | *seruo Conchubrano* to the space before | *Amen* inclusive there are 444 letters and spaces between words, a way of concluding the preceding forty-four chapters of the Life.

The text of *Audite sancta studia* contains from | *Deum deorum omnium* to *Qui regnas in secula seculorum Amen* | inclusive exactly 4444

---

[75] Lapidge and Sharpe (1985), 308, 1303, 1304. Howlett (2005b), 17–8.

letters and spaces between words, which introduce the *Vita Sancte Monenne* of 44 chapters and the signature passage of Conchubranus of 444 letters and spaces between words. This may be related to the traditional date ascribed to Patrick's foundation of Armagh, AD 444, and to *Annus I* of the *Annales Cambriae*, AD 444, and the year in which according to the *Vita Sancti Reguli* Saint Rule is supposed to have met King Ungus at Rig Monaid in Saint Andrews, the event occurring at the 444th word *peruenit | ad uerticem montis regis id est Rig Mund*, in AD 444.[76]

The text of *Audite facta* contains from | *Audite* to *in secula seculorum Amen* | inclusive 282 words and 1666 letters, MDCLXVI, a number that exhibits the descending value of Roman numerals.

Cormac's Polyphonic Colophon

MS London, British Library, Additional 36929 is an Irish Gallican psalter, written probably during the middle or second half of the twelfth century, its Psalms divided into three groups of fifty.[77] On folio 59r, after the canticles which follow the first group of fifty Psalms, the *Hymnus Trium Puerorum* and the *Canticum Isaie*, below the words *Domine saluum me fac: et psalmos nostros cantabimus cunctis | diebus uitae nostrae in domo domini*, the last of which echo Ps 22:6, the scribe wrote a polyphonic colophon:

> Cormacus scripsit hoc ψ salterium ora pro eo.
> Qui legis hec ora pro sese qualibet hora.
>
> Cormac wrote this Psalter; pray for him.
> You who read these things, pray for himself [perhaps meaning 'yourself'] in every hour.

Here are two lines of dactylic hexameter verse, which both exhibit correct quantities, with two elisions in the first line in *psalterium ora* and *pro eo*, with nothing extraordinary in the second. Both verses exhibit internal vowel rhymes in the third and the sixth feet. In the first compare the *o* in *hoc* with that in *eo*. In the second compare the *i*, *e*, and *ora* in *legis hec ora* with those in *qualibet hora*. The second verse also exhibits internal alliteration on *qu* and *l* in *qui legis* and *qualibet*. Both verses exhibit chiastic disposition of sounds. In the first, around the central *o* in *hoc* in the third foot, compare the *ps-t* of *scripsit* with that of

---

[76] Howlett (2000), 76–85.
[77] Howlett (1995c).

psalterium, the *ri* of *scripsit* with that of *psalterium*, the *u* of *Cormacus* with that of *psalterium*, and the *or-a* of *Cormacus* with that of *ora*. The first and last of these chiastically disposed sounds in the first verse are echoed at the centre of the second verse. Around the central *a* of *ora* in the third foot, compare the *o* of *ora* with that of *pro*, the *r* of *ora* with that of *pro*, the *e* of *hec* with that of *sese*, the *s* of *legis* with that of *sese*, the *i* of *legis* with that of *qualibet*, and the *e* of *legis* with that of *qualibet*. The first verse is linked to the second by parallelism of *Cormacus scripsit hoc* with *qui legis hec* and of *ora pro eo* with *ora pro sese*.

Cormac refers to himself and his work in places determined by calculation of sesquioctave ratio 9:8. The fifteen words of the colophon divide by sesquioctave ratio at 8 and 7, at the end of the first verse, *eo*. The twenty-nine syllables divide by sesquioctave ratio at 15 and 14, at *eo*. The seventy letters divide by sesquioctave ratio at 37 and 33, at *eo*. The fifteen words divide by one-ninth at 1.67, the twenty-nine syllables at 3, and the seventy letters at 8. The first word names the writer *Cormacus* in three syllables and eight letters. Let us note also the references to Cormac's work. From *hoc psalterium* to *qui legis hec* inclusive there are eight words, fourteen syllables, and thirty-two letters.

The reader is invited to pray for Cormac at *ora*, the fifth word from the beginning of the colophon, and for himself at *ora*, the fifth word from the end. Between *ora* and *ora* there are five words. Before the former *ora* there are ten syllables. From the latter *ora* inclusive to the end there are ten syllables. From the first syllable of the former *ora* to the first syllable of the latter *ora* inclusive there are ten syllables.

These verses are set to a polyphonic composition in three vertically arranged four-lined staves drawn in red ink, with red guilloche separating the stave of the upper voice from that of the middle voice and that of the middle voice from that of the lower voice. Bar lines in red ink make divisions of notation coincide with word divisions. Cormac the composer refers to himself at the symmetrical centre of his polyphonic music. In the upper voice the central twenty-eighth note of fifty-five is the last of *eo*. In the middle voice the central twenty-third note of forty-five is the last of *eo*. In the lower voice the central fifteenth note of twenty-nine is the last of *eo*.

The music in the first half of the first verse is identical with that in the first half of the second verse in all three voices. As the lower voice has one note for each syllable, the fifth note of the first verse, required for the second syllable of *scripsit*, though obscured by abrasion, is confirmed by parallelism with the fifth note of the second verse. Both verses of the lower voice are exactly parallel except for the antepenultimate note of the first, which has no pair in the second. Both verses of the middle voice are exactly parallel except for the tenth note of the first, which has no pair in the second. Both verses of the upper voice are exactly parallel except for the penultimate note of the first, which has no parallel in the second. The voices begin on DDG in both verses and end in unison on GGG in both verses. The only other points at which the voices sing in unison, AAA and AAA, coincide with the rhymes in the third foot of each of the hexameters, the latter sung on note A on the vowel *a* in *ora*.

Conclusion

Having considered four models from Late Antiquity, three texts from the first century of Hiberno-Latin computistic learning, and examples of computistic phenomena in Hiberno-Latin literature from the fifth century to the twelfth under twenty-three headings, one for every letter of the Latin alphabet, let us conclude in a sentence of fifty-two words. We might even add a coda to achieve a total of 365 characters, one for each day of a solar year.

DÁIBHÍ Ó CRÓINÍN

THE CONTINUITY OF IRISH COMPUTISTICAL TRADITION

Abstract

It is well known that the study of computus in Ireland in the sixth and seventh centuries was at a level not equaled anywhere else in Europe, with the possible exception of Visigothic Spain. Not so well known, however, is the fact that computistics continued to thrive in Ireland, not only into the eighth and ninth centuries, but well beyond that. In fact, the eleventh and twelfth centuries saw a high-point of scholarly activity, in the related fields of chronology and chronography, both in Latin and in the vernacular. The best known Irish scholar of the period, Marianus Scottus of Fulda and Mainz, established a pattern for computistical and chronographical studies for centuries to come. This paper presents some of the evidence for that *Blütezeit*.

Keywords

Adam & Eve (sons of), Ambrose of Milan's *De Noe et Arca*, *Anonymus ad Cuimnanum*, apocrypha, Book of Glendalough, Dublittir Ua hUathgaile, Flann Mainistrech (Monasterboice), (Ps-) Hilary, *Tractatus Hilarii in septem epistolas catholicas*, Hugo of St Victor's *De Arca Noe morali liber*, Irish Computus, *Lebar Bretnach*, *Lebar Gabála Érenn*, Marianus Scottus (Mainz), Máel Brigte Ua Máelsuanaig, *Saltair na Rann*, *Sex Aetates Mundi*.

World History was a preoccupation of Irish scholars, writing in both Latin and the vernacular (and sometimes both), during the eleventh and

---

[1] The best-known representative of the tradition in Latin is Marianus Scottus (†1082), originally of Moville (Co. Down), and subsequently exiled in Fulda and Mainz; see von den Brinken (1961). For his Irish milieu, see Mac Carthy (1892) and Ó Cuív (1990).

twelfth centuries.¹ This is reflected not only in the proliferation of translations of Roman texts, for example, Lucan's *Pharsalia* and Dares Phrygius' Siege of Troy,² but also in a marked upsurge of interest in both biblical and pseudo-biblical literature. A parallel development saw the composition (in the Irish language, in both prose and verse) of indigenous texts relating Ireland's role in that scheme of World History.³ These were works of biblical history and Irish pseudo-history, such as *Saltair na Rann* ('The Psalter of Verses'),⁴ *Lebar Gabála Érenn* (the so-called 'Book of Invasions'),⁵ and *Lebar Bretnach* (a Latin version of Nennius' *Historia Brittonum*).⁶ One such work in the biblical genre is the Irish *Sex Aetates Mundi* (hereafter SAM), compiled probably c.1080.⁷

The modern scholar who contributed most to the codicological and structural analysis of this corpus, the late Hans Oskamp, once suggested that 'of all these tracts [SAM, *Lebar Gabála, Lebar Bretnach*] *Sex Aetates Mundi* undoubtedly came first in all the synchronistic compilations of the eleventh and twelfth centuries'.⁸ In my 1983 edition of the text I argued that the recension in the best manuscript, Oxford, Bodleian Library, Rawlinson B 502, 40v–45r, was probably the work of Dublittir Ua hUathgaile, resident *fer légind* (scholar) of Killeshin (Co.

---

² Stokes (1881); idem (1884); idem (1909); Calder (1907); see Harris (1998). The paper by Ní Shéaghdha (1984) (delivered as a Statutory Lecture of the School of Celtic Studies in the Dublin Institute for Advanced Studies) is a very general treatment of the subject. See now Murray (2006) and literature there cited.

³ The best example of the vernacular Irish tradition was Flann Mainistrech, scholar of Monasterboice (Co. Louth); see Mac Airt (1953–4); idem (1955–6); idem (1958–9). Flann composed a parallel series of historical poems on the subject of native Irish history; see Mac Neill (1913a). However, there is no adequate modern survey of the literature produced by Flann (†1056) and his contemporaries, Eochaid Ua Flainn [Ua Flannacáin] (†1004), Gilla Cóemáin (fl.1072), Tanaide Ua Máelchonaire (fl.?), Gilla In Choimded Ua Cormaic (fl.1118). For a useful general survey of the Irish background, see Ó Cuív (1963) and Byrne (1974). The contemporary production of Welsh translations of some of these texts is a phenomenon that would be worth investigating, especially to see if there was any link between these Irish and Welsh literary projects. A useful starting-point is Jones (1968).

⁴ See Stokes (1883).

⁵ Macalister and Mac Neill (1916); Macalister (1938–56). See the very useful summary of recent researches on this text in Carey (1993) and for the scholarly background, Carey (1994).

⁶ Todd and Herbert (1848); Hogan (1895); van Hamel (1932). There is a substantial modern literature devoted to this text, conveniently summarized by Dumville (1975–6).

⁷ Ó Cróinín (1983); see Tristram (1985). For a survey of the manuscript evidence from this period, see Oskamp (1974).

⁸ Oskamp (1975), 114. See also Oskamp (1968). For additional supporting evidence in favour of Oskamp's theory, see Ó Cróinín (1998).

Laois), whose floruit can be placed probably *c*.1100 (and perhaps even slightly later). He very likely succeeded Conchobar Ua hUadgaile (†1082 according to the Annals of the Four Masters) in that position, and is listed among the witnesses to the charter in the Book of Durrow (MS Dublin, Trinity College, A.4.15 (57), 248v), which records the cession to the Columban community and monastery in Durrow (Co. Offaly) of a small *erles* ('enclosure') called Ind Ednán, in recompense for an *erles* of theirs which had been granted by the monastery of Killeshin (Glenn Uissen) to the Dál Cais. Analysis of the identifiable names in the Durrow charter revealed that the document could be dated to *c*.1110 x *c*.1116, with outside limits of *c*.1082 x *c*.1120.[9] Dublittir and his brother or cousin Dúnchad are both mentioned in the witness-list, and the Uí Uathgaile genealogy is preserved in Rawl. B 502. This manuscript was written in a south-Leinster scriptorium (possibly Killeshin, though Glendalough too has been proposed),[10] *c*.1130,[11] and contains a miscellany of biblical, genealogical, legal, and literary texts in Irish.

The rough floruit that I had assigned to Dublittir (the last quarter of the eleventh century and the first quarter of the twelfth) would accord well with the language of SAM, which is Middle Irish of a consistent standard throughout.[12] It would also mean that the copy of the text in Rawl. B 502 must be very close in date to the author/compiler/translator's own time, which would add extra weight to the statement in that manuscript that the work was Dublittir's, since an explicit attribution by name at the beginning of the work (§1: *tintúd Duiblitrech huí hUathgaile*

---

[9] Ó Cróinín (1983), 41–8.

[10] For modern arguments in favour of identifying Rawl. B 502 with the 'lost' manuscript long known as the Book of Glendalough, see Ó Riain, (1981) and idem (1982); for counter-arguments, see Breatnach (1997). The debate is still ongoing.

[11] See Byrne (1979). The article by Bhreathnach (1994) has nothing new to offer, but is a usefull summary of the evidence.

[12] Prof. Máire Herbert (1986), in a review of my edition, seems to me to have hopelessly misunderstood the evidence for origins and transmission. Her reference to a text 'made ostensibly [*sic*] in the light of critical conventions' (p 102) is clearly intended as a criticism of my editorial methodology, which she describes thus: 'This apparently accepts from the outset that the text is amenable to general procedures of textual criticism. Therefore it surveys the evidence of the principal witnesses, sets out their relationships, and makes editorial decisions accordingly'. That a text created by a scholar *c*.1100 could not be reconstructed from MS witnesses dating from *c*.1100–1120 is a nihilistic conclusion and a counsel of despair. I treated all the wider textual matters relating to the other recensions of SAM in my M. Phil. dissertation (UCD 1977), but some of that discussion was omitted from the published edition.

forsin Pandecht Cirine tria Góedeilg in so sís) to one either still alive or recently deceased, and resident in the monastery where the manuscript was written (if – as I am inclined to believe – it was produced in Killeshin) would strike a very odd note indeed if it were anything but true.[13]

SAM is a mixed prose-verse synopsis of Old and New Testament history based on the well-known theme of the Six Ages of the world popularized by patristic authors such as Saint Augustine. The germ of the idea was to be found in the apocryphal Epistle of Barnabas: 'On the Sabbath', which treated of a created universe lasting six days. By uniting hexaemeron, chiliasm, and eschatology with the statement in 2 Pet 3:7-8 that 'one day is with the Lord as a thousand years', to declare that '[. . .] in six days, that is in six thousand years, everything will be completed', the millennarian mind produced a concept that lasted a thousand years. Since Genesis determined the chronographical pattern of creation, and since the books of Kings and Chronicles set the historiographical pattern, early Christian commentators naturally felt free to draw on the vast corpus of patristic and post-patristic literature on the subject that came into existence during the first four centuries of the Church. In the composition of the Irish *Sex Aetates*, therefore, it is no coincidence that its author drew for the most part on Genesis and Kings for detailed genealogies and biblical history, coupled with an allegorical approach to both Old and New Testament events that allowed him also a free hand to use extra-canonical and apocryphal texts to fill out his story.[14]

The most remarkable feature of SAM, indeed, is the frequency with which its author has drawn on apocryphal works, several of which are unique to our text. For example, he gives (§§9–11) the names of Adam and Eve's fifty-two sons, in prose and in verse mnemonic form, with the additional detail – ascribed to Jerome – that the sons numbered fifty-two and the daughters seventy-two (*Hironimus: Tot filii Adae quot sunt*

---

[13] Bhreathnach (1994), 45, states: 'Rather than interpret this expanded [sic] version as a genuine ascription, it could be inferred that it is the work of an enthusiastic copyist or compiler willingly ascribing more than his fair share to Dublittir, author of "Rédig dam" (but not of the *Sex Aetates*) and a *fer léigind* of his own monastery, Killeshin.' Comment is superfluous. She teeters on the edge of opting for Killeshin as place of origin for the MS, but her nerve failed her.

[14] The apocryphal sections were collected from my edition and published in Herbert and McNamara (1989).

dominici dies in anno, id est .lii. Tot filiae Adae quot sunt lingae, id est .lxxii.).[15]

SAM §§36–37 give the names of Lot's daughters in the characteristic Irish *tres linguae sacrae* form: *Iachether* and *Iachereth* in Hebrew, *Chiros* and *Miros* in Greek, *Sadibilis* and *Adherens* in Latin. These too are unique both in Hiberno-Latin and vernacular Irish literature, and though similar material does occur elsewhere in medieval Latin literature outside Ireland (and in Jewish tradition), the SAM details are not found in any other text.[16] The names of other biblical wives are given who likewise are not mentioned in the canonical books of the Old Testament. In the New Testament section of the work, SAM §66 gives a list of the names of the seventy-two disciples of Christ, another example of the 'names for the nameless' syndrome which, while characteristic also of the early Christian centuries, was a distinctive feature of Hiberno-Latin literature from an early date.[17]

Of more immediate interest to us, however, from a computistical point of view, is another section of SAM (§19), whose origins are likewise to be found in the apocryphal literature of the early church, and at the same time vividly illustrates the continuity of the Irish (Hiberno-Latin) exegetical and computistical tradition from the seventh century to the eleventh. This text concerns the symbolism of Noah's Ark, which was an important theme in medieval Latin exegesis generally, following on the example of commentaries such as Ambrose of Milan's *De Noe et Arca*. Our section, however, has close parallels in a seventh-century Hiberno-Latin commentary, the *Tractatus Hilarii in septem epistolas catholicas*,[18] composed (according to its editor, Robert McNally) at a date between 690 and 708. Taking the SAM text first (§19):

Cestnaigthir *hic* do thomus 7 do figuir 7 do eterchert na háircce: *Arca graece, Secreta latine interpretatur* .i. deirrit. *Longuitudo hautem arcae*

---

[15] For similar details from other contemporary Irish texts, see Ó Cróinín (1983), 142–3. In the context of my title, 'The continuity of Irish computistical tradition', I might point out that this detail concerning the numbers of Adam and Eve's sons and daughters is at least as old as the eighth century, as proved by its occurrence in the text of the Hiberno-Latin grammar known as the *Anonymus ad Cuimnanum*, where it appears in the prologue (ed. by Bischoff and Löfstedt in CCSL 133D, 2: *Apocripha enim ferunt Adam habuisse filios LII numero dominicarum totius anni dierum et filias tot, quot linguae hominum sunt*).

[16] For discussion of this material, see Ó Cróinín (1983), 154–5.

[17] For discussion, see Ó Cróinín (1983), 17–74.

[18] CCSL 108B, 51–124.

perseuerantiam perfectionis significat .i. feidligthetu na fuirbthitad tóirnes a fat. *Latitudo eius latitudinem caritatis significat* .i. forleithi na deisirce tóirnes a lethet. *Altitudo eius altitudinem spei significat* .i. arddi 7 áithi na freiscisen fil ó duini co Dia tóirnes a hailt .i. a hairddi, *id est, unum cubitum, id est, uniuersalem catholicam fidem significat* .i. oénchubat a hairdi .i. oéndatu 7 huilidetu inna hirsi catholicda fon huili ndomun tóirnes in sin.

'It is enquired of here concerning the dimensions of the Ark, and of its shape and symbolical meaning. It is interpreted as 'arca' in Greek and 'secreta' in Latin, i.e. hidden. The length of the Ark, however, signifies the lasting nature of perfection, i.e. its length signifies the perseverance of perfection. Its width signifies the breadth of charity, i.e. its breadth signifies the extent of charity. Its height signifies the loftiness of hope, i.e. the height signifies the rarity and sharpness of the anticipation that man has of God, i.e. its height, i.e. one cubit, i.e. the unity and universality [at once] of the Catholic faith throughout the whole world is signified by that.'[19]

Compare the following passage from the seventh-century *Tractatus Hilarii*:

Trecenta in longitudine habebat [arca] cubita, id est perseuerantiam in fide Trinitatis; quinquaginta in latitudine, id est caritatis longitudo [sic, rectius latitudo?]; triginta in altitudine, id est tres leges ad caelum ascendentes, id est tristige.[20]

It may be noted, by way of parenthesis, that, besides the connections with other Hiberno-Latin texts remarked on by McNally, there is a close parallel also between our SAM text and Hugo of St Victor's *De Arca Noe morali liber* I:[21] *longitudo in fide Trinitatis trecentos cubitos habeat, in latitudine charitatis quinquaginta, in altitudine spei quae in Christo est triginta.* It was long ago pointed out by Bernhard Bischoff[22] that Hugo's work may have been known to and used by the twelfth-century Armagh biblical commentator Maél Brigte Ua Maélsuanaig (floruit *c.*1138).[23]

[19] Text and translation: Ó Cróinín (1983), 71, 113.

[20] CCSL 108B, 91.

[21] PL 176, 634.

[22] Bischoff (1954), rev. ed. in Bischoff (1966), here p 270 n 143. There is an English translation by O'Grady (1976).

[23] For Maél Brigte's gospel commentary, Rittmueller (1983); addenda and corrigenda in eadem (1984).

One final example from SAM (§20) will illustrate the nature of the work's dependence on earlier, non-Irish apocryphal sources. I give here first the Irish text, with a brief explanation of its purpose, then the Latin source, which I recently discovered:

> *Interrogatio hic.*, Cia haés ésci 7 cia lathi sechtmaine 7 cia lathi mís gréne do-chuas isind áirc 7 tícht essi? Ocus cia bliadain, hi commun nó i n-emblesim? Ní annsa: hi sechtmaid fichet <.i. Maii> do-chuas inti 7 tícht essi, 7 is sí tarrasair for Sléib Armeniae. *Terticim Kalendas* Iúin do-chuas inti 7 sechtmad Id Maii tícht essi, ar aí lathi mís gréne. Can asa thucthar sin? Ní annsa: ar is oénlathi déc fil etir sechtmad Id 7 *treticim*, 7 iss ed ón beres bliadain gréne do bliadain éscae, ar it cethri lathi cóicat <.i. ar tríb cétaib> [fil] i ssuidiu 7 is ed ón fil etir treticim 7 sechtmad Id hi cind bliadna.
>
> Oéndiden *hautem* do-chuas inti. Can asa thucthar sin? Ní annsa: ar is i n-oéndiden tarmthecht Ádaim. Cubaid dano ciambad inti na-tíssed dígal forsin mbith. Cest: cia lathi sechtmaine tícht essi? Ní annsa: .i. Mairt. Can asa thucthar sin? Ní annsa; ar is hi *treticim Kalendarum* Iúin do-chuas inti. Secht lá déc 7 bliadain de-side ro-buí Noé, sind áirc. Hi sechtmaid déc dano mís Maii ro-thinscan in díliu ferthain. Sé cét bliadan dano aés Noé tan tánic in díliu.

> 'Here is a question: At what age of the moon, and on what day of the week, and on what day of the solar month was the Ark entered and left? And what kind of year [was it], 'common' or 'embolismic'? Not difficult! It was entered and departed from on the 27th of May, when the Ark happened upon Mount Armenia. It was entered on the 13th Kalends of June, and it was left on the 7th Ides of May, as regards the day of the solar month. Whence is that derived? Not difficult! For there are eleven days between the 7th Ides and the 13th, and that, moreover, is the amount by which a solar year surpasses a lunar year. For there are 354 days in that [= the lunar year], and that is what separates the 13th and the 7th Ides at the end of a year.
>
> It was entered, then, on a Friday. Whence is that derived? Not difficult! For Adam's transgression was on a Friday. It is fitting, therefore, that it should be on that day that retribution should come upon the world. It is asked, on what day of the week was the Ark left? Not difficult! i.e., on a Tuesday. Whence is that arrived? Not difficult! For it was entered on the 13th Kalends of June. Hence Noe was in the Ark for a year and seventeen days. Moreover, on the 17th of May the deluge began to pour. Noe, then, was 600 years old when the deluge came here.'[24]

---

[24] Text and translation: Ó Cróinín (1983), 72, 114. For commentary, see ibidem, 145–7.

	Year	Month	Day
Noah enters Ark	600	2	10
Flood begins	600	2	17
Ark settles	600	7	17
Mountains appear	600	10	1
Raven sent out	600	11	11
Noah opens Ark	601	1	1
Noah leaves Ark	601	2	27
–	600	2	10
Time in Ark	1		17

Figure 1 The Genesis account of the Flood and the time spent by Noah in the ark as outlined in SAM (Ó Cróinín (1983), 146).

The problem here set by the question (*interrogatio*) has its origins in the ambiguities of the Genesis account of the Flood, especially Gen 7:11 and 8:3–4. According to Gen 8:3–4, the Flood began exactly with the beginning of the calendar year 600 in the life of Noah, and ceased exactly with the end of that same year, which was understood to have been a *bliadain éscae*, a lunar year of 354 days. According to the other version of the story, however, the Flood commenced 47 days after the first day of the calendar year, i.e. on the 17th day of the second month (Gen 7:11), and ended on the 27th day of the second month in the following year (as in Gen 8:14). This second version, therefore, has added eleven days (reckoning inclusively) to the lunar year of 354 days, thereby clearly suggesting that the year of the Flood should be reckoned rather as a solar year (*bliadain gréne*) of 365 days. The SAM text, however, states explicitly that the Flood occurred in a leap-year (of 366 days), as *Figure 1* shows:

If the Flood occurred in a leap-year, then that year had 366 days; $366+17=383$; $383-7\times54=5$. Accordingly, if the first of these 383 days (rekoning inclusively) was a Friday (*Oéndiden do-chuas inti*), then the last would have occurred five weekdays later (reckoning inclusively), so it would have been a Tuesday.

Thus the text is a combination of computistical calculation and exegetical interpretation, a very common feature of medieval biblical commentary, and one that reflects the nature of its medieval source. What is of particular interest to us in the present context of my theme is the newly-discovered source; the parallels with our SAM text will be immediately obvious:

Dicunt Septuaginta quod ab initio mundi usque ad diluuio fuerunt anni (e)duo milia clxii; secundum Hebreos duo milia xxx.i. Item ibi: in anno endecadi factum est diluuium. Kal. Ian. fuerat, secunda feria, et fuit bissextus, et luna vii. in ipsis. Kal. Apr. fuerat .ii. feria, luna viii. Kal. Mai. iiii. feria fuerat, luna viii. et introiuit Noe in arcam secunda feria, luna xxvii / cum vi. epactas xxviii. / et xiii. Kal. Iun. recte secunda feria intrauit, quia in illo die firmamentum fuit factum, et exiuit de arca .vi. feria, et recte .vi. feria, quia Adam in illo die uenit in mundum quando exiuit de paradyso. Et Kal. Ian. fuerat .iii. feria, luna xviii. Et Kal. Apr. fuerat iii feria, luna xviiii. Et Kal. Maii / cum .i. epacta et Kal. Apr. [. . .] / v. feria fuerat, luna xx. Et exiuit de arca vii. Id. Maii, luna xxviii. articulo diei, id est in vi. hora [corr. < id], in v. feria, luna xxvii. In tertio decimo Kal. Iun., mense secundo, primo anno post sexcesimos Noe annos, prima luna, Maii xx.mo vii die, in vii. Id., vi. feria, in ipso die uenit de arca. Compleuit totum annum secundum lunam in arca, a xxma viii.ua luna Maii, id est xvi/i/i Kal. Iun., usque in xxviii luna Maii, id est vii. Idus Maii; remanent autem .xi. dies de anno solis, id est a vi. Id. Maii usque in xiii Kal. Iun.

Xv cubitis diluuium altior quam mons Olimphy, qui ex/c/elsior est quam nubes pluuiales. Alii dicunt quod ille non numeratur cum cetera, quia in uertice eius non est crassa terra. Uenit ad uesperum, id est diei lunae, .i. vi. hora diei.[25]

This computistical-exegetical text is preserved in MS Padua, Biblioteca Antoniana, I 27 (*saec.* ix/x), 123v.[26] The Padua codex is a miscellany, made up principally – though not exclusively – of computistical and astronomical texts. Thus, while it contains Hrabanus Maurus, *De computo*; excerpts from Bede's *De temporibus* and *De temporum ratione*; various computistical *argumenta* and tables, etc., it also preserves a number of important early Italian historical sources, including the *Translatio Sancti Benedicti* and the *Chronicon Langobardorum* (of which this is the best copy). On the face of it, therefore, the manuscript appears to have only an incidental Irish connection, but a closer analysis of the contents reveals a different picture.

On folios 66r–71v is a copy of the text known in the early middle ages as *Disputatio Cori et Praetextati*, which is an epitome (probably

---

[25] I have not troubled to correct the various numbers in the text, which have been corrupted in transcription. Discussion of them would take us too far from my present topic. Words and sections between slashes are added above the line in the MS by the scribe-corrector.

[26] For a comprehensive description of the manuscript's contents, see Appendix II.

Irish, and probably of the seventh century) of Macrobius' *Saturnalia* I 11–15.²⁷ This is followed immediately (folios 71v–75v) by Anatolius, *De ratione paschali*, which was long believed to be one of the so-called 'Irish Forgeries' associated with the Paschal controversy of the sixth and seventh centuries.²⁸ Following this again (folios 76r–77v) is the only surviving copy of the long-lost *latercus* (otherwise known as the Irish 84-year Easter table), which disappeared completely after the Columban community of Iona abandoned its use in 716, until it was rediscovered in 1985.²⁹

Immediately after the 84-year Easter table in turn is an acephalous lemmatized commentary on the *argumenta* of Dionysius Exiguus that is also demonstrably Irish in origin (folios 77v–78r).³⁰ These computistical texts, well-known in seventh-century Irish circles, clearly suggest a direct access to Irish materials, some of them very early. Padua I 27 also contains (folios 119r–123r) excerpts from the *Collectio canonum Hibernensis* collection of Irish canons, possibly another indication of a direct connection with Ireland.³¹ Our text on the subject of Noah's Ark is found immediately following that collection of excerpts from the *Hibernensis* (folio 123v).

The sentence immediately following the excerpt on Noah's Ark in Padua I 27 reads: *Item in ipso questionario. Interrogatio: Quomodo diuisum est orbis a filiis et nepotibus Noe?*, etc.; this also has a connection with SAM, for §6 of that work reads: *Cinnas ro-gablaigset clanna Noé, et cia lín cenél n-écsamail i tarrasatar?* ('How did the family of Noah divide, and in how many distinct races did they remain?'), which is followed at SAM §24 by an account of Noah's sons Sem,

---

²⁷ I have had an edition in preparation for far longer than I care to admit! There is a dense discussion of the manuscript transmission in Arweiler (2000), though the author has not realized the exclusively Irish nature of the text transmission, and has an incomplete list of MS copies (5 only of the 7 known to me): Oxford, Bodleian Library, Bodley 309, 101r–105v; Padua, Biblioteca Antoniana, I 27, 66r–71v; Genève, Bibliothèque Publique et Universitaire, 50, 163r–164v; Köln, Dombibliothek, 83², 204r–205v; and Wien, Österreichische Nationalbibliothek, ser. nov. 37, 3r–7v. The additional MSS are: Paris, Bibliothèque Nationale, Nouvelle acquisition latine 1613 (originally part of Tours, Bibliothèque Municipale, 334), 41v–46r, and Vatican, Biblioteca Apostolica, Reg. Lat. 586, 117v–125r.

²⁸ See Krusch (1880), 311–27; see Mc Carthy and Breen (2003).

²⁹ See Mc Carthy and Ó Cróinín (1987–8), repr. in Ó Cróinín (2003).

³⁰ See Ó Cróinín (2008).

³¹ The late Prof. Maurice Sheehy, who at one time was preparing an edition of the *Hibernensis*, informed me that the Padua excerpts did not belong to any known recension of that text; for these excerpts see Appendix I.

Cham and Japheth (with the names of their wives, Olla, Oliua, and Olibana!) and their respective territories, very like what follows in the Padua text:

> *Responsio: Sem, ut estimatur, Asia; Cham Africam, et Iaphet Eurupam sortitus est. De Iaphet natae sunt .xv. gentes; de Cham .xxx.; de Sem .xxvii., simul .lxxii., de quibus orti sunt gentes .lxxii. Delatet Deus Iaphet et habitet in tabernaculis Sem. Ubi haec benedictio completa est? Responsio: Id est, quando Romani, qui de genere Iaphet sunt, obtinuerunt Iudeam, quod de gente Sem sunt. Item Sem accepit Asiam, et orientem. Cham, Africam et meridiem. Iaphet, cum partem Asiae et Eurupam et occidentem.*

Compare the following in SAM §24:

> *Ro-thuisim Noé trí macu ria ndílinn, conid uadib-side génatar na dá cenél sechtmogat iar ndílinn. Is amlaid so hautem ro-sílsat na cenéla .i. secht cenéla fichet díb ó Sém, trícha immorro ó Chám, a cóic déc ó Iaféth; conid dá cenél sechtmogat sin uli 7 conid dá bérla sechtmogat tuctha dóib post ic tairmesc in Tuir Nemruaid.*

> 'Noah begot three sons before the deluge, and it is from them that the seventy-two races derived after the deluge. This, then, is how those races gave issue: twenty-seven of them from Sem; thirty from Cham, and fifteen from Japheth. So that there are seventy-two races in all, and they were given seventy-two languages afterwards, at the stopping of [the building of] Nimrod's Tower.'[32]

The *Questionario* format suggests that our Padua text (and the very similar material in SAM) derives ultimately from some such question-and-answer collection; that would explain the *Interrogatio*-format of the SAM version. The more detailed account in the Padua manuscript would appear to present an earlier (seventh-century?) stage of the transmission, but the SAM text is nonetheless very good. Whether it is derived independently from a miscellany of biblical-computistical classroom material (as the *Interrogatio* formula might suggest), or formed an integral part of a longer Latin text similar in content to the Irish SAM, which was then translated into Irish, is impossible to tell. It is equally difficult to establish whether Dublittir Ua hUathgaile's contribution to this enterprise involved only translation (*tintúd*) or something more. He was equally at home in both Latin and Irish, as

---

[32] Text and translation: Ó Cróinín (1983), 73, 115.

the mixed nature of the SAM text clearly indicates, and would have been quite capable of collecting different excerpts from earlier Hiberno-Latin sources and combining them into the text that we now have.[33]

For the present, however, we cannot do more than point to the existence in the Padua manuscript of material in Latin which is preserved also in the Irish text of SAM. That the Latin *Quaestionario*-format was earlier in date than the text in SAM, and represents an earlier stage of the Irish text history, would appear to be certain. One thing, however, is clear from this new evidence: apart from its exceptional qualities as a translation, and as an uncommonly rich source of biblical apocrypha, some of it unique, the Irish *Sex Aetates Mundi*, in terms of its sources, is another example of the kind of continuity best exemplified by the exactly contemporary Marianus Scottus of Mainz and Fulda, whose autograph manuscript World Chronicle (MS Vatican, Biblioteca Apostolica, Pal. Lat. 830)[34] is prefaced by several of the so-called 'Irish Paschal Forgeries'. In both cases we are witness to the continuing tradition of biblical, computistical, and grammatical studies[35] in Ireland from the seventh century down to the eleventh and twelfth, and the preservation, even in those parts of the country where the vernacular came under pressure from the Anglo-Norman occupiers, of materials that represent the seamless garment of the Irish learned tradition.[36]

---

[33] In the acephalous text of SAM in *Leabhar na hUidhre* (MS Dublin, Royal Irish Academy, 23 E 25) most of what was originally Latin has been rendered into Irish; this, and other considerations, led me to the view (which I still hold) that the LU recension is an inferior one. The fact that it originally contained a shorter recension than that in Rawl. B 502 does not, in my view, necessarily mean that its text represents an earlier stage in the transmission.

[34] For Marianus and his sources, besides the article by von den Brinken (1961), see also Kenney (1966), 614–6, No. 443.

[35] The notion – as silly as it is widespread in current 'scholarship' – that knowledge of Latin withered away in Ireland after the 10th century is belied, e.g., by the sophisticated treatment of Boethian metrics in the 12th-century Irish manuscript Firenze, Biblioteca Medicea Laurenziana, Plut. 78.19, 1v–3v; see Oskamp (1977); Ó Néill (2005).

[36] Even in the vernacular tradition, the poem printed in Appendix III is clear evidence for the continuity of computistical studies in Irish schools where instruction was in Irish. For the most spectacular example of all – 'a verse mnemonic in Irish for finding the lune of the day was being recited in West Cork that presupposed the lunar calendar of the Celtic Easter cycle, which had been abandoned over 1200 years earlier'! – see Holford-Strevens (1999), 871; Mc Carthy (1993), 212; the original source of the mnemonic is O'Leary (*c.*1920), 1.

We may leave the final word on this subject to Bernhard Bischoff: '[E]s gab in Irland im XI.–XII. Jahrhundert noch klösterliche Schulen, in denen ebensolche Artes-Studien wie in guten festländischen und englischen Schulen betrieben wurden.'[37]

APPENDICES

Appendix I: The *Hibernensis* excerpts in MS Padua, Biblioteca Antoniana, I 27, 119r–123r

INCIPIUNT EXEMPLARIA DIUERSA ET UTILIA SANCTORUM PATRUM AD PROUECTVM ANIMARUM UEL DE CONGRUA SI QUIS EA RETINERE UOLUERIT MENTE DEUOTA

The passages excerpted are as follows (Wasserschleben (1874)'s pagination; in the transcription proper, square brackets mark text that is not in Wasserschleben's *Collectio*; arrow-brackets indicate omitted letters / text that I have restored from Wasserschleben; a full edition of these excerpts, including commentary, is in preparation):

XXVII 10–17 (88–91); [+ excerpt < Greg.Mag., *Moralia* —]; XLIV 20 (179); LIII 4 (214); LII 3 (211–2); [+ excerpt < Isidore + Ezek. —]; LII 5 (212); [+ excerpt *De peccato* —]; LIV 1–4 (215–6); LIV 8 (216–7); [+ excerpt < Jerome —]; I 1 (3); I 3 (3–4); [+ *Ordo ep. quadripart. e.* —]; I 6 (5); [+ excerpt on *Corepiscopi* —]; I 10 (7–8); I 21 (11); II 1 (12); II 2 (13); II 11 (15); II 13 (16); II 16 (17); II 22 (18); II 24 (19); II 26 (19); III 1–2 (20); III 5 (21); IV 1 (22); IV 4 (23); XIII 1 (38); XIII 7 (40); XV 2 (42–3); XV 8 (45); XVI 2 (45–6); XVI 7 (45); XVI 9 (48); XXI 16 (68); [+ excerpt < Jerome —]; XXI 10 (65–6); XXII 2 (74); XXII 3–4 (74); XXVII 19 (92); XXVIII 12 (97); XXXIX 1–2 (147); XXIX 3 (99); XXIX 9 (102); XXIX 2 (99); XXIX 1 (98); XXIX 4 (99–100); XXX 2 (103) XXX 3 (103); XXXIII 1–2 (118); XXXIII 3 (118–9); XXXIV 4 (122–3); XXXVII 4 (132); XXXVII 25 (138); XXXVII 11/13 (134); [XXXVII 14 (135) om.]; XXXVII 10 (134); XXXVII 15 (135); XXXVII 39 (140–1); XXXVII 21 (136) [XXXVII 17 (135) om.]; XLII 1–4 (161–3) — but without Paul, Nechias, Pelagius — XLII 10 (164); XLIII 7 (173–4).

---

[37] Bieler and Bischoff (1956), 220.

INCIPIUNT EXEMPLARIA DIUERSA ET UTILIA SANCTORUM PATRUM AD PROUECTUM ANIMARUM UEL DE CONGRUA SI QUIS EA RETINERE UOLUERIT MENTE DEUOTA DE EO QUOD QUATTUOR MODIS IUDICAT DEUS OMNE PECCATUM; IN PRIMIS DE MAGNORUM PECCATURUM MAGNIS UINDICTIS.

[XXVII 10] Initium omnis peccati superbia est, per quam diabolus <factus> est, qui<a> bonus per naturam a Deo conditus, hi<n>c in sempiternum damnatur [cum omnibus suis]. Adam primus homo diuine mandatum transgressione morte dam[p]natur. Kain primus homicida vii uindictas soluit. Delamech primo adultero et homicida septuagies vii.es uindicatur. Iniquitatem mundi dilui/um/ tersit. Uanam gloriam turrem aedificantium linguarum confusio dam[p]nauit. E<t> [pro] scelere in naturali ignis Sodomam et Gomorram conbussit. Egyptus crudeliter blebem captiuans adficitur .x. plagis et in mari* rubro nouissime mergitur. Chore quoque indignam deo hostiam inmolans cum suis alienum igne crematur. Aperta est /terra, et reliqua/ <et deglutiuit Dathan et operuit super synagogam Abiron>. Dauid populum adnumerante adogantiae causa uirorum excercitus sui .lxx. /milia/ occisa sunt. Duo filii Aaron ignem alienum offerentes moriuntur iuxta hostias. Dcc Hebreorum pro mandati transgressione in heremo perierunt. In crucis uindicta .xi./es/ centena milia Hebreorum gladio fameque in Hierosolimis perieunt et c. milia publice uenundata sunt.

[XXVII 11] <ITEM> DE MINORUM PECCATORUM GRAUIBUS UINDICTIS IN LEGE. In sabbato homo ligna colligens coram omni populo occisus est. Item <in lege> qui maledixerit patre uel matri morte moriatur. IN REGUM <libris> Ozias arcam subleuans boue cadente moritur. Pueros xlii ad Heliseum dicentes: ascende calue duo ursi deuorauerunt eos. IN DEUTERONOMIO. Moyses propter unius uerbi paruam infidelitatem terram repromissionis non intrauit. Zacharias <pater Iohannis> contradicen/s/ angelo in uno sermone .viiii.uem mensibus mutilauit. Christus dicit: qui dixerit fratri suo racha siue fatue, et reliqua <reus erit concilio et gehennae ignis>. Ananias quoque et saphyra de propria pecunia mentientes moriuntur. Caupo percutiens Thomam moritur. Qui <enim> transgreditur unum mandatum, et reliqua <factus est omnium reus.>

[XXVII 1] ITEM DE MAGNIS PECCATIS LEUIBUS UINDICTIS. Daniel dicit ad Nabuchodonosor: Redime peccata tua elemosynis tuis o rex. SALOMON. Sicut aqua estinguit ignem ita elemonysa, et reliqua <extinguit peccatum>. DOMINUS PER EZECHIEL. Si dixero iniquo morte morieris, si auertat se ab iniquitatibus suis uita uiuet et non morietur. ITEM. In quacumque die conuersus fuerit peccatos a malitia sua omnes iniquitates illius obliuioni tradentur. <Christus dicit: Remittuntur illi peccata multa, quia dilexit multum.> IN AEPISTOLIS. Dic prius iniquitates tuas et saluaueris [salvus eris]. [PROPHETA. Dixi pronuntiabo, et reliqua.] <ITEM> APOSTOLOS. Qui furabatur, iam non furetur, sed laboret, et reliqua <manibus, ut habeat, unde tribuat indigentibus.> IACOBUS. Confitemini alterutrum peccata, et cetera <vestra et salvemini.> HIERONIMUS. Dominus paciens et multae misricordiae <ut> uitam aeternam penitentibus breui et diligenti tribuat.

[XXVII 13] <ITEM> DE TARDIS UINDICTIS IN PECCATORES. Dominus in Adam post annos dcccc/es/ xxx. uindicat. Kain fratricida non subito interficitur, sed per septem generationes uixit. Lamech <parricida> post multum tempus consumitur. Saul <similiter peccans> non cito moritur, de quo Dominus ait: Penitet me, ordinasse Saul in regem. Salomone peccante in filio sui uindicatur. Peccata filiorum Iacob in tertia et quarta generatione uindicantur. Dominus ad Moysen de Chanabeis dicit: Nondum conplete sunt iniquitates eorum, IN PROPHETIS. Non uult Deus mortem peccatoris, et reliqua. <sed ut se convertat et vivat in aternum.> HIERONIMUS. Minatur Dominus peccatores percutere, sed per multos annos ex<s>pectat, sicut a passione [sua] Dominus <-i nostri Jesu Christi> xl annos ex<sp>ectauit Iudeos. PELAGIUS AIT: Plerique [homines] <etiam> contra se calumniantur, cum in praesenti non reddat <-et> Deus [malum], non intelligentes, quodsi ita fieret, nullus pene hominum remansisset, nec unquam de iniustis fierent <justi.>

[XXVII 14] [ITEM] DE UELOCIBUS UINDICTIS IN PECCATORES. IN LEGE. Cito Maria lepra percutitur. Cito duo filii Aaron morte finiuntur. <In regum libris> Cito Ozias arcam tangens moritur. <In novo> Cito Iudas Schariod pena consumitur. Cito Anna <Ananaia> et Saphira moriuntur. Cito Olimphus Arrianus episcopus in balneis santam trinitatem blasphemans conbustus uisibiliter moritur. Arrius <ipse> Christi fidei contrarius apud Constantinopolim uisceribus fusis cito interiit.

[XXVII 15] DE NOTATIONE DIUERSITATIS IUDITII DEI IN UTRISQUE. AUGUSTINUS AIT: Notanda est diuersitas/ 119ᵛ) i-udicii Dei in utrisque. In uno atrocitas ueritatis alacri<ta>sque misericordiae Dei ne addat peccata peccatis. In altero pacienciae modus et penitenteae* (corr > -iae) expectatio putatur. Siue quia haec uita quasi uapor aut somnium, siue umb<r>a in comparatione uitae futurae. ITEM. Non requirendum est cur Dominus aliquando tarde, aliquando cito, aliquando grauiter, aliquando leuiter iudicat, Salomone dicente: Noli alta <multum> sapere, et altiora te ne requiras. PAULUS AIT: Inuestigabiles illius et inscrutabilia sunt iuditia eius. GREGORIUS. Quis occulta iudicia Dei sciat? Ea quae in diuino examine compraehendere non possumus, timere magis quam discutere debemus.

[XXVII 16] [ITEM] DE QUATTUOR MODIS QUIBUS NON CITO UINDICAT DEUS. AUGUSTINUS: Quattuor modis cito non <non cito> uindicat Deus: primo ut pacientia eius probetur, ut <inde> dicitur: Paciens Dominus et multae misericordiae; secundo, ut sit homo inexcusabilis, ut apostolus dicit <inde dicitur>: Inexcusabilis <er>o homo; tertio, quia merita martyrum orant pro eis, ut Stephanus orauit; quarto, ut peccata peccatis addantur, ut in fine mundi cum diabolo [sine fne] puniantur. HINC IN LEGE DICITUR Noua peccata ueteribus peccatis addam. <Hi sunt uasa irae.> ET IN APOCALIPSYN IOHANNES: Adhuc qui <in> sordibus suis sordescat <qui in sordibus sunt.>

[XXVII 17] [ITEM] DE TRIBUS UINDICTIS QUIBUS CITO UINDICAT <iudicat> DEUS. AUGUSTINUS: Tribus quoque modis cito uindicat Deus: primo, ad probationem penitentiae et ad timorem incutiendum aliis; secundo, ne addant peccata, in poena, <ut et> cito consummatur. Inde Paulus ait: Dedi hominem huiusmodo <-i>, et reliqua. Et ut Thomas cito in cauponem uindicat, ut Dominus illi in futuro misereatur; tertio, quia animae sanctorum sub ara /Dei/ clamant: uindica sanguinem nostrum. [GREGORIUS IN MORALIBUS IOB.] Cum tempus uitae a diuina nobis praesentia sit procul dubio praefixum, querendum ualde est qua ratione dicit quod ininqui ex praesenti seculo ante tempus proprium subtrahantur. Omnipotens enim Deus et si plerumque mutat sententiam nunquam ex hac uita quisque subtrahitur quo ex diuina potentia ante tempora praescitur. Sed sciendum quia creans et ordinans nos omnipotens

Deus iuxta singulorum merita disponit et terminum, ut ualde malus breuiter uiuat ne multis bene agentibus noceat. Et iterum bonus diutius in hac uita subsistat, ut multis boni operis adiutor exsistat. Vel rursum malus longius differatur in uita ut praua adhuc opera augeat, et quorum temptatione purgati iusti uerius uiuant. Et uel certe iterum bonus citius subtrahatur, ne si hic diu uixerit eius innocentiam malicia corrumpat. Sciendum tamen quod benignitatis Dei est peccatoribus spatium penitentiae largiri.]

[XLIV 20] DE <nomine> BASILICA<e et ejus scissura.> SYNODUS HIBERNENSIS: Basilion grece rex latine, hinc et basilica regalis [domus dicitur], quia in primis temporibus reges tantum sepeliebantur [*space here in MS*] [inde] nomen sortita est, nam ceteri homines siue igni siue aceruo lapidum conditi sunt. <Item: Nemo alienus libertatem scindendi basilicam sine principis permissione habet, et si hoc ausus fuerit, reddet secundum dignitatem ejusdem loci.>

[LIII 4] DE TRIBUS NECESSARIAS HOMINIS. HIERONIMUS. Unusquisque tria necessaria habet, seruus, cans, bos; seruus ad operandum, canis ad custodiendum bos, ad arandum; seruus utatur bonis domini sui; canis de micis omnium pascatur, bos de herbis et frumento et paleis pascatur.

[LII 3] <De quinque causis, quibus tonsus est Petrus. Romani dicunt quod> BEATO PETRO PRO QUINQUE CAUSIS TONSUS EST. Prima, <ut praedixi,> ut adsimilaret spineam Christi coronam; secunda, ut clerici a laicis in tonsura discretionem haberent, et sicuti in habitu, ita operibus discererentur; tertia, ut sacerdotes ueteris testamenti reprobaret suspiciendo tonsuram in illo loco capitis, ubi columba super capud Christi descendit; quarta, ut derisiones gannituram in regno Romano suscepturus propter Deum sustineret; quinta, ut a Symone mago Christianorum discerneret tonsuram, in cuius capite cesaries ab aure ad aurem tonsa anteriore parte cum ante magi in fronte circum habebant.

[LII 4] <De sacramento coronae in anima habendo.> HESIDORUS DICIT: [Tonsurae aeclesiasticae usus a Nazereis incipit quod crinite seruato post uite magnae continentiae capud radebant, et capillos in igne mittebant, ut deuotionem Domino consecrarent. Inde apostoli hoc exemplum imitantes quasi diuini [*space here in MS*] mancipati

Domino consecraret crine praecioso. EZECHIEL DICITUR. Tu fili hominis/ 120ʳ) adsume tibi gladium acutum et duces (?) per caput tuum [*space here in MS*].

[LII 5] <De corona sacerdotium et regnum significante.> ESIDORUS: In tonso capite sacerdotium et regnum ecclesiae figuratur in inferiore et superiore circuli corona, thiara enim apud ueteres in capite sacerdotum erat. Unde Petrus ait: Vos estis genus electum, regale sacerdotium.

[DE TRIBUS MODUS LIBERIS A PECCATO ALIQUO SINODUS. Tres sunt peccata aliena]

Appendix II: Descriptions of MS Padua, Biblioteca Antoniana, I 27

Bibl.: Giuseppe Abate & Giovanni Luisetto, *Codici e manoscritti della Biblioteca Antoniana col catalogo delle miniature* [Fonti e studi per la storia del santo a Padova], 2 vols. (Vicenza 1975) i 28–33 ['XIII. Raccolta de esempli di SS. Padri'] — Patrick McGurk, *Catalogue of astrological and mythological illuminated manuscripts of the Latin middle ages,* 4: Astrological manuscripts in Italian libraries (other than Rome) (London 1966) 64–72 — Wesley M. Stevens, *Hrabanus Maurus 'On reckoning'* (Ph.D. diss., Emory University, Atlanta, Georgia 1968) [University Mincrofilms Inc., Ann Arbor, Michigan 1980] 178–88 — Luigi Guidaldi, *I più antichi codici della Biblioteca Antoniana di Padova (codici del sec. IX)* (Padova 1930) 21–8 (with plates) — Beniamoni Pagnin, 'Un presunto calendario bolognese nel codice Antoniano 27,' *Il Santo* 4/2 (Padova 1932) 316–22 — Augusto Campana, 'Veronensia,' *Miscellania Giovanni Mercati* [Studi e Testi 122] (Vatican City 1946) 1, 57–91 — A. Campagna, 'Nota Pacifichiana,' *Atti dell' Accademia d'agricoltura, scienze e lettere di Verona* 6/7 (1955–6) 135–49 — B. Pagnin, 'La provenienzia del Codice Antoniano 27 e del *Chronicon Regum Langobardorum* in esse contenuto,' *Miscellanea in onore di Roberto Cessi* (Rome 1958) 1, 29–41 — Daniel Mc Carthy and Dáibhí Ó Cróinín, 'The "lost" Irish 84-year Easter table rediscovered,' *Peritia* 6–7 (1987–8) 227–42, repr. in Dáibhí Ó Cróinín, *Early Irish history and chronology* (Dublin 2003) 58–75.

Contents of Manuscript (§§ in references to Isidore, *Etymologiae* refer to Lindsay's 1911 edition; ll. in references to Bede's computistical texts refer to Jones' 1943 edition):

(1) 1r–40v: Hrabanus Maurus, *De computo* — (2) 40v–41r: *De sex aetatibus mundi* (= Bede, *De temporibus* 16) — (3) 41r–43v: *Cursus et ordo temporum* — (4) 43v–44r: *De athomis momentis punctis et horis* — (5) 44r–45v: Isidore (?), *De diebus et uocabulis dierum* — (6) 45v–46v: Isidore, *De mensibus et nomina mensium* (= Isidore, *Etymologiae* V 33, §§1–2) — (7) 46v: *Argumentum ad septuagesimum et quadragesimum et Pascha* — (8) 46v–47r: *Argumentum ad LXXmum* — (9) 47r: *Ad XLmam argumentum* — (10) 47r–47v: *Item ad XL* — (11) 47v–48r: *Argumentum de Pascha inueniendum* — (12) 48r–49r: *Argumentum* (= PL 90, 722–3) — (13) 49r: *Ratio de bissexto* (= Bobbio Computus 40: PL 129, 1295–6) — (14) 49v–50r: *Interrogatio de bissexto* (= PL 90, 723) — (15) 50r–54r: *Interrogationes* (from *mundus* to *dies*) — (16) 54r–56r: *Uersus metrici Bedae quos inueni in alio loco* — (17) 56r–56v: *Uersus heroici de XII signis* (= MGH Poetae 4,2, 693) — (18) 56v–57r: *Uersus de nominibus uentorum* (= Baehrens (1883), 383) — (19) 57r–57v: *Uersus metrici de regularibus lunarum* (= van Wijk (1936), 119–20) — (20) 57v: *Uersus metrici de regularibus kalendarum* — (21) 57v: *Hec sunt opera quae fecit Deus in principio* — (22) 57v–59r: *Inquisitio et consideratio de septuagesima et sexagesima*, etc. — (23) 59r–63r: *Rationem humani corporis singulorumque membrorum differentiam Lactantius siue plerique auctorum ita diffinierunt* (= Isidore, *Differentiae* II 17, 19–23, 25, 24, 26–28) — (24) 63r–63v: *De ponderibus* — (25) 63v–64r: *De mensuris* — (26) 64r: *De sacrificiis ueteris testamenti* — (27) 64r–64v: *De cursu que cantatur in aecclesia dei* — (28) 64v–66r: *Translatio Sti Benedicti* (= Morin (1902)) — (29) 66r–71v: *Disputatio Cori et Praetextati* (= Macrobius, *Saturnalia* I 11–15) — (30) 71v–75v: Anatolius, *De ratione paschali* (= Breen and Mc Carthy (2003), 44–53; facsimile ibidem, 54–62) — (31) 76r–77v: *latercus* (Irish 84-year Easter table) (= Mc Carthy and Ó Cróinín (1987–88), repr. in Ó Cróinín (2003), 68–9; revised Mc Carthy (1993), 218–9; facsimile Warntjes (2007), 80–2) — (32) 77v–78r: Commentary on Dionysiac *Argumenta* (= Ó Cróinín (2008), 264–72) — (33) 78v–85r: Dionysiac Easter table (AD 532–1063) — (34) 85v: *Computus Bedae seu aliis doctoribus* (= PL 90, 799–800) — (35) 85v–89v: Tables of concurrents, termini of feasts, etc. (= PL 90, 715–6) — (36) 90r–90v: *Uersus cicli anniuersalis* (= PL 90, 860) — (37) 90v: *Uersus septem dierum* (= Baehrens (1883), 353) — (38)

90v: *Uersus de XII signis* (= Riese (1869–70), ii 92, nr. 640, 4–12, 1–3) — (39) 90v: *Uersus de singulis mensibus* (= Riese (1869–70), ii 91–2, nr. 639) — (40) 90v–91r: On the zodiac (= Gillert (1880), 251) — (41) 91ʳ: *Item* (= PL 94, 638) — (42) 91r–91v: *De duodecim signis* (= Isidore, *Etymologiae* III 71, §§22–32*) — (43) 91v–92r: *Sol igitur primum annum* (= PL 129, 1342–5*) — (44) 92r–93r: *Annus solaris habet IIIIor tempora* (Ps. Bede = Monachi ordinis S. Benedicti abbatiae Montis Casini (1873), 88–9) — (45) 93r: *De stellis que planetae uocantur* (= Isidore, *Etymologiae* III 71, §§20–21) — (46) 93r–94r: *De sideribus uel stellis* (= Isidore, *Etymologiae* III 71, §§4–19) — (47) 94r: *Planetae stelle sunt [Rerumque retro in II folio rerumque ut supra]* (= Isidore, *Etymologiae* III 71, §§20, 33–37) — (48) 94r: *Haec stellae cometes uocantur* — (49) 94: *Indictio dicta est* — (50) 94v: *Pro quibus mysteriis canitur missa* — (51) 94v: *De septem gradibus ecclesiasticis* (= Wilmart (1923), 311) — (52) 95v: *Quo ordine duodecim signa in caelo consistant* (= Bede, *De temporum ratione* 16, ll. 23–33, with movable chart) — (53) 95v: *Dimensio celestium* — (54) 96r: Diagram of *horologium uiatorum* (= PL 90, 954) — (55) 96r: *Spera caeli quater senis* (= MGH Poetae 4,2, 692) — (56) 96r: *De lune cursu per signa* (= Bede, *De temporum ratione* 19) — (58) 96v: *Ex libro questionum Alcuini* — (59) 97r: Table illustrating 1ˢᵗ text on 96v (*Cursus lune*: Bede, *De temporum ratione* 19) — (60) 97v: *Aetas lunae in alphabetis distincta* (= Bede, *De temporum ratione* 23) — (61) 98r: *De aetate lunae si quis computare non potest* (= Bede, *De temporum ratione* 23) — (62) 98r–98v: *De tribus annis circuli decennouenalis quo argumentum generale lunae cum ceteris seruare querunt* (= Bede, *De temporum ratione* 20, ll. 25–55) — (63) 98v: Table of zodiac signs & months (= PL 90, 754) — (64) 99r: Table: *Aetas lunae in alphabetis distincta* (= PL 90, 754–6) — (65) 99v: lunar table (= PL 90, 802C) + *Ecce quinque uocales* — (66) 99v–100r: *Quot punctos siue horas unaquaequae luna per unam quamque noctem luceat* — (67) 100v: *Signa duodecim zodiaci* — (68) 100v: *De signis duodecim* (= PL 90, 357–9*) — (69) 101r: *De circulo magno Paschae* (= Bede, *De temporum ratione* 45) — (70) 101r: *Luna Ianuarii media nocte accenditur* — (71) 101r–101v: *De annis communibus et embolismis* (= Isidore, *Etymologiae* VI 17, §§21–24) — (72) 101v: *Item* — (73) 101v: *Item Beda* (= Bede, *De temporum ratione* 36, ll. 18–24) — (74) 101v: *De bissexto* (Bede, *De temporibus* 10, ll. 1–7) — (75) 101v: *De ratione bissexti* — (76) 101v: *Annus autem ciuilem id est solarem* (= Bede, *De temporum ratione* 36, ll. 38–39) — (77) 101v–102r: *Annus lunaris quadrifariae accipitur* (= Bede, *De temporum ratione* 36, ll. 7–25) — (78) 102r: *Item* — (79) 102r:

Extracts from canons: *Canones Carthaginenses*, etc., on martyrs — (80) 102v: blank — (81) 103r–108v: Calendar (see Borst (2001), 116–8) — (82) 109r–114r: Tables for calculating quadragesima, septuagesima, etc. — (83) 114v: *Argumenta de luna* — (84) 115r: *Saltus dictus est a saliendo* — (85) 115v: *De nominibus digitorum* (= Isidore, *Etymologiae* XI 1, §§70–72) — (86) 115v: *Item de numero* (= Bede, *De temporum ratione* 1, ll. 22–49) — (87) 115v–116r: *Romana Computatio* (= Jones (1939), 107) — (88) 116r–118r: Diagrams illustrating finger-counting — (89) 118v: *De septem miraculis manufactis* (= PL 90, 966; cf. Jones (1939), 89–90) — (90) 118v: *Ianuarius aquarium* — (91) 119r–123r: *Collectio canonum Hibernensis* (= Appendix I) — (92) 123r: Computistical notes on Noe's Ark (= p 332 above) — (93) 123r: *Chronicon Langobardorum* (cf. Pagnin (1958)) — (94) 124r–124v: Excerpts from Fathers — (95) 125r: Rota of months, days & regulars — (96) 125v: Rota of concurrents — (97) 126r: Rota for day on which year begins — (98) 126v: Rota for Easter & other movable feasts — (99) 127r–127v: Computistical *argumenta* — (100) 127v: Rota for indictions — (101) 127v–129v, 130r: More computistical *argumenta* — (102) 128r: Rota of zodiac signs — (103) 128v: Blank — (104) 130v–133r: Ps-Bede, *De stellis celorum* (Maass (1898), 582–94) — (105) 130v: *helix arcturus maior habet autem in capite stellas obscuras VII* — (106) 133r: *eo quod contraria sit cani.*

Appendix III: An Irish Poem on the Computus (MS Oxford, Bodleian Library, Rawlison B 512, 52r–v)[1]

1. A Loingsig, a haés mac nEirc,
   atfes, at fer condeirc,
   in fetar cethardha cain
   do bith i ngach aénbliadain?
2. Dá griantairisim glanda,
   dá equinoctus amra
   cecha bliadna — buan in bágh —
   de reir ríaghla na Rómán.
3. Griantairisim geimridh gluair,
   in .xii. Enáir fuair;

---

[1] First published by Grosjean (1931). I have made minimal editorial changes. The Loingsech addressed by the poet has not been identified.

      in .xii. Iúil, iar tain,
      griantairisim int shamhraidh.

4. Equinoctus erraigh éim
      in .xii. April féil;
      equinoctus fogmair fil
      in .xii. Octimbir.

5. Fil cethrar amra aili
      for aénlaithi sechtmaine,
      co tí lá brátha dia mbrúdh —
      nochan fhágar an athgudh.

6. Aili uathad — érim tric —
      fris riaglaidter ind Init;
      cethramad déac — deagda fásc —
      fris riaglaidtar in Mór-Cháscc.

7. Cethramad — nát legait liúin —
      uathad amra esca Iúin;
      domnach ina diaid — ní dís —
      is é domnach Cingcigis.

8. Sechtmad déac éscai Iúil,
      is é side fil fo diúig;
      domnach iar sin — ségdo fásc —
      is é domnach na Sam-Chásc.

9. Secht n-aés for Init i fuss,
      .vii n-aés forsin Inid Corgus,
      .vii. n-aés for cella, cia be,
      .vii. n-aés Cásc Esérghe.

10. Secht n-aés forsin Min-Cáisc máir,
      a .vii. forsin Fresgabáil;
      a secht for Cinquagis cain,
      secht n-aés for Cáisc int shamhraidh.

11. Bad eol duit, a Loingsig láin —
      bid do lebar it leth-láim! —
      ó .iii. co nómaid — núall nglé —
      .vii. n-aésa na Inite.

12. Ó .xxiiii. aéin
      co n-ice .xxx. tondbain,
      it é sin secht n-aés — ro-clos —
      do bith forsin Corgos.

13. Ótha .xii. n-dil
      co .xviii. — ni denim —

.vii. n-aés co[í]dchi, cin chati,
bís for o[í]dchi Caplaiti.
14. Ó .v.x. — derb ro-clecht —
co n-ice aén find fichet,
secht n-aés na Cásca Móiri
iar n-ugtarás chanóine.
15. Óthá aili .xx. finn
co hochtmad .xx., foidim —
ní baés do neoch dia mbí fásc —
it é .vii. n-aés na Min-Chásc.
16. Ó cóiced .xx. co prím —
cen imroll, cen imirím —
.vii. n-aésa sin — ségdo in dáil —
bíte forsin Fresgabáil.
17. Ó cóiced uathad — ní bréc —
co n-ice oín delbda déc,
.vii. n-aés do grés — grím cen geis —
bít for domnach Ceingcigeis.
18. Ó .xviii. — derbh ro-dlecht —
cosin .xxviiii.,
.vii. n-aés na Sam-Chásca sin,
bat mebuir lat, a Loingsigh!
A Loingsig!

1. O Loingsech, of the people of Erc! You should know — you are a man of vision — what are the four fair things that are found in every year?
2. [There are] two clear solstices, two famous equinoxes in every year — long-lasting the affection — according to the rules of the Romans.
3. The solstice of the clear winter, on the 12th of cold January, and on the 12th of July — after a time — the summer solstice.
4. The eminent spring equinox [occurs] on the famous 12th of April; the autumn equinox is on the 12th of October.
5. There are four famous other things that occur on the same weekday, until the Day of Judgement presses upon them — let it not be gainsaid!
6. There is one single day — a happy knowledge — by which Lent is regulated; the fourteenth day — goodly information — by which Low Sunday is regulated.

7. The famous fourth day of the moon in June — let no one be remiss! — the Sunday after that — not two — that is Quingagesima.
8. The seventeenth day of the moon in July, that is in the end; the Sunday after that — a happy knowledge — that is the Sunday of Summer-Easter.
9. There are seven ages [of the moon] for the beginning [*Initium*] of Lent, here, seven ages for the Lent of Quadragesima; seven ages for *cella*, whatever it be, and seven ages for the Easter of the Resurrection.
10. There are seven [lunar] ages for the great Low Sunday; seven for the Ascension; seven for fair Quinquagesima, seven for Summer Easter.
11. You should know, Loingsech — let your book be in your hand! — from the 3rd to the 9th — a bright acclamation — are the seven [lunar] days of Lent.
12. From the 24th alone to the 30th at the end, those are the seven [lunar] ages —it has been heard — that exist for Quadragesima.
13. From the dear 12th to the 18th — no little thing — [those are] the seven [lunar] days always, without dignity (?) — that are found on the eve of Maundy Thursday.
14. From the 15th — a practised certainty — until the fair 21st, [those are] the seven [lunar] ages of great Easter, according to the authority of the canon.
15. From the fair 21st until the 28th — a messenger — it is no foolishness for him who has knowledge — [are] the seven [lunar] ages of Low Sunday.
16. From the 25th to the 1st, without confusion, without miscalculation, those are the seven [lunar] ages — happy the gathering! — that occur on the Ascension.
17. From the 5th — no lie! — until the comely 11th, [those are] the seven ages, usually — a deed without taboo — that occur on Quinquagesima Sunday.
18. From the 17th — it was laid down with certainty — until the 29th, [those are] the seven [lunar] ages for *Sam-Cháisc* — remember them, Loingsech!

BIBLIOGRAPHY

Abels, R. (1983) 'The council of Whitby: A study of early Anglo-Saxon politics,' *Journal of British Studies* 23, 1–25.

Alberigo, G. (1994) *Les conciles œcuméniques. Les décrets, tome II, vol. 1: De Nicée à Latran V*, Paris.

Allenbach, J. (1995) *Biblia Patristica. Index des citations et allusions bibliques dans la littérature patristique, vol. 6: Hilaire de Poitiers, Ambroise de Milan, Ambrosiaster*, repr. 2001, Paris.

Anderson, M. (1973) *Kings and Kingship in early Scotland*, Edinburgh.

Arweiler, A. (2000) 'Zu Text und Überlieferung einer gekürzten Fassung von Macrobius' Saturnalia I,12,2–I,15,20,' *Zeitschrift für Papyrologie und Epigraphik* 131, 45–57.

Baehrens, E. (1883) *Poetae Latini Minores, vol. 5*, Leipzig.

Baillie, M. (1994) 'Dendrochronology raises questions about the nature of the AD 536 dust-veil event,' *The Holocene* 4, 212–7.

Baker, P.S. and M. Lapidge (1995) *Byrhtferth's Enchiridion*, Oxford.

Balbi, J. (1460) *Catholicon*, Mainz, repr. 1971, Westmead.

Bayet, J. (1973) *Littérature latine*, Paris.

Bayliss, M. and M. Lapidge (1998) *Collectanea Pseudo-Bedae*, Dublin.

Beaucamp, J., R. Bondoux, J. Lefort, M.F. Rouan and I. Sorlin (1979) 'Temps et histoire I: Le prologue de la Chronique pascale,' *Travaux et Mémoires. Centre de Recherche d'Histoire et Civilisation de Byzance* 7, 223–301.

Bergmann, W. (1985) *Innovationen im Quadrivium des 10. und 11. Jahrhunderts. Studien zur Einführung von Astrolab und Abakus im lateinischen Mittelalter*, Stuttgart.

Bhreathnach, E. (1994) 'Killeshin, an Irish monastery surveyed,' *Cambrian Medieval Celtic Studies* 27, 33–47.

Bieler, L. (1961) *Irland: Wegbereiter des Mittelalters*, Lausanne.

— (1975) *The Irish Penitentials*, Dublin.

— (1979) *The Patrician texts in the Book of Armagh*, Dublin.

Bieler, L. and B. Bischoff (1956) 'Fragmente zweier frühmittelalterlicher Schulbücher aus Glendalough,' *Celtica* 3, 211–20.

Bisagni, J. and I. Warntjes (2008) 'The Early Old Irish material in the newly discovered *Computus Einsidlensis* (c.AD 700),' *Ériu* 58, 77–105.

Bischoff, B. (1932) 'Regensburger Beiträge zur mittelalterlichen Dramatik und Ikonographie,' *Historische Vierteljahrschrift* 27, 509–22, repr. in Bischoff (1967), 156–68.

— (1954) 'Wendepunkte in der Geschichte der lateinischen Exegese im Frühmittelalter,' *SE* 6, 189–281, repr. in Bischoff (1966), 205–73.

— (1966) *Mittelalterliche Studien. Ausgewählte Aufsätze zur Schriftkunde und Literaturgeschichte, Band 1*, Stuttgart.

— (1967) *Mittelalterliche Studien. Ausgewählte Aufsätze zur Schriftkunde und Literaturgeschichte, Band 2*, Stuttgart.

— (1998) *Katalog der festländischen Handschriften des neunten Jahrhunderts, Teil 1: Aachen – Lambach*, Wiesbaden.
— (2004) *Katalog der festländischen Handschriften des neunten Jahrhunderts, Teil 2: Laon – Paderborn*, Wiesbaden.
Blackburn, B. and L. Holford-Strevens (1999) *The Oxford companion to the year*, Oxford.
Böhne, W. (1965) 'Beginn und Dauer der römischen Fastenzeit im sechsten Jahrhundert,' *ZKG* 77, 224–37.
Borst, A. (1994) *Das Buch der Naturgeschichte. Plinius und seine Leser im Zeitalter des Pergaments*, Heidelberg.
— (1998) *Die karolingische Kalenderreform*, Hannover.
— (2001) *Der karolingische Reichskalender und seine Überlieferung bis ins 12. Jahrhundert*, 3 vols., Hannover.
— (2004) *Der Streit um den karolingischen Kalender*, Hannover.
— (2006) *Schriften zur Komputistik im Frankenreich von 721 bis 818*, 3 vols., Hannover.
Botte, B. (1980) *Ambroise de Milan: Des sacrements. Des mystères. Explication du Symbole*, 2nd ed., Paris.
Breatnach, C. (1997) 'Rawlinson B 502, Lebar Glinne dá Locha and Saltair na rann,' *Éigse* 30, 109–23.
Brosset, M.F. (1870) *Deux historiens arméniens: Kiracos de Gantzac, XIIIe siècle, Histoire d'Arménie; Oukhtanès d'Ourha, Xe siècle, Histoire en trois parties*, 2 vols., St. Petersburg.
Brooks, G.E.W. (1902–4) *The Sixth Book of the Select Letters of Severus, Patriarch of Antioch, in the Syriac Version of Anathasius of Nisibis*, 2 vols., London.
Brown, P. (1998) *Pouvoir et persuasion dans l'Antiquité tardive. Vers un Empire chrétien*, Paris.
Bruckner, A. (1936) *Scriptoria medii aevi Helvetica. Denkmäler schweizerischer Schreibkunst des Mittelalters, Band 2: Schreibschulen der Diözese Konstanz: St. Gallen 1*, Genf.
Brunhölzl, F. (1975) *Geschichte der lateinischen Literatur des Mittelalters, Band 1: Von Cassiodor bis zum Ausklang der karolingischen Erneuerung*, München.
Bucherius, A. (1634) *De doctrina temporum commentarius in Victorium Aquitanum*, Antwerpen.
Butzmann, H. (1964) *Die Weissenburger Handschriften*, Frankfurt.
Byrne, F.J. (1974) 'Senchas: the nature of Gaelic historical tradition,' *Historical Studies* 9, 137–59.
— (1979) *A thousand years of Irish script: an exhibition of Irish manuscripts in Oxford Libraries*, Oxford.
Calder, G. (1907) *Imtheachta Æniasa. The Irish Aeneid*, London.
— (1917) *Auraicept na n-éces. The scholar's primer*, Edinburgh, repr. 1995, Dublin.

Capitale, M. (1990) *Milano capitale dell'impero romano 286–402 d.C.*, Milano.
Carey, J. (1993) *A new introduction to Lebor Gabála Érenn*, London.
— (1994) *The Irish national origin-legend: synthetic pseudohistory*, Cambridge.
Chadwick, H. (1981) *Boethius: The consolations of music, logic, theology and philosophy*, Oxford.
Chaîne, M. (1925) *La chronologie des temps chrétiens de l'Égypte et de l'Éthiope*, Paris.
Charles-Edwards, T.M. (1998) 'The construction of the *Hibernensis*,' *Peritia* 12, 209–37.
Chevalier, J. and A. Gheerbrant (1969) *Dictionnaire des symboles*, repr. 2004, Paris.
Colgrave, B. (1927) *The life of Bishop Wilfrid by Eddius Stephanus*, Cambridge.
— (1940) *Two Lives of Saint Cuthbert*, Cambridge.
Colgrave, B. and R.A.B. Mynors (1969) *Bede's Ecclesiastical History of the English people*, Oxford.
Collingwood, R.G., R.P. Wright, S.S. Frere and R.S.O. Tomlin (1965) *The Roman inscriptions of Britain, vol. 1: Inscriptions on stone*, new ed. 1995, Gloucester.
Cordoliani, A. (1942) 'Études de comput II – Un texte espagnol de comput du VIIIe (?) siècle,' *Bibliothèque de l'école des chartes* 103, 65–8.
— (1943) 'Les traités de comput du Haut Moyen Age (526–1003),' *Archivum Latinitas Medii Aevi* 17, 51–72.
— (1945–6) 'Les computistes insulaires et les écrits pseudo-alexandrins,' *Bibliothèque de l'école des chartes* 106, 1–34.
— (1955a) 'Les manuscrits de comput ecclésiastique de l'Abbaye de Saint Gall du VIIIe au XIIe siècle,' *ZSK* 49, 161–200.
— (1955b) 'L'évolution du comput ecclésiastique à Saint Gall du VIIIe au XIe siècle,' *ZSK* 49, 288–323.
— (1958) 'Textes de comput espagnol du VII siècle: Le Computus Cottonianus,' *Hispania Sacra* 11, 125–36.
D'Arbois de Jubainville, H. (1866) 'Gloses irlandaises du neuvième siècle,' *Bibliothèque de l'école des chartes* 12, 509–10.
De Angelis, V. (1977–80) *Papiae Elementarium: Littera A*, 3 vols., Milano.
— (1867) 'Gloses irlandaises du neuvième siècle,' *Bibliothèque de l'école des chartes* 13, 471–5.
Declercq, G. (2000) *Anno Domini. The origins of the Christian era*, Turnhout.
— (2000a) *Anno Domini: Les origines de l'ère chrétienne*, Turnhout.
— (2002) 'Dionysius Exiguus and the introduction of the Christian era,' *SE* 41, 165–246.
Dindorf, L. (1832) *Chronicon Paschale*, 2 vols., Bonn.
Draak, M. (1957) *Construe Marks in Hiberno-latin manuscripts*, Amsterdam.
Du Cange, C. (1688) *Pascalion, siue Chronicon Paschale*, Paris.
Duchesne, L. (1880a) 'Review of Bruno Krusch, *Der 84jährige Ostercyclus und seine Quellen*,' *Revue critique d'histoire et de littéraire*, nouvelle série 10, 145–8.

— (1880b) 'Review of Bruno Krusch, *Der 84jährige Ostercyclus und seine Quellen*,' *Bulletin critique de littérature, d'histoire et de théologie* 1, 243–6.

— (1886–92) *Le liber pontificalis: Texte, introduction et commentaire*, 2 vols., Paris.

Dulaurier, É. (1859) *Recherches sur la chronologie arménienne technique et historique*, Paris.

Dumville, D.N. (1975–6) 'The textual history of Lebor Bretnach: a preliminary study,' *Éigse* 16, 255–73.

— (1977–9) 'Ulster heroes in the early Irish annals: A caveat,' *Éigse* 17, 47–54.

— (1981) 'English libraries before 1066: Use and abuse of the manuscript evidence,' in M.W. Herren, *Insular Latin studies*, Toronto, 153–78.

— (1983) 'Motes and beams: Two Insular computistical manuscripts', *Peritia* 2, 248–56.

Duval, Y.-M. (1969) 'La "manœuvre" frauduleuse de Rimini. À la recherche du Liber aduersus Vrsacium et Valentem,' in *Hilaire et son temps, Actes du colloque de Poitiers 29 septembre–3 octobre 1968 à l'occasion du XVIe centenaire de sa mort*, Paris, 51–103.

— (1973) *Le livre de Jonas dans la littérature chrétienne grecque et latine. Sources et influences du Commentaire sur Jonas de saint Jérôme*, 2 vols., Paris.

Elrington, C.R. (1847–64) *The whole works of Most Rev. James Ussher D.D.*, 17 vols., Dublin.

Englisch, B. (1994) *Die Artes liberales im frühen Mittelalter (5.–9. Jh.). Das Quadrivium und der Komputus als Indikatoren für Kontinuität und Erneuerung der exakten Wissenschaften zwischen Antike und Mittelalter*, Stuttgart.

— (2002) *Zeiterfassung und Kalenderproblematik in der frühen Karolingerzeit. Das Kalendarium der Hs. Köln DB 83–2 und die Synode von Soissons 744*, Stuttgart.

Esposito, M. (1907) 'An unpublished astronomical treatise by the Irish monk Dicuil,' *PRIA* 26 C, 378–446, repr. in Esposito (1990) VII.

— (1990) *Irish books and learning in mediaeval Europe*, ed. M. Lapidge, Aldershot.

Etter, H.F. (1988) *Die Zürcher Stadtheiligen Felix und Regula. Legenden, Reliquien, Geschichte und ihre Botschaft im Licht moderner Forschung*, Zürich.

Faris, M.J. (1976) *The bishop's synod: The first synod of St. Patrick – A symposium with text, translation and commentary*, Liverpool.

Favale, A. (1958) *Teofilo d'Alessandria (345c.–412). Scritti, Viata et Dottrina*, Turino.

Favier, J. (1886) 'Manuscripts de la Bibliothèque de Nancy,' *Catalogue general des manuscripts des bibliothèques publiques de France, Departments – Tome IV*, Paris, 121–299.

Fleckenstein, J. (1953) *Die Bildungsreform Karls des Großen als Verwirklichung der norma rectitudinis*, Bigge.

Floëri, F. and P. Nautin (1957) *Homélie pascales, 3: Une homélie anatolienne sur la date de Pâques en l'an 387*, repr. 2004, Paris.

Foley, W.T. and A.G. Holder (1999) *Bede: A Biblical Miscellany*, Liverpool.

Fontaine, J. (1960) *Isidore de Seville, Traité de la nature*, Bordeaux.

— (1992) *Ambroise de Milan, Hymnes*, Paris.

Folkerts, M. (1978) 'Die älteste mathematische Aufgabensammlung in lateinischer Sprache: Die Alkuin zugeschriebenen *Propositiones ad acuendos iuvenes*. Überlieferung, Inhalt, kritische Edition,' *Denkschriften der österreichischen Akademie der Wissenschaften, mathematisch-naturwissenschaftliche Klasse* 116, 6. Abhandlung, 15–76.

Folkerts, M. und H. Gericke (1993) 'Die Alkuin zugeschriebenen *Propositiones ad acuendos iuvenes* (Aufgaben der Schärfung des Geistes der Jugend),' in P.L. Butzer and D. Lohrmann, *Science in Western and Eastern civilization in Carolingian times*, Basel, 283–358.

Freeman, A.M. (1924) 'The annals in Cotton MS Titus A xxv,' *Revue Celtique* 41, 301–30.

— (1925) 'The annals in Cotton MS Titus A xxv,' *Revue Celtique* 42, 283–305.

— (1926) 'The annals in Cotton MS Titus A xxv,' *Revue Celtique* 43, 358–84.

— (1927) 'The annals in Cotton MS Titus A xxv,' *Revue Celtique* 44, 336–61.

Gaidoz, H. (1867) 'Note on the Irish glosses recently found in the library of Nancy,' *PRIA* 10 C, 70–1.

Gesenius, W., E. Kautzsch and A.E. Cowley (1910) *Gesenius' Hebrew Grammar*, Oxford.

Gillert, K. (1880) 'Lateinische Handschrifen in St. Petersburg,' *Neues Archiv* 5, 241–65.

Ginzel, F.K. (1914) *Handuch der mathematischen und technischen Chronologie, Band 3*, Leipzig.

Grabowski, K. and D. Dumville (1984) *Chronicles and annals of mediaeval Ireland and Wales*, Woodbridge.

Grosjean, P. (1931) 'A poem on the computus,' in J. Fraser, P. Grosjean and J.G. O'Keeffe, *Irish Texts, vol. 1*, London, 52–4.

Grotefend, H. (1982) *Taschenbuch der Zeitrechnung des deutschen Mittelalters und der Neuzeit*, 12th repr., Hannover.

Grumel, V. (1958) *La Chronologie*, Paris.

— (1960) 'Le problème de la date pascale aux IIIe et IVe siècles,' *Revue des études byzantines* 18, 163–78.

Gryson, R. (1968) *Le prêtre selon saint Ambroise*, Louvain.

Gy, P.-M. (1955) 'Semaine sainte et triduum pascal,' *Maison-Dieu* 41, 7–15.

Hägermann, D. (2000) *Karl der Große, Herrscher des Abendlandes*, München.

— (2003) *Karl der Große*, Reinbeck.

Harris, J.R. (1998) *Adaptations of Roman epic in medieval Ireland. Three studies in the interplay of erudition and oral tradition*, Lewiston.

Hefele, C.J. (1907) *Histoire des conciles d'après les documents originaux, tome I, partie 1*, Paris.

Helm, R. (1956) *Eusebius Werke, vol. 7: Die Chronik des Hieronymus*, 2nd repr., Berlin.
Hennessy, W.M. (1887) *Annála Uladh – The Annals of Ulster, vol. 1: A.D. 431–1056*, Dublin.
Herbert, M. (1986) 'The Irish Sex Aetates Mundi: First editions,' *Cambridge Medieval Celtic Studies* 11, 97–112.
Herbert, M. and M. McNamara (1989) *Irish biblical apocrypha*, Edinburgh.
Hess, H. (1958) *The Canons of the Council of Sardica A.D. 343*, Oxford.
— (2002) *The Early Development of Canon Law and the Council of Serdica*, Oxford.
Hogan, E. (1895) *The Irish Nennius from L. na Huidre and homilies and legends from L. Brecc*, Dublin.
Holder, A. (1889) *Inventio sanctae crucis*, Leipzig.
Holford-Strevens, L. (2005) *The history of time: A very short introduction*, Oxford.
Howlett, D. (1993) 'Two works of Saint Columban,' *Mittellateinisches Jahrbuch* 28, 27–46.
— (1994a) *Liber epistolarum Sancti Patricii episcopi: The book of letters of Saint Patrick the bishop*, Dublin.
— (1994b) 'The earliest Irish writers at home and abroad,' *Peritia* 8, 1–17.
— (1995a) *The Celtic Latin tradition of Biblical style*, Dublin.
— (1995b) 'Five Experiments in Textual Reconstruction and Analysis,' *Peritia* 9, 1–50.
— (1995c) 'The polyphonic colophon to Cormac's Psalter,' *Peritia* 9, 81–91.
— (1996a) 'Seven studies in seventh-century texts,' *Peritia* 10, 1–70.
— (1996b) *The English origins of Old French literature*, Dublin.
— (1996c) '*Rubisca*: An edition, translation, and commentary,' *Peritia* 10, 71–90.
— (1997a) *British books in Biblical style*, Dublin.
— (1997b) 'Insular Latin writers' rhythms,' *Peritia* 11, 53–116.
— (1998a) *Cambro-Latin compositions: Their competence and craftsmanship*, Dublin.
— (1998b) 'Hellenic learning in Insular Latin,' *Peritia* 12, 54–78.
— (1998c) 'Arithmetic rhythms in Latin letters,' *Archivum Latinitatis Medii Aevi* 56, 193–225.
— (1998d) '*Synodus prima Sancti Patricii*: An exercise in textual reconstruction,' *Peritia* 12, 238–53.
— (1999a) *Sealed from within: Self-authenticating Insular charters*, Dublin.
— (1999b) 'Dicuill on the islands of the north,' *Peritia* 13, 127–34.
— (2000) *Caledonian Craftsmanship: The Scottish Latin Tradition*, Dublin.
— (2001a) 'Hiberno-Latin syllabic poems in the Book of Cerne,' *Peritia* 15, 1–21.
— (2001b) 'Further manuscripts of *Ailerani Sapientis Canon Euangeliorum*,' *Peritia* 15, 22–6.

— (2002) 'A miracle of Maedóc,' *Peritia* 16, 85–93.
— (2003–4a) 'Numerical punctilio in Patrick's *Confessio*,' *Peritia* 17–8, 150–3.
— (2003–4b) 'The prologue to the *Collectio canonum Hibernensis*,' *Peritia* 17–8, 144–9.
— (2005a) *Insular Inscriptions*, Dublin.
— (2005b) 'Three poems about Monenna,' *Peritia* 19, 1–19.
— (2006a) 'On the new edition of Anatolius *De ratione paschali*,' *Peritia* 20, forthcoming.
— (2006b) 'Computus in the *Collectanea Pseudo-Bedae*,' *Peritia* 20, forthcoming.
— (2006c) *Muirchú moccu Machténi's "Vita Sancti Patricii" Life of Saint Patrick*, Dublin.
— (2006d) 'Gematria, number and name in Anglo-Norman,' *French Studies Bulletin* 101, forthcoming.
— (2007a) 'A Triad of Texts about Saint David,' in J.W. Evans and J.M. Wooding, *St David of Wales: Cult, Church and Nation*, Woodbridge, 253–73.
— (2007b) 'Text and form of the Cambro-Latin sequence *Arbor eterna*,' *Archivum Latinitatis Medii Aevi* 65, forthcoming.
Howlett, D. and C. Thomas (2003) '*Vita Sancti Paterni*, The Life of Saint Padarn and the original *Miniu*,' *Trivium* 33, 1–103.
Huber, W. (1969) *Passa und Ostern. Untersuchungen zur Osterfeier der alten Kirche*, Berlin.
Hunt, R.W. and A.G. Watson (1999) *Bodleian library quarto catalogues, vol. IX: Digby manuscripts*, Oxford.
Hunter Blair, P. (1970) *The world of Bede*, London.
Hussey, J.M. (1966) *Cambridge Medieval History, vol. IV: The Byzantine Empire, part I: Byzantium and its neighbours*, Cambridge.
Ideler, L. (1825–6) *Handbuch der mathematischen und technischen Chronologie*, 2 vols., Berlin.
Inglebert, H. (1996) *Les Romains chrétiens face à l'histoire de Rome. Histoire, christianisme et romanités en Occident dans l'Antiquité tardive (IIIe–Ve siècles)*, Paris.
Ireland, C. (1996) 'Aldfrith of Northumbria and the learning of a *sapiens*,' in K.A. Klar, E.E. Sweetser and C. Thomas, *A Celtic Florilegium*, Lawrence, 73–7.
Jaffé, P. and W. Wattenbach (1874) *Ecclesiae Metropolitanae Coloniensis Codices Manuscripti*, Berlin.
James, E. (1984) 'Bede and the tonsure question,' *Peritia* 3, 85–98.
Jan, W. (1718) *Historia cycli dionysiani cum argumentis paschalibus et aliis eo spectantibus*, Wittenberg.
Jeanjean, B. and B. Lançon (2006) *Saint Jérôme, Chronique: Continuation de la Chronique d'Eusèbe, années 326–378*, Rennes.
Jones, C.W. (1934) 'The Victorian and Dionysiac paschal tables in the West,' *Speculum* 9, 408–21, repr. in Jones (1994) VIII.

— (1937) 'The "Lost" Sirmond Manuscript of Bede's Computus,' *English Historical Review* 51, 204–19, repr. in Jones (1994) X.
— (1939) *Bedae pseudepigrapha. Scientific writings falsely attributed to Bede*, Ithaca.
— (1943) *Bedae opera de temporibus*, Cambridge.
— (1943a) 'A Legend of St. Pachomius,' *Speculum* 18, 198–210, repr. in Jones (1994) VII.
— (1994) *Bede, the schools, and the computus*, ed. W.M. Stevens, Aldershot.
Jones, T. (1968) 'Historical writing in medieval Wales,' *Scottish Studies* 12, 15–27.
Jounel, P. (1983) 'L'année,' in A.-G. Martimor, *L'église en prière. Introduction à la liturgie, 4: La liturgie et le temps*, Paris, 43–166.
Kaltenbrunner, F. (1876) *Die Vorgeschichte der Gregorianischen Kalenderreform*, Wien.
Karst, J. (1911) *Eusebius Werke, Bd. 5: Die Chronik. Aus dem Armenischen übersetzt*, Leipzig.
Kasten, B. (1986) *Adalhard von Corbie. Die Biographie eines karolingischen Politikers und Klostervorstehers*, Düsseldorf.
Kenney, J.F. (1966) *The sources for the early history of Ireland: Ecclesiastical: An introduction and guide*, 2nd ed., Dublin.
Kent, R.G. (1951) *Varro, on the Latin Language*, 2 vols., Cambridge.
Krusch, B. (1880) *Studien zur christlich-mittelalterlichen Chronologie. Der 84jährige Ostercyclus und seine Quellen*, Leipzig.
— (1884) 'Die Einführung des griechischen Paschalritus im Abendlande,' *Neues Archiv der Gesellschaft für ältere deutsche Geschichtskunde* 9, 99–169.
— (1910) 'Das älteste Lehrbuch der dionysianischen Zeitrechnung,' in *Mélanges offerts à Émile Chatelain*, Paris, 232–42.
— (1926) 'Ein Bericht der päpstlichen Kanzlei an Papst Johannes I. von 526 und die Oxforder HS. Digby 63 von 814,' in A. Brackman, *Papsttum und Kaisertum*, München, 48–58.
— (1938) 'Studien zur christlich-mittelalterlichen Chronologie: Die Entstehung unserer heutigen Zeitrechnung,' *Abhandlungen der Preußischen Akademie der Wissenschaften Jahrgang 1937, phil.-hist. Klasse 8*, Berlin.
Labourt, J. (2002-3) *Saint Jérôme, Correspondance*, 8 vols., Paris.
Lagrange, M.-J. (1930) *L'Évangile de Jésus-Christ*, Paris.
Langlois, C.V. and C. Seignobos (1898) *Introduction aux études historiques*, repr. 1992, Paris.
Lapidge, M. (2006) *The Anglo-Saxon library*, Oxford.
Lapidge, M. and M.W. Herren (1979) *Aldhelm. The prose works*, Cambridge.
Lapidge, M. and M. Winterbottom (1991) *Wulfstan of Winchester: The life of St Æthelwold*, Oxford.

Lapidge, M. and R. Sharpe (1985) *A bibliography of Celtic Latin literature: 400–1200*, Dublin.

Law, V. (1982) *The Insular Latin grammarians*, Suffolk.

— (1995) *Wisdom, authority and grammar in the seventh century: Decoding Virgilius Maro Grammaticus*, Cambridge.

Lejbowicz, M. (2006) 'Des tables pascales aux tables astronomique et retour. Formation et réception du comput patristique,' *Methodos* 6, 1–67 (http://methodos.revues.org).

Lehmann P. (1912), 'Cassiodorstudien,' *Philologus* 71 (1912), 278–99.

— (1959) 'Cassiodorstudien,' in P. Lehmann, *Erforschung des Mittelalters, Band 2*, Stuttgart, 38–108 (repr. from *Philologus* 71 (1912), 278–99; 72 (1913), 503–17; 73 (1914), 253–73; 74 (1918), 351–8).

Lindsay, W.M. (1911) *Isidori Hispalensis episcopi etymologiarum sive originum libri XX*, 2 vols., Oxford.

— (1915) *Notae Latinae*, Cambridge.

Lionard, P. (1961) 'Early Irish grave slabs,' *PRIA* 61 C, 95–169.

Lohrmann, D. (1993) 'Alcuins Korrespondenz mit Karl dem Großen über Kalender und Astronomie,' in P.L. Butzer and D. Lohrmann, *Science in Western and Eastern civilization in Carolingian times*, Basel, 79–114.

Maass, E. (1898) *Commentariorum in Aratum reliquiae*, Berlin.

Mac Airt, S. (1951) *The Annals of Inisfallen (MS. Rawlinson B. 503)*, Dublin.

— (1953–4) 'Middle-Irish poem on world-kingship,' *Études Celtiques* 6, 255–80.

— (1955–6) 'Middle-Irish poem on world-kingship,' *Études Celtiques* 7, 18–45.

— (1958–9) 'Middle-Irish poem on world-kingship,' *Études Celtiques* 8, 98–119, 284–97.

Mac Airt, S. and G. Mac Niocaill (1983) *The Annals of Ulster (to A.D. 1131)*, Dublin.

Macalister, R.A.S. (1938–56) *Lebor Gabála Érenn*, 5 vols., Dublin.

Macalister, R.A.S. and E. Mac Neill (1916) *Leabhar Gabhála: The Book of Conquests of Ireland*, Dublin.

Mac Carthy, B. (1892) *The codex Palatino-Vaticanus No. 830*, Dublin.

— (1901) *Annals of Ulster, vol. 4: Introduction and Index*, Dublin.

Mac Neill, E. (1913a) 'Poems by Flann Mainistrech on the dynasties of Ailech, Mide and Brega,' *Archivium Hibernicum* 2, 37–99.

Mac Neill, E. (1913b) 'The authorship and structure of the "Annals of Tigernach",' *Ériu* 7, 30–120.

Magnou-Nortier, É. (2002) *Le Code Théodosien Livre XVI et sa réception au Moyen Âge. Texte latin de l'édition Mommsen et traduction française*, Paris.

Maraval, P. (1997) *Le christianisme de Constantin à la conquête arabe*, Paris.

Martin, A. (1996) *Athanase d'Alexandrie et l'Église d'Égypte au IVe siècle (328–373)*, Roma.

Martin, A. and M. Albert (1985) *Histoire "Acéphale" et Index syriaque des Lettres festales d'Athanase d'Alexandrie*, Paris.
Mayeur, J.-M., C. Pietri, L. Pietri, A. Vauchez and M. Venard (1995) *Histoire du christianisme des origines à nos jours. Tome II: Naissance d'une chrétienté (250–430)*, Paris.
Mayr-Harting, H. (1991) *The coming of Christianity to Anglo-Saxon England*, 3rd repr., London.
Mayr, J. (1955) 'Der Computus Ecclesiasticus,' *ZKT* 77, 301–30.
Mc Carthy, D.P. (1993) 'Easter principles and a fifth-century lunar cycle used in the British Isles,' *Journal for the History of Astronomy* 24, 204–24.
— (1994) 'The origin of the *latercus* Paschal cycle of the Insular Celtic churches,' *Cambrian Medieval Celtic Studies* 28, 25–49.
— (1996) 'The lunar and paschal tables of *De ratione paschali* attributed to Anatolius of Laodicea,' *Archive for History of Exact Sciences* 49, 285–320.
— (1998) 'The status of the pre-Patrician Irish Annals,' *Peritia* 12, 98–152.
— (2003) 'The Emergence of *Anno Domini*,' in G. Moreno-Riaño and G. Jaritz, *Time and eternity. The medieval discourse*, Turnhout, 31–53.
— 'Chronological synchronisation of the Irish Annals,' www.cs.tcd.ie/Dan.McCarthy/chronology/synchronisms/annals-chron.
— 'The chronology of St. Colum Cille,' www.celt.dias.ie/publications/tionol/dmcc01.pdf.
Mc Carthy, D.P. and A. Breen (1997) 'Astronomical observations in the Irish annals and their motivation,' *Peritia* 11, 1–43.
— (2003) *The ante-Nicene Christian Pasch: De ratione paschali – The Paschal tract of Anatolius, bishop of Laodicea*, Dublin.
Mc Carthy, D.P. and D. Ó Cróinín (1987–8) 'The "lost" Irish 84-year Easter cycle rediscovered,' *Peritia* 6–7, 227–42, repr. in D. Ó Cróinín (2003), 58–75.
McCormick, M. (1975) *Les annals du haut moyen âge*, Turnhout.
McLynn, N.B. (1994) *Ambrose of Milan: Church and Court in a Christian Capital*, Berkeley.
Meersseman, G.G. and E. Adda (1966) *Manuale di computo con ritmo mnemotecnico dell'arcidiacono Pacifico di Verona*, Padova.
Meslin, M. (1967) *Les ariens d'occident (335–430)*, Paris.
Miller, M. (1991) 'The chronological structure of the Sixth Age in the Rawlinson Fragment of the 'Irish World-Chronicle',' *Celtica* 22, 79–111.
Mittler, E. (1986) *Bibliotheca Palatina*, Heidelberg.
Moloney, M. (1964) 'Beccán's hermitage in Aherlow: The riddle of the slabs,' *North Munster Antiquarian Journal* 9, No. 3, 99–107.
Monachi ordinis S. Benedicti abbatiae Montis Casini (1873) 'Florilegium Casinense,' in idem, *Bibliotheca Casinensis seu codicum manuscriptorum, vol. 1*, Montecassino.
Morin, G. (1902) 'La Translation de S. Benoît et la Chronique à l'époque de Leno,' *RB* 19, 337–56.

Morris, J. (1972) 'The Chronicle of Eusebius: Irish fragments,' *Bulletin of the Institute of Classical Studies* 19, 80–93.

Muratori, L.A. (1713) *Anecdota, quae ex Ambrosianae Bibliothecae codicibus, tomus 3*, Padova.

Murphy, F.X. (1945) *Rufinus of Aquileia (345–411). His Life and Works*, Washington.

Murray, K. (2006) *Translations from Classical literature: Imtheachta Æniasa and Stair Ercuil ocus a Bás*, London.

Nauroy, G. (1974) 'La méthode de composition d'Ambroise et la structure du De Iacob et uita beata,' in Y.-M. Duval, *Ambroise de Milan. XVIe centenaire de son élection épiscopale*, Paris, 115–53, repr. in Nauroy (2003), 301–54.

— (2003) *Ambroise de Milan. Écriture et esthétique d'une exégèse pastorale. Quatorze études*, Bern.

Neale, J.M. (1850–73) *A history of the Holy Eastern Church*, 5 vols., London.

Neugebauer, O. (1979) *Ethiopic astronomy and computus*, Wien.

— (1982) 'On the Computus paschalis of "Cassiodorus",' *Centaurus* 25, 292–302.

Newton, R.R. (1972) *Medieval chronicles and the rotation of the earth*, London.

Neyrand, L. (2003) *Eusèbe de Césarée. Histoire ecclésiastique*, Paris.

Nicolet, C. (2003) *La fabrique d'une nation: La France entre Rome et les Germains*, Paris.

Ní Shéaghdha, N. (1984) 'Translations and adaptations into Irish,' *Celtica* 16, 107–24.

Ochsenbein, P. (2000) 'Der erste bekannte Schreiber im Kloster St. Gallen: Presbyter Winitharius,' *Helvetia archeologica* 124, 146–57.

O'Conor, C. (1814–26) *Rerum Hibernicarum Scriptores Veteres*, 4 vols., Buckingham.

Ó Cróinín, D. (1982a) 'Mo-Sinnu moccu Min and the computus of Bangor,' *Peritia* 1, 281–95, repr. in Ó Cróinín (2003), 35–47.

— (1982b) 'Review of M.W. Herren, *Insular Latin studies*,' *Peritia* 1, 404–9.

— (1982c) 'A seventh-century Irish computus from the circle of Cummianus,' *PRIA* 82 C, 405–30, repr. in Ó Cróinín (2003), 99–130.

— (1983) *The Irish Sex Aetates Mundi*, Dublin.

— (1983a) 'Early Irish annals from Easter-tables: A case restated,' *Peritia* 2, 74–86, repr. in Ó Cróinín (2003), 76–86.

— (1983b) 'New heresy for old: Pelagianism in Ireland and the papal letter of 640,' *Speculum* 60, 503–16, repr. in Ó Cróinín (2003), 87–98.

— (1983c) 'The Irish provenance of Bede's Computus,' *Peritia* 2, 229–47, repr. in Ó Cróinín (2003), 173–90.

— (1983d) 'Sticks and stones – A reply,' *Peritia* 2, 257–60.

— (1984) 'Rath Melsigi, Willibrord, and the earliest Echternach manuscripts,' *Peritia* 3, 17–49, repr. in Ó Cróinín (2003), 145–65.

— (1986) 'New light on Palladius,' *Peritia* 5, 276–83, repr. in Ó Cróinín (2003), 28–34.

— (1993) 'The Irish as mediators of antique culture,' in P.L. Butzer and D. Lohrmann, *Science in Western and Eastern civilization in Carolingian times*, Basel, 41–51.
— (1998) 'Lebar Buide Meic Murchada,' in T.C. Barnard, D. Ó Cróinín and K. Simms, *A miracle of learning. Essays for William O'Sullivan*, Aldershot, 40–51.
— (2000) 'Who was Palladius "first bishop of the Irish"?,' *Peritia* 14, 205–37.
— (2003) *Early Irish history and chronology*, Dublin.
— (2003a) 'Bede's Irish computus,' in Ó Cróinín (2003), 201–12.
— (2008) 'Dionysius Exiguus in the classroom,' in J.W. Dauben, S. Kirschner, A. Kühne, P. Kunitzsch and R.P. Lorch, *Mathematics celestial and terrestrial. Festschrift für Menso Folkerts*, Stuttgart, 253–74.
Ó Cuív, B. (1963) 'Literary creation and Irish historical tradition,' *Proceedings of the British Academy* 49, 233–62.
— (1990) 'The Irish marginalia in Codex Palatino-Vaticanus no. 830,' *Éigse* 24, 45–67.
O'Grady, C. (1976) 'Turning-points in the history of Latin exegesis in the early Irish Church: AD 650–800,' in M. McNamara, *Biblical studies: the medieval Irish contribution*, Dublin, 73–160.
O'Leary, P. (c.1920) *Irish numerals and how to use them*, Dublin.
Ohashi, M. (2003) '"Sexta aetas continet annos praeteritos DCCVIIII" (Bede, *De temporibus*, 22): A scribal error?,' in G. Jaritz and G. Moreno-Riaño, *Time and Eternity. The Medieval Discourse*, Turnhout, 55–61.
— (2005) 'Theory and History: An interpretation of the paschal controversy in Bede's *Historia ecclesiastica*,' in S. Lebecq, M. Perrin and O. Szerwiniack, *Bède le Vénérable. Entre tradition et postérité*, Lille, 177–85.
Ó Néill, P. (2005) 'Irish glosses in a twelfth-century copy of Boethius's *Consolatio Philosophiae*,' *Ériu* 55, 1–47.
Ó Riain, P. (1981) 'The Book of Glendalough or Rawlinson B 502,' *Éigse* 18, 161–76.
— (1982) 'NLI G 2, f. 3 and the Book of Glendalough,' *ZCP* 39, 29–32.
Oroz Reta, J., M.-A. Marcos Casquero and M.C. Díaz y Díaz (1993–4) *San Isidoro de Sevilla: Etimologías*, 2 vols., 2nd repr., Madrid.
Orth, P., M. Mersiowsky and A. Mentzel-Reuters (2004) *Théodore Mommsen et le Moyen Âge, Catalogue de l'exposition préparée par les Monumenta Germaniae Historica à l'occasion du centième anniversaire de la mort de Théodore Mommsen et organisée par l'Institut historique allemand à Paris, du 27 avril au 28 mai 2004*, Paris.
Oskamp, H.P.A. (1968) 'On the author of Sex Aetates Mundi,' *Studia Celtica* 3, 127–40.

— (1974) 'Ierse Handschriften — 1000 tot 1300. Verslagen en beschouwingen,' *Nederlandse Organisatie voor Zuiver Wetenschappelijk Onderzoek, Jaarboek 1974*, 91–5.

— (1975) 'The Yellow Book of Lecan proper,' *Ériu* 26, 102–14.

— (1977) 'A schoolteacher's hand in a Florentine manuscript,' *Scriptorium* 31, 191–7.

Pagnin, B. (1958) 'La provenienzia del Codice Antoniano 27 e del *Chronicon Regum Langobardorum* in esse contenuto,' in *Miscellanea in onore di Roberto Cessi, vol. 1*, Roma, 29–41.

Palanque, J.R. (1933) *Saint Ambroise et l'Empire romain: contribution à l'histoire des rapports de l'église et de l'état à la fin du quatrième siècle*, Paris.

Pallarès, J.G. (1984) 'Astronomía en el Computus Cottonianus,' *Faventia* 6, 73–89, repr. in Pallarès (1999), 7–20.

— (1988–9) 'Los textos latinos de cómputo de los mss. Paris, Bibl. Nat., NAL, 2169 y Léon, Bibl. de la Catedral, N. 8: Una edition,' *Analecta Sacra Tarraconensia* 61–2, 373–410, repr. in Pallarès (1999), 63–92.

— (1989) 'El Computus Cottonianus en los mss. Londres, B.M., Cotton Caligula A XV; Paris, B.N., NAL 2169 y León, Archivo de la Catedral, N. 8: Un nuevo enfoque de la cuestión,' *Actas del VII Congreso Español de Estudios Clásicos, vol. 3*, Madrid, 501–6, repr. in Pallarès (1999), 57–62.

— (1994) 'Hacia una nueva edición de los *argumenta paschalia* de Dionisio el Exiguo,' *Hispania Sacra* 49, 13–31, repr. in Pallarès (1999), 93–109.

— (1999) *Studia Chronologica. Estudios sobre manuscritos latinos de cómputo*, Madrid.

Papias (1496) *Elementarium doctrinae rudimentum*, Venezia, repr. 1966, Turino.

Parkes, M.B. (1987) 'The contribution of Insular scribes of the seventh and eighth centuries to the "Grammar of legibility",' in A. Maierù, *Grafia e interpunzione del latino nel medioevo*, Roma, 15–29, repr. in Parkes (1991), 1–18.

— (1991) *Scribes, scripts, and readers: Studies in the communication and presentation of medieval texts*, London.

Paul, A. (2000) 'Genèse et avènement des 'Écritures' chrétiennes,' in J.-M. Mayeur, C. and L. Pietri, A. Vauchez and M. Venard, *Histoire du christianisme, tom. 1: Le Nouveau Peuple (des origines à 250)*, Paris, 673–737.

Peden, A. (2003) *Abbo of Fleury and Ramsey: Commentary on the calculus of Victorius of Aquitaine*, Oxford.

Pedersen, O. (1978) 'Astronomy,' in D.C. Lindberg, *Science in the Middle Ages*, Chicago, 303–37.

— (1983) 'The ecclesiastical calendar and the life of the church,' in G.V. Coyne, *Gregorian reform of the calendar*, Vatican, 17–74.

Pietri, C. (1976) *Roma Christiana: Recherches sur l'Eglise de Rome, son organisation, sa politique, son idéologie de Miltiade à Sixte III (311–440)*, 2 vols., Roma.

Piganiol, A. (1972) *L'empire chrétien (325–395)*, 2nd ed., Paris.

Pillonel-Wyrsch, R.-P. (2004) *Le calcul de la date de Pâques au Moyen Âge*, Fribourg.

Pitra, J.B. (1852) *Spicilegium Solesmense I*, Paris, repr. 1962, Graz.

Plummer, C. (1896) *Venerabilis Baedae Opera Historica*, 2 vols., Oxford.

Poole, R.L. (1918a) 'The earliest use of the Easter cycle of Dionysius Exiguus, part 1,' *English Historical Review* 33, 57–62.

— (1918b) 'The earliest use of the Easter cycle of Dionysius Exiguus, part 2,' *English Historical Review* 33, 210–3.

Puech, H.-C. (1979) *Sur le manichéisme et autres essais*, Paris.

Radner, J.N. (1978) *Fragmentary annals of Ireland*, Dublin.

Richard, M. (1939) 'Les écrits de Théophile d'Alexandrie,' *Le Muséon* 52, 33–50.

— (1974) 'Le comput pascal par octaétéris,' *Le Muséon* 87, 307–39.

Riese, A. (1869–70) *Anthologia latina sive poesis latinae supplementum. Pars prior: carmina in codicibus scripta*, 2 vols., Leipzig.

Rittmueller, J. (1983) 'The gospel commentary of Máel Brigte ua Máelsuanaig and its Hiberno-Latin background,' *Peritia* 2, 185–214.

— (1984) 'Afterword: the gospel of Máel Brigte,' *Peritia* 3, 215–8.

Rodgers, R.H. (1975) *Rutilius Taurus Aemilianus Palladius, Opus agriculturae de veterinaria medicina de insitione*, Leipzig.

Rousseau, P. (1999) *Pachomius: The Making of a community in fourth-century Egypt*, Berkeley.

Rühl, F. (1897) *Chronologie des Mittelalters und der Neuzeit*, Berlin.

Saenger, P. (1997) *Space between words: The origins of silent reading*, Stanford.

Sansterre, J.-M. (1972) 'Eusèbe de Césarée et la naissance de la théorie "césaropapiste",' *Byzantion* 42, 131–95, 532–94.

Savage, J.J. (1928) 'An Old-Irish gloss in Cod. Laur. XLV, 14,' *ZCP* 17, 371–2.

Savon, H. (1977) *Saint Ambroise devant l'exégèse de Philon le Juif*, 2 vols., Paris.

— (1993) 'Les recherches sur Saint Ambroise en Allemagne et en France de 1870 à 1930,' in J. Fontaine, R. Herzog and K. Pollman, *Patristique et Antiquité tardive en Allemagne et en France de 1870 à 1930. Influences et échanges. Actes du Colloque franco-allemand de Chantilly (25–27 octobre 1991)*, Paris, 111–28.

— (1995) 'Saint Ambroise a-t-il imité le recueil de lettres de Pline le Jeune?,' *RÉA* 41, 3–17.

— (1997) *Ambroise de Milan (340–397)*, Paris.

Senebier, J. (1779) *Catalogue raisonné des manuscrits conservés dans la Bibliothèque de la Ville et République de Genève*, Genève.

Schieffer, T. (1954) *Winfrid-Bonifatius und die christliche Grundlegung Europas*, Freiburg.

Schmid, J. (1904) *Die Osterfestberechnung auf den britischen Inseln vom Anfang des 4. bis zum Ende des 8. Jahrhunderts*, Regensburg.
— (1907) *Die Osterfestberechnung in der abendländischen Kirche. Vom ersten allgemeinen Konzil zu Nikäa bis zum Ende des VIII. Jahrhunderts*, Freiburg.
Schove, D.J. (1984) *Chronology of eclipses and comets AD 1–1000*, Woodbridge.
Schuba, L. (1992) *Die Quadriviums-Handschriften der Codices Palatini Latini in der Vatikanischen Bibliothek*, Wiesbaden.
Schwartz, E. (1905) 'Christliche und jüdische Ostertafeln,' *Abhandlungen der königlichen Gesellschaft der Wissenschaften zu Göttingen, philologisch-historische Klasse, Band 8, Nr. 6*, Berlin.
Schwartz, E. and T. Mommsen (1903–9) *Eusebius Werke, 2. Band: Die Kirchengeschichte*, 3 vols., Leipzig.
Sharpe, R. (1995) *Adomnán of Iona – Life of St Columba*, London.
Simonetti, M. (1975) *La crisi ariana nel IV secolo*, Roma.
Singmaster, D. (1998) 'The history of some of Alcuin's *Propositiones*,' in P.L. Butzer, H.T. Jongen and W. Oberschelp, *Karl der Große und sein Nachwirken: 1200 Jahre Kultur und Wissenschaft in Europa, vol. 1*, Turnhout, 11–29.
Smyth, A.P. (1972) 'The earliest Irish annals: Their first contemporary entries, and the earliest centres of recording,' *PRIA* 72 C, 1–48.
Stancliffe, C. (2003) *Bede, Wilfrid, and the Irish*, Jarrow.
Stern, L.C. (1910) *Epistolae Beati Pauli glosatae glosa interlineali: Irisch-lateinischer Codex der Würzburger Universitätsbibliothek*, Halle.
Stevens, W.M. (1981) 'Scientific instruction in early Insular schools,' in M.W. Herren, *Insular Latin studies*, Toronto, 83–111.
— (1985) *Bede's Scientific Achievement*, Jarrow.
Stevenson, J. (1995) *The "Laterculus Malalianus" and the school of Archbishop Theodore*, Cambridge.
Stokes, W. (1881) *Togail Troi. The destruction of Troy*, Calcutta.
— (1883) *Saltair na Rann: A collection of 162 Early Middle Irish poems*, Oxford.
— (1884) 'The destruction of Troy,' in W. Stokes and E. Windisch, *Irische Texte, 2. Serie*, Leipzig, 1–142.
— (1909) *In Cath Catharda. The Civil War of the Romans. An Irish version of Lucan's Pharsalia*, Leipzig.
— (1993) *The Annals of Tigernach*, 2 vols., Felinfach (repr. from *Revue Celtique* 16 (1895) 374–419; 17 (1896) 6–33, 119–263, 337–420; 18 (1897) 9–59, 150–97, 267–303).
Stokes, W. and J. Strachan (1901–3) *Thesaurus Palaeohibernicus: A collection of Old Irish glosses, scholia, prose, and verse*, 2 vols., Cambridge, repr. 1987, Dublin.
Strobel, A. (1977) *Ursprung und Geschichte des frühchristlichen Osterkalenders*, Berlin.

— (1984) *Texte zur Geschichte des frühchristlichen Osterkalenders*, Münster.
Ström, H. (1939) *Old English personal names in Bede's History: An etymological-phonological investigation*, Lund.
Springsfeld, K. (2002) *Alkuins Einfluß auf die Komputistik zur Zeit Karls des Großen*, Stuttgart.
— (2003) 'Rechnen,' in W. Dreßen, G. Minkenberg and A.C. Oellers, *Ex-Oriente, Isaak und der weiße Elefant, Band 1: Die Reise des Isaak, Bagdad*, Aachen, 224–33.
Teres, G. (1984) 'Time computations and Dionysius Exiguus,' *Journal for the History of Astronomy* 15, 177–88.
Testard, M. (2002) *Ambroise: De officiis*, Paris.
Thomas, C. (1981) *Christianity in Roman Britain to AD 500*, Berkeley.
Thorndike, L. and P. Kibre (1963) *A Catalogue of incipits of medieval scientific writings in Latin*, London.
Tierney, J.J. (1967) *Dicuili Liber de mensura orbis terrae*, Dublin.
Todd, J.H. and A. Herbert (1848) *The Irish version of the Historia Britonum of Nennius*, Dublin.
Tremp, E., K. Schmuki and T. Flury (2004) *Karl der Große und seine Gelehrten. Zum 1200. Todesjahr Alkuins (gest. 804)*, St. Gallen.
Tristram, H.L.C. (1985) *Sex Aetates Mundi. Die Weltzeitalter bei den Angelsachsen und Iren*, Heidelberg.
Ussher, J. (1639) *Britannicarum Ecclesiarum Antiquitates*, Dublin, repr. in C.R. Elrington (1847–64), vols. v–vi.
van de Vyver, A. (1931) 'Cassiodore et son oeuvre,' *Speculum* 6, 244–92.
— (1957) 'L'évolution du comput Alexandrin et Romain du IIIe au Ve siècle,' *RHE* 52, 5–25.
van der Hagen, J. (1734) *Observationes in veterum patrum et pontificium, prologos et epistolas paschales, aliosque antiquos de ratione paschali scriptores*, Amsterdam.
van Euw, A. (1998a) 'Kat. Nr. 23: Beda Venerabilis: Naturlehre, historiographische und zeitrechnerische Werke (Dom Hs. 103),' in J.M. Plotzek and U. Surmann, *Glaube und Wissen im Mittelalter – Die Kölner Dombibliothek*, München, 129–35.
— (1998b) 'Kat. Nr. 24: Kompendium der Zeitrechnung, Naturlehre und Himmelskunde (Dom Hs. 83^{II}),' in J.M. Plotzek and U. Surmann, *Glaube und Wissen im Mittelalter – Die Kölner Dombibliothek*, München, 136–56.
van Hamel, A.G. (1932) *Lebor Bretnach, the Irish version of the Historia Britonum ascribed to Nennius*, Dublin.
van Wijk, W.E. (1936) *Le nombre d'or. Étude de chronologie technique suivie du texte de la Massa compoti d'Alexandre de Villedieu*, La Haye.
Verbist, P. (2003) 'Abbo of Fleury and the computational accuracy of the Christian era,' in G. Jaritz and G. Moreno-Riaño, *Time and Eternity. The Medieval Discourse*, Turnhout, 63–80.

Vogel, C. (1981) *Introduction aux sources de l'histoire du culte chrétien au Moyen Âge*, Spoleto.
von den Brinken, A.-D. (1957) *Studien zur lateinischen Weltchronistik bis in das Zeitalter Ottos von Freising*, Düsseldorf.
— (1960) 'Die Welt- und Inkarnationsära bei Heimo von St. Jakob. Kritik an der christlichen Zeitrechnung durch Bamberger Komputisten in der ersten Hälfte des 12. Jahrhunderts,' *Deutsches Archiv für Erforschung des Mittelalters* 16, 155–94.
— (1961) 'Marianus Scottus. Unter besonderer Berücksichtigung der nicht veröffentlichten Teile seiner Chronik,' *Deutsches Archiv für Erforschung des Mittelalters* 17, 191–238.
von Oppolzer, T. (1887) *Canon der Finsternisse*, Wien.
Waitz, G. (1840) *Über das Leben und die Lehre des Ulfila. Bruchstücke eines ungedruckten Werkes aus dem Ende des 4. Jahrhunderts*, Hannover.
Walker, G.S.M. (1957) *Sancti Columbani opera*, Dublin.
Wallace-Hadrill, J.M. (1988) *Bede's ecclesiastical history of the English people: A historical commentary*, Oxford.
Wallis, F. (1999) *Bede: The reckoning of time*, Liverpool.
Walsh, M. and D. Ó Cróinín (1988) *Cummian's letter De controversia paschali together with a related Irish computistical tract De Ratione Conputandi*, Toronto.
Warntjes, I. (2005) 'A newly discovered Irish computus: *Computus Einsidlensis*,' *Peritia* 19, 61–4.
— (2007) 'The Munich Computus and the 84 (14)-year Easter reckoning,' *PRIA* 107 C, 31–85.
Wasserschleben, H. (1874) *Die irische Kanonensammlung*, Giessen.
Weber Jones, L. (1932) *The script of Cologne from Hildebald to Hermann*, Cambridge.
Williams, D.H. (1995) *Ambrose of Milan and the end of the Arian-Nicene conflicts*, Oxford.
Willis, J. (1983) *Martianus Capella*, Leipzig.
Wilmart, A. (1923) 'Les ordres du Christ,' *Revue des sciences religieuse* 3, 305–27.
— (1933) 'Un nouveau texte du faux concile de Césarée sur le comput pascal,' in idem, *Analecta Reginensia. Extraits des manuscrits latins de la reine Christine conservés au Vatican*, Vatican, 19–27.
Winterbottom, M. (1978) *Gildas. The Ruin of Britain and other works*, London.
Zeller, K. (1991) '*Cassianus natione Scythia*, ein Südgallier,' *WS* 104, 161–8.
Zelzer, M. (1978) 'Zum Osterfestbrief des heiligen Ambrosius und zur römischen Osterfestberechnung des 4. Jahrhunderts,' *WS* 91, 187–204.
Zimmer, H. (1881) *Glossae Hibernicae*, Berlin.
Zinner, E. (1925) *Verzeichnis der astronomischen Handschriften des deutschen Kulturgebietes*, München.

INDICES

Index of Sources

Vulgate

Genesis (Gen): 161, 166, 180, 327, 331
  1:5: 275, 281
  7:11: 331
  8:3–4: 331
Exodus (Ex):
  6:18–20: 167
  12:2: 22, 36–7
  12:5–8, 11–14, 29, 31, 33, 34, 39: 24, 37, 39
  12:40–41: 167
  16: 24, 39
  17:1–7: 24, 39
Leviticus (Lev):
  23:5: 22, 37–9, 225
  25:10: 225
Numbers (Num):
  9:3: 22, 36–7
  28:16: 22, 38
Deuteronomy (Dt):
  16:1: 22, 37
Joshua (Jos):
  12–14: 285
Judges (Jds):
  3:7–11: 265
Samuel: 180
Kings: 161, 180, 327
1 Kings (1 Kgs):
  6:1: 167
2 Kings (2 Kgs):
  16:2: 172
  17:6: 172
  18:10–11: 171–2, 174–5, 177
  22:1–23:30: 244
Chronicles: 327
Ezra: 180
Nehemia: 180
Job: 180
Psalms:
  22:6: 321
  88:29: 17
  88:38: 20, 36
  117:24: 17, 23, 36–7
  118: 317
  118:126: 17, 36
  140:3: 24, 39
Qoheleth (Qo):
  3:1: 17, 36
Isaiah (Is):
  1:3: 17, 36
  49:8: 17, 37
  54:13: 35
Jeremiah (Jer):
  8:7: 17, 36
  25:1: 177
  31:22–24: 35
Matthew (Mt):
  5:17: 22, 37
  26:2: 131
Luke (Lk):
  13:32: 36
  22:7–22: 36, 39
  27:7–12: 36
John (Jn):
  1:17: 22, 37
  2:9: 38
  2:18: 24, 39
  2:19: 24, 38
  4:14: 36
  6:45: 35
  17:1: 21, 36
  21:11: 290
Acts of the Apostles (Act):
  2:8–11: 152
Romans (Rom):
  8:16–17: 282
1 Corinthians (1 Cor):
  5:7: 38
  5:8: 24, 39
  10:2–4: 24, 39
2 Corinthians (2 Cor):
  2:15: 24, 39

3:6–17: 39
6:2: 17, 37
Galatians (Gal):
  2:2: 153
  4:10–11: 36
Colossians (Col)
  3:11: 152
  4:3: 39
  4:6: 24, 39

2 Peter (2 Pet):
  3:7–8: 327
1 John (1 Jn):
  2:18: 24, 39
Apocalypse (Apoc):
  13:18: 265
  21:12: 294
  21:17: 294

Authors and Texts

Admonitio generalis: 243–4
  Praefatio: 244
  71: 244
Acts of the Council of Ceasarea: 74, 137
Adomnán: 159, 183–8, 259
  De locis sanctis: 159, 185–8, 260, 291
    I 1: 291–4
  Vita Columbae: 186
    II 46: 185
Æthelstan:
  Charter of 928: 314
Aileranus Sapiens: 259, 284
  Canon Euangeliorum: 284–5
  Interpretatio Mystica: 284
Alcuin (?): 113, 132, 185, 205, 209, 232, 238–41, 244–9, 254, 257–8, 308, 314, 355
  Calculatio: 91, 204, 217, 234
  Carmen LXXII: 210
  Epistola 126: 247–8
  Epistola 145: 248
  De bissexto: 93, 221, 235, 240, 246–7
  De saltu lunae: 240, 246–7
  Propositiones ad acuendos juvenes: 240–1
  Ratio de luna: 239–40, 246–7
Aldhelm:
  Epistola ad Gerontium: 197

Ambrose of Milan: 1–2, 4–5, 7–9, 14, 21–2, 27, 30–5
  De Iacob et uita beata: 16
  De officiis: 20
    I 26: 20
    I 49: 16
  De Noe et Arca: 324, 328
  De sacramentis:
    I 4: 21
  Epistula 70: 8
  Epistula extra collectionem 13: 1–39
  Hymni: 25
Anatolian Homily of 387: 13
Anatolius (?): 137, 259, 274
  De ratione paschali: 128, 137, 179, 181, 253, 262–4, 333, 342
    10: 220
Annales Bertiniani: 203
Annales Cambriae: 321
Annales Lundenses: 203
Annales Prumienses: 203
Annals of Boyle: 162–4, 166, 178
Annals of Inisfallen: 162–4, 166, 170–6, 178, 181, 278
Annals of Tigernach: 162–4, 166, 170–6, 178–81, 187–8
Annals of Ulster: 162–4, 183–4, 259, 287–9, 300
Anonymus ad Cuimnanum: 289, 324
  Prologue: 328

Athanasius: 1, 27–32, 132, 135, 142
  Festal index: 28–31
Audite facta: 321
Audite omnes amantes Deum:
  282–3, 290, 307
Audite sancta studia: 320–1
Auraicept na n-éces: 259, 265
Augustinus Hibernicus: 259, 285
  De mirabilibus sacrae scripturae:
    260, 272
  II 4: 285–6
Barnabas:
  Epistle:
    15: 327
Bede: 40, 54, 58, 96, 143, 159–60,
  186–8, 190–2, 205, 208–9,
  238–9, 246, 249, 251–4,
  257–9, 291,
  De temporibus: 121, 192–5, 332
    1–16: 193
    10: 343
    13: 149
    14: 77–8, 93, 190–203, 222
    16: 342
    17–22 (*Chronica minora*):
      159–203
  De temporum ratione: 65, 115–6,
    123–4, 126, 137, 194, 197, 199,
    208–9, 221, 245, 253–4, 332
    1: 265, 344
    6: 249–50
    11: 256
    16: 343
    19: 137, 343
    20: 70, 148, 230, 343
    21: 70, 233
    22: 220
    23: 137, 233, 343
    24: 97, 103
    30: 92,
    36: 343
    39: 220–1
    43: 250–1

    45: 223, 230, 250, 343
    47: 77, 191, 197–202
    49: 77, 191
    50: 149
    51: 197, 201
    52: 77, 191
    54: 77, 191
    56: 97
    57: 77
    58: 77, 191
    66 (*Chronica maiora*): 159–89,
      191, 199
  Easter table: 208–9, 242, 245–6,
    255, 258
  Epistola ad Pleguinam: 187,
    191–2
  Historia abbatum:
    15: 200
  Historia ecclesiastica gentis
    Anglorum: 160, 182, 189,
    191, 197
    Prologue: 277
    II 19: 149, 152
    III 3–5: 183
    III 4: 186
    III 8: 154
    III 21: 156
    III 22: 156
    III 24: 156
    III 25: 143–58
    III 29: 149
    IV 26: 186
    V 15: 185, 187
    V 16–17: 185, 187
    V 21: 197, 201
    V 24: 181
  Vita Cuthberti:
    24: 185
Bobbio Computus (MS Milano,
  Biblioteca Ambrosiana, H 150
  inf): 114, 116, 126
  prefix: 98
  3: 70

21: 70
22: 70
29: 92
39: 93
40: 93, 342
45: 92
65: 103
89: 125
146: 129
153–155: 77
Bonifatius:
  Epistola 75: 252–3
  Epistola 76: 252–3
Brían mac Con Catha:
  Rubisca: 312–4
Calendar: 238–258
Canones lunarium decemnovennalium circulorum: 43, 76
Cassiodorus:
  Institutiones: 67
Chronicon Langobardorum: 332, 344
Chronicon paschale: 7
Chronograph of 354:
  Cyclus paschalis: 30, 32–3
Codex Theodosianus:
  XVI 1: 13
  XVI 5: 13, 22
Cogitosus: 259
  Vita Sanctae Brigitae: 260, 294
    32: 204–7
Collectanea Pseudo-Bedae: 259, 301–3
Collectio Canonum Hibernensis: 333, 336–41, 344
  Prologus: 303–5
  XXVII 1: 338
  XXVII 10: 337
  XXVII 11: 337
  XXVII 13: 338
  XXVII 14: 338
  XXVII 15: 339
  XXVII 16: 339
  XXVII 17: 339–40

XLIV 20: 340
LII 3: 340
LII 4: 340–1
LII 5: 341
LIII 4: 340
Cologne Prologue: 114, 131
Columbanus: 259, 261, 283, 288
  Carmen de mundi transitu: 260, 283
  Epistola ad Gregorium Papam: 134, 251
    4: 145
  Precamur Patrem: 260, 283, 288
Columbanus of Saint-Trond:
  Ad Fidolium: 314
Commentary on the argumenta of Dionysius Exiguus: 333, 342
Comp. Col.: 217, 219, 253
Computus Carthaginiensis:
  II 9: 5
Computus Cottonianus of 688/9 (MS London, British Library, Cotton Caligula A XV, 73r–80r): 40, 43, 53–6, 58, 60, 62–3, 70, 73–8, 81, 91, 96–7, 100–1, 105
Computus Digbaeanus of 675 (MS Oxford, Bodleian Library, Digby 63, 72v–79r): 40–105
Computus Einsidlensis (MS Einsiedeln, Stiftsbibliothek, 321 (647), 82–125): 92–3, 125, 221
Computus Oxoniensis (MS Oxford, Bodleian Library, Bodley 309, 62r–v; Computus Hibernicus in Graff's article – not to be confused with Comp. Hib. of Borst's editions, which is the Munich Computus here): 122–3, 125–6, 134, 139–40, 268–79
Computus paschalis of 562: 40, 43–5, 57–9, 62–3, 67–9, 77–8, 80, 82, 84–5, 87–8, 91, 94.

Computus Rhenanus of 775 (MS Köln, Dombibliothek, 103, 184v–190v and MS Wolfenbüttel, Herzog-August-Bibliothek, Weißenburg 91, 169r–173v): 40, 43, 70, 75–7, 101–2.
Computus Parisinus of 819/20 (MS Paris, Bibliothèque Nationale, Nouvelle acquisition latine 1615, 154r–155r): 68–9, 72, 79–84, 87–9, 91, 95, 107–9 (facsimile).
Conchubranus: 259, 320
  Vita Sancte Monenne: 260, 320–1
Concilium Arelatense a. 314:
  §1: 131
Concilium Aurelianense a. 541:
  §1: 41, 149, 196, 243
Concilium Chalcedonense:
  Expositio fidei: 26
Concilium Constantinopolitanum I:
  Expositio fidei: 26
  Canonum II: 14
  Canonum III: 14
Concilium Nicaenum I: 19
Continuationes Chronicarum Fredegarii:
  16: 196
  24: 251
Cormac: 259
  Polyphonic colophon: 321–3
Cú Chuimne: 259
  Cantemus in omni die: 259, 306–8
Cummian: 133, 136
  De controuersia paschali: 118, 124, 129, 131–2, 136, 152, 196–7, 266–8, 277, 288
Cummianus Longus (?): 259, 288
  Celebra Iuda: 260, 288
  Commentary on the Gospels: 288
  De figuris Apostolorum: 260, 288–9
  Penitentiale: 288
Cyril of Alexandria (?): 42, 112, 124, 128, 131–2, 134–6, 142, 251, 253
  Epistola: 123–7, 131, 135, 137
  Prologus: 138
Dares Phrygius:
  Acta diurna belli Troiani: 325
De computo dialogus: 118
De divisionibus temporum: 112–26, 133–6
De mirabilibus sacrae scripturae: vide Augustinus Hibernicus
De ratione computi liber: 254
De ratione conputandi (Brussels Computus): 60, 126, 145, 260, 268, 288
  13: 118, 121, 124, 126
  26: 132
  71: 125
  73: 148
  103: 149
  106: 132
De sollemnitatibus paschae: 137–8, 253
Dial. Burg.: 196
  14: 93, 221
Dial. Neustr.:
  8: 70
  21: 70
Dicuil:
  Liber de astronomia:
    I 5: 101
    I 7: 232
  Liber de Mensura Orbis Terrae: 310–2
Dionysius Exiguus: 2, 40–2, 123–4, 131–2, 134–6, 142–3, 181, 190–1, 200, 202, 205–8, 249, 253, 274
  Argumenta: 40–111, 249

I: 43, 45–6, 54–62, 64–9, 73–4, 76, 78–84, 87–9, 93–5, 137, 191, 222
II: 43, 45–6, 54–62, 64–9, 73–4, 76, 78–80, 82, 84, 87–8, 93–4, 137, 222
III: 45–7, 50, 54–62, 64–70, 73–4, 76, 78–80, 82–4, 87–9, 137, 93–5, 149
IV: 45, 47–8, 50, 54–62, 64–9, 73–4, 76, 78–80, 82–4, 87–9, 93–5, 137, 222
V: 43, 45, 48, 54–62, 64–9, 73–4, 76, 78–80, 82, 84, 87–8, 93–4, 137
VI: 45, 48, 54–62, 64–9, 73–4, 76, 78–80, 82, 84, 87–8, 93–4, 137
VII: 48, 54–62, 65–9, 73–4, 76, 78, 80, 84, 87–9, 93–5, 137
VIII: 45, 48, 54–62, 64–9, 73–4, 76, 78–80, 82, 84, 87–8, 93–4, 137
IX: 45, 48–9, 54–62, 64–9, 73–5, 79–80, 84–9, 93–5, 137
X: 45, 49–50, 54–62, 64–9, 73–5, 79–80, 84–5, 87–9, 93–5, 137
XI: 45, 50, 54–62, 64, 67, 71–4, 79–80, 89–91, 93, 95–7, 217
XII: 50, 54–62, 64, 67, 71–4, 79–80, 89–91, 93, 95–7, 217
XIII: 51, 54–62, 64, 67, 71–5, 79–80, 89–91, 93, 95–105, 217
XIV: 51–62, 64–7, 69, 71, 73–4, 76, 79–80, 89, 91–3, 95, 97, 137, 217
XV: 53–62, 67, 79–80, 89, 92, 95, 97, 217
XVI: 45, 53–62, 67, 74–5, 79–80, 89, 92–4, 97, 217, 221
Ciclus: 42, 90, 92, 122, 181, 190–1, 199–203, 207–9, 242, 249
Epistola ad Bonifatium et Bonum: 42, 137
Epistola ad Petronius (Prologus; Libellus): 6, 42, 90, 134–5, 137, 208, 250
Disputatio Cori et Praetextati: 138, 332–3, 342
Disputatio Morini: 45, 112–3, 117, 119, 125–37, 140–2
Einhard:
  Vita Karoli Magni:
    25: 239
Eucherius of Lyon:
  Formulae spiritalis intelligentiae: 205–6, 237
  Instructiones: 237
Eusebius: 132, 135, 137, 259, 285
  Chronicon: 137, 161–2, 165–6, 169, 171–82
  Historia ecclesiastica: 179
    V 24: 23
    X 5: 130
Flann Mainistrech: 324–5
Fragmentary annals of Ireland:
  166: 154
Fragmentum Nanciacense (MS Nancy, Bibliothèque Municipale, 317 (356)): 40, 43, 69–72, 74, 79–80, 89–90, 95–105, 110–1 (facsimile).
Gennadius: 137
  De viris illustribus: 138
Gildas: 259, 261–2
  De poenitentia:
    1: 261
  De excidio Britanniae:
    23,2: 261

Gregory of Tours:
   Historiarum libri decem:
      V 17: 149, 254
      X 23: 149, 196
Gregory the Great: 259, 262
   Moralia in Job: 286
Harley Glossary: 314
Hilarus:
   Epistola ad Victorium: 138, 253
Hisperica Famina: 314
Hrabanus Maurus: 40, 209, 248
   De computo: 78, 332, 342
      10: 121
      17: 248
      43: 103
      62: 77
      67: 77
      69: 77
      70: 70
      72: 77
      73: 70
      78: 77
      79: 98
      83: 150, 209
      90: 77
Hugo of St Victor:
   De Arca Noe morali liber: 324
      1: 329
Imtheachta Aeniasa: 325
In Cath Catharda: 325
Irish Annals: 159–89
Irish poem on the computus: 344–7
Isidore of Seville: 118, 120, 217, 259, 268
   Allegoriae: 235
   Chronica maiora: 165–8, 176, 183–4
   De natura rerum: 168, 253
      VII 4: 224
      VII 5: 224
      IX: 117
   Differentiae: 205, 235
      II 4–16: 237
      II 17: 342
      II 19–28: 342
   Etymologiae: 138, 268, 278, 342
      III 4: 257
      III 29: 117
      III 58: 252
      III 71: 343
      V 29: 120
      V 33: 342
      V 38–9 (*Chronica minora*): 168–9, 172, 174
      VI 17: 149, 223, 343
      VI 18: 225, 236
      XI 1: 344
      XIII 1: 117
      XVI 25–26: 253
   Liber numerorum:
      15: 121
Jerome: 4–6, 137, 159, 179–80, 259, 327
   Chronicon: 5, 138, 160–1, 165–8, 171, 173–5, 181, 268, 278
   Epistola 96: 9
   Epistola 98: 9
   Epistola 100: 9
   In Danielem: 178, 180, 188
John Malalas:
   Chronographia: 192
Joseph Scottus: 259, 308
   Abbreuiatio commentarii Hieronimi in Isaia: 308,
   Carmen figuratum: 260, 308–10
Josephus:
   Antiquitates: 178, 180
Julius Africanus: 178
Laidcenn mac Baíth Bannaig: 259, 286, 314
   Egloga: 260, 286–7
   Lorica: 260, 287–8
Laterculus Malalianus: 192
Latercus (84 (14)-year Easter reckoning): 128, 133, 143–58, 181–2, 188, 210, 333, 342
Lebar Bretnach: 324–5
Lebar Gabála Érenn: 324–5

Lect. comp.: 76, 245, 253
   I 3: 70
   I 4: 70
   I 11: 220–1
   IIII 1–7: 77–8
   V 1: 77–8
   V 5: 98
   V 9: 102
   VI 6: 77, 85–6
Leo I: 253
   Epistola ad Marcianum: 138
Leo monachus: 253
Lib. ann.: 76, 218–9, 240
   2: 77
   3: 77
   6: 77
   8: 77
   14: 223
   15: 77
   19: 217–8
   20: 77
   24: 77
   25: 77
   40: 102
   42: 98
   44: 103
   45: 103
Lib. calc.: 245
   17–23: 77
   31: 98
   32: 102
   82: 92
Lib. comp.: 76
   I 3: 92–3
   I 10: 217–8
   I 11: 217–8
   II 1: 224
   II 7–12b: 77
   III 13: 77
   II 14: 220
   II 15: 70
   II 18: 220
   IIII 2: 102, 226
   IIII 3: 98
   IIII 5: 103
   IIII 19: 217–8
   IIII 28a: 218–9
Liber pontificalis: 178
Lucan:
   Pharsalia: 325
Macrobius:
   Saturnalia:
      I 11–15: *vide Disputatio Cori et Praetextati*
Máel Brigte Ua Máelsuanaig: 324
   Gospel commentary: 329
Marcellinus:
   Chronicon: 183–4
Marianus Scottus: 126, 324
   Chronicon: 335
Martianus Capella: 259
   De nuptiis Philologiae et Mercurii: 268
      II: 265
      VIII: 252
Maximus Confessor: 40, 56, 90, 97
   Computus ecclesiasticus:
      I 11: 98
      I 12: 98, 100, 102
      I 19: 92
      I 27: 99
      II 1: 98
      II 5: 98, 100, 102
      II 27: 149
      III 10: 145
Memoria abbatum nostrorum: 260, 290
ps-Methodius:
   Revelationes: 206, 237
Mo Sinu maccu Min: 122, 259, 266, 294
Muirchú moccu Macthéni: 259
   Vita Sancti Patricii: 260, 289–90, 297
      10: 297–301
Munich Computus (MS München, Bayerische Staatsbibliothek, Clm 14456,

8r–46r): 92–3, 114, 116, 125–6, 221
Nennius:
  Historia Brittonum: 325
Orationes Moucani: 314
Ordo Romanus: 205–6, 236
Pachomian legend: 34–5, 122–5, 134–5,
Pacificus of Verona:
  Computus:
    §6: 70
    §14: 70
    §208–9: 77
    §213: 77
    §216: 77
    §219: 77
    §249: 77
    §250: 77
    §309: 77
    §§337–342: 103
Palladius:
  De agricultura: 218
Paschasinus: 250–1
  Epistola ad Leonem: 42, 137, 253
Passio sanctorum Felicis et Regulae: 205, 237
Patrick: 259, 261, 268, 289–90, 297, 300, 321
  Epistola ad milites Corotici: 260, 281
  Confessio: 260, 281–2
Périgueux table: 216–7
Philocalus: 233
Prosper of Aquitaine: 259
  Chronicon: 300
Proterius: 2, 8–9
  Epistola ad Leonem: 6, 42, 137, 253
Quaest. Austr.: 196
Rhygyfarch ap Sulien:
  Vita Sancti Dauid:
    §2: 302
Romana computatio: 344

Rufinus: 159, 179–82, 188–9
  Translation of Eusebius' Historia ecclesiastica: 179–80, 188
Saltair na Rann: 325
Sex Aetates Mundi: 324–36
  §1: 326–7
  §6: 333
  §§9–11: 327–8
  §19: 328–9
  §20: 330
  §24: 333–4
  §§36–37: 328
  §66: 328
Stephen of Ripon:
  Vita Wilfridi:
    10: 156–7
Suggestio Bonifati primiceri: 74, 85–6, 124, 137
Sulpicius Severus:
  Epistola ad Rufinum: 181
Supputatio Romana: 30–1, 33, 100, 102–3
Synodus episcoporum: 260–1, 279–80
Theophilus: 1–2, 7–9, 11, 13–4, 21–2, 34, 39, 124, 132, 134–6, 142
  Epistola: 13
  Prologus: 7, 138
Tírechán: 259, 289
  Collectanea: 259, 289–90
Togail Troi: 325
Tractatus Hilarii in septem epistolas catholicas: 324, 328–9
Translatio Sancti Benedicti: 332, 342
Varro:
  De lingua Latina:
    VI 2: 117
Versiculi familiae Benchuir: 260, 290–1
Versus de annis a principio: 260, 283–4
Victorius of Aquitaine: 41, 122, 134–6, 138, 143, 149, 190, 196, 221, 238, 243, 255, 274

Prologus: 41, 132, 138, 142, 253
   1: 197
   4: 156
   7: 222
   9: 222, 224
Calculus: 66, 138
Cyclos: 41, 124, 133, 138, 145, 149, 153, 196–203, 224, 243, 251–2
Vita Ceolfridi Abbatis:
   20: 200

Vita Cuthberti:
   I 1: 197
   III 6: 185
Vita Sancti Maedoci: 260, 315–9
Vita Sancti Reguli: 321
Willibrord:
   Calendar: 233
   Easter table: 70
Wulfstan of Winchester:
   Vita S. Æthelwoldi:
      12: 150

Manuscripts

Angers, Bibliothèque Municipale, 44 (48): 288
Basel, Universitätsbibliothek, F III 15k: 63, 77, 114, 116, 118–9, 150
Bern, Burgerbibliothek, 417: 86, 116, 118–9, 126
Bern, Burgerbibliothek, 610: 114, 116,
Bern, Burgerbibliothek, 645: 221, 255
Berlin, Staatsbibliothek, lat. Fol. 307:
Berlin, Staatsbibliothek, Phillipps 1831: 193, 201
Berlin, Staatsbibliothek, Phillipps 1832: 193, 202
Berlin, Staatsbibliothek, Phillipps 1833: 218–9
Berlin, Staatsbibliothek, Phillipps 1869: 254, 256
Besançon, Bibliothèque Municipale, 186: 114, 116, 126
Bruxelles, Bibliothèque Royale, 5301–20: 154
Bruxelles, Bibliothèque Royale, 5413–22: 127
Bruxelles, Bibliothèque Royale, 7672–4: 315
Cambridge, Corpus Christi College, 279: 279
Dijon, Bibliothèque Publique, 448: 114, 116, 118–9
Dublin, Primate Marsh's Library, Z3.1.5 (formerly V.3.4): 315, 317
Dublin, Royal Irish Academy, 23 E 25 (*Leabhar na hUidhre*): 335
Dublin, Trinity College, A.4.15 (57) (Book of Durrow): 326
Dublin, Trinity College, E.3.11 (175): 315
Einsiedeln, Stiftsbibliothek, 167: 194
Einsiedeln, Stiftsbibliothek, 321 (647): *vide Computus Einsidlensis*
Firenze, Biblioteca Medicea Laurenziana, Plut. 78.19: 335
Genève, Bibliothèque Publique et Universitaire, 50: 114–6, 118–9, 124, 126–7, 333
Karlsruhe, Badische Landesbibliothek, Aug. CLXXI: 67
Karlsruhe, Badische Landesbibliothek, Aug. CLXVII: 86
Köln, Diözesan- und Dombibliothek, 83^2: 11, 86, 114, 116, 217, 219, 245, 253, 333
Köln, Diözesan- und Dombibliothek, 102: 76
Köln, Diözesan- und Dombibliothek, 103: 76, 233, *et vide Computus Rhenanus*.
Köln, Diözesan- und Dombibliothek, 138: 236

Leiden, Universiteitsbibliotheek, Scaliger 28: 114, 116, 118
Léon, Biblioteca de la Catedral, N. 8: 78
London, British Library, Additional 36929: 321
London, British Library, Cotton Caligula A XV: 73, 104, 114, 116, 127, *et vide Computus Cottonianus*
London, British Library, Cotton Cleopatra A II: 320
London, British Library, Cotton Vespasian A XIV: 315
London, British Library, Harley 3017: 86
London, British Library, Royal 13 A XI: 218–9
Madrid, Biblioteca Nacional, 3307: 218–9, 226
Milano, Biblioteca Ambrosiana, D 17 inf: 67
Milano, Biblioteca Ambrosiana, F 60 sup: 265
Milano, Biblioteca Ambrosiana, H 150 inf: *vide Bobbio Computus*
Montpellier, Bibliothèque de la Faculté de Médecine, 157: 116.
München, Bayerische Staatsbibliothek, Clm 14456: 70, 91, *et vide Munich Computus*
München, Bayerische Staatsbibliothek, Clm 14725: 43, 116
München, Bayerische Staatsbibliothek, Clm 14746: 193
München, Bayerische Staatsbibliothek, Clm 21557: 195
Nancy, Bibliothèque Municipale, 317 (356): *vide Fragmentum Nanciacense*
Oxford, Bodleian Library, Auc. F. 3.14: 194
Oxford, Bodleian Library, Auc. F. Infra I.2: 194–5
Oxford, Bodleian Library, Bodley 309 (Sirmond MS): 40, 57, 61, 64–5, 112–42, 198, 277, 333, *et vide Computus Oxoniensis*
Oxford, Bodleian Library, Digby 63: 40, 43, 45, 64, 83, 114, 116, 127, *et vide Computus Digbaeanus*
Oxford, Bodleian Library, Hatton 42: 303
Oxford, Bodleian Library, Rawlinson B 485: 315
Oxford, Bodleian Library, Rawlinson B 502: 163, 170, 172, 187–8, 325–6, 335
Oxford, Bodleian Library, Rawlinson B 503: 170
Oxford, Bodleian Library, Rawlinson B 505: 315
Oxford, Bodleian Library, Rawlinson B 512: *vide Irish poem on the computus*
Padua, Biblioteca Antoniana, I 27: 67, 77–8, 209–10, 331–44
Paris, Bibliothèque Nationale, Lat. 2200: 67
Paris, Bibliothèque Nationale, Lat. 4860: 91, 127, 194
Paris, Bibliothèque Nationale, Lat. 10837: 70
Paris, Bibliothèque Nationale, Lat. 16361: 114, 116, 126–7
Paris, Bibliothèque Nationale, Nouvelle acquisition latine 1613: 116, 333
Paris, Bibliothèque Nationale, Nouvelle acquisition latine 1615: 68, 116, 194, *et vide Computus Parisinus*
Paris, Bibliohèque Nationale, Nouvelle acquisition latine 2169: 78
Schaffhausen, Stadtbibliothek, Ministerialis 61: 121

St. Gallen, Stiftsbibliothek, 110: 201
St. Gallen, Stiftsbibliothek, 184: 217–8
St. Gallen, Stiftsbibliothek, 225: 204–37
St. Gallen, Stiftsbibliothek, 248: 86, 195
St. Gallen, Stiftsbibliothek, 250: 195
St. Gallen, Stiftsbibliothek, 251: 194, 209, 224
St. Gallen, Stiftsbibliothek, 397: 218
St. Gallen, Stiftsbibliothek, 878: 194
St. Gallen, Stiftsbibliothek, 902: 224
Strasbourg, Bibliothèque Nationale et Universitaire, 326: 116
Tours, Bibliothèque Municipale, 334: 127, 333
Valenciennes, Bibliothèque Municipale, 404 (386): 232
Vatican, Biblioteca Apostolica, Pal. Lat. 830: 126, 335
Vatican, Biblioteca Apostolica, Pal. Lat. 1447: 57, 64–6, 90, 106, 121
Vatican, Biblioteca Apostolica, Pal. Lat. 1448: 57, 61, 64–6, 90, 121, 193, 209–10
Vatican, Biblioteca Apostolica, Reg. Lat. 586: 114, 127, 131, 333
Vatican, Biblioteca Apostolica, Reg. Lat. 1260: 63, 77, 90
Vatican, Biblioteca Apostolica, Reg. Lat. 2077: 30, 32–3
Vatican, Biblioteca Apostolica, Rossi. 247: 116, 126
Vatican, Biblioteca Apostolica, Urb. Lat. 290: 119, 126
Vatican, Biblioteca Apostolica, Vat. Lat. 642: 114, 116
Vatican, Biblioteca Apostolica, Vat. Lat. 5755: 56–7, 64–7
Wien, Österreichische Nationalbibliothek, 458 (*olim* Salisburgensis 174): 291
Wien, Österreichische Nationalbibliothek, ser. nov. 37: 333
Würzburg, Universitätsbibliothek, M. p. misc. F. 5a: 67
Würzburg, Universitätsbibliothek, M.p.th. f. 61: 266
Wolfenbüttel, Herzog-August-Bibliothek, Weißenburg 91: 75–6, *et vide Computus Rhenanus*